Agroecological Practic
for Sustainable Agricult

Principles, Applications, and Making the Trans

Agroecological Practices for Sustainable Agriculture

Principles, Applications, and Making the Transition

Editor

Alexander Wezel

ISARA-Lyon, France

World Scientific

NEW JERSEY • LONDON • SINGAPORE • BEIJING • SHANGHAI • HONG KONG • TAIPEI • CHENNAI • TOKYO

Published by

World Scientific Publishing Europe Ltd.
57 Shelton Street, Covent Garden, London WC2H 9HE
Head office: 5 Toh Tuck Link, Singapore 596224
USA office: 27 Warren Street, Suite 401-402, Hackensack, NJ 07601

Library of Congress Cataloging-in-Publication Data
Names: Wezel, Alexander, editor.
Title: Agroecological practices for sustainable agriculture : principles, applications, and making the transition / edited by: Alexander Wezel (ISARA-Lyon, France).
Description: New Jersey : World Scientific, 2017.
Identifiers: LCCN 2016056719 | ISBN 9781786343055 (hc : alk. paper)
Subjects: LCSH: Agricultural ecology. | Sustainable agriculture.
Classification: LCC S589.7 .A36952 2017 | DDC 577.5/5--dc23
LC record available at https://lccn.loc.gov/2016056719

British Library Cataloguing-in-Publication Data
A catalogue record for this book is available from the British Library.

Desk Editors: Kalpana Bharanikumar/Mary Simpson

Typeset by Stallion Press
Email: enquiries@stallionpress.com

Printed in Singapore

Preface

"But what are agroecological practices?" This was the question, raised four years ago by my colleagues when we decided to create a new teaching module called "Agroecological Cropping Practices," within our international MSc Agroecology programme. Upon an initial search for definitions and clearer descriptions, we found that we could not adequately respond to this question. Attempting to define what agroecological practices could be, we conducted a thorough literature review and discussed with other colleagues and practitioners, their perspectives on which practices qualified as agroecological practices. Although during this initial step we limited ourselves to cropping practices under temperate climates, it still remained a significant and time-consuming effort to evaluate, define, and summarise the breadth of agroecological practices included in this first paper.

In the years since our initial effort, other authors have evaluated and discussed the criteria for agroecological practices in cropping or livestock systems. As such, interesting recent advancements have emerged concerning definitions and developments of agroecological practices. Under the framework of agroecology, "practice" is considered one of the three major interpretations of the term agroecology, along with agroecology as a scientific discipline and as a movement. Each of these interpretations have a common goal: to develop and design sustainable agriculture and food systems. The question of which practices constitute agroecological practices is of great importance to farmers, professionals in agriculture, and other concerned stakeholders, including policy makers. The Food and Agriculture Organisation of the United Nations (FAO) highlighted the need to identify agroecological practices during the International Symposium on Agroecology for Food Security and Nutrition, hosted for the first time in September 2014 in Rome, Italy. Moreover, concrete policies now exist that aim to promote agroecology and advocate for the implementation of agroecological practices. The French government initiative designed to promote agroecology, proposed in December 2012, serves as an example of such a policy, from which evolved the establishment of a law, *la loi d'avenir pour l'agriculture, l'alimentation et la forêt* (Law for the Future of Agriculture, Food, and the Forest).

Therefore, the objective of this book is to illustrate and summarise recent advances concerning the development and implementation of agroecological practices. The first few chapters provide examples of agroecological cropping practices, followed by chapters related to agroecological practices in livestock systems, and closing with chapters on current methods to teach agroecological practices in higher education. Following these discussions, a last chapter addresses the potential for expanded implementation of agroecological practices and policy.

Chapter 1 by Clara Nicholls, Miguel Altieri, and Luis Vazquez illustrates the underlying agroecological principles that need to be considered for the conversion of farming systems to more sustainable models. They argue that modern agroecosystems require systemic change, and that redesigned farming systems will not emerge from simply implementing a set of practices, but rather from the application of existing, well-defined agroecological principles. These principles can be applied by way of various practices and strategies, and each will have unique effects on productivity, stability, and resiliency within the farm system.

In Chapter 2, Alexander Wezel and Erin Silva address agroecological cropping practices. The chapter describes the state of the art regarding agroecological management, along with the diversity of concepts and interpretations included in the designation of agroecological practices. Subsequently, an overview of agroecological cropping practices, including their constraints and potential applications, is provided, along with specific examples.

The following five chapters provide more detailed insight into agroecological soil, nutrient, weed, and pest management.

Matthew Ryan and Joséphine Peigné review relevant agroecological principles and their integration into management strategies for regenerating soil health and improving crop production. As soil management strongly intertwines with other aspects of crop and livestock production, a systems approach is employed for this assessment, providing an overview of applied agroecological soil management while considering profitability, resilience, and environmental impact over the short- and long-terms.

Jennifer Blesh and Ryan Galt illustrate different strategies for transitioning to agroecological nutrient management practices in the U.S. Corn Belt. They explore farmers' approaches to agroecological soil fertility management, drawing upon an interdisciplinary case study of working farms in the U.S. Corn Belt. They analyse the social and ecological processes occurring on farms that affect transitions, as well as the role of institutions and organisations, and the political and economic context.

In Chapter 5, Richard Smith and David Mortensen analyse the role of the cropping system in defining weed management challenges and opportunities. They illustrate the need to apply an expanded assembly framework in order to further develop ecologically-based weed management strategies. They highlight the higher-order constraints that operate in agroecosystems, whose limitations must be understood and considered to revise weed management toward more environmentally-benign practices. Further, they characterise cropping systems along axes of disturbance intensity and frequency, illuminating the nature of the filtering processes that occur in these systems and the phenological, reproductive, and growth traits of weeds most likely to be constrained by these filters.

The subsequent chapter addresses the potential integration of GM crops and traits into agroecological practises. Heikki Hokkanen and Ingeborg Menzler-Hokkanen review emerging integrated pest management (IPM) options for GM crops and traits. They emphasise the need for flexible, system-appropriate strategies for the development of new management options with GM crops, tailoring them according to the conditions of the target environments and specific agroecosystems. As the optimal IPM toolbox for any region depends on which tools best work together to optimise plant and ecosystem health and performance, GM crops providing pest and disease resistance may be considered as one tool in the IPM toolbox. Thus, a key question posed in this chapter asks: are GM crops just basic IPM tools, or do they have additional features providing added value?

In Chapter 7, Ingeborg Menzler-Hokkanen and Heikki Hokkanen discuss a specific biocontrol practice, entomovectoring. They describe how pollinators, in this case honey bees, can be used as an entomovector to control diseases in strawberry production. In this system, honey bees transport the biocontrol agent from their hives to the strawberry flowers, efficiently and effectively preventing the development of grey mould on later maturing strawberry fruits.

Steve Gliessman, Rose Cohen, and Heather Putnam examine in Chapter 8 agroecology as a platform for participatory knowledge generation and exchange, serving as a foundation upon which to restructure our food systems. Using a model for assessing agroecological transformation and change, they evaluate a case study based on a long-term partnership between an international non-profit organisation, local organisations in Nicaragua and Mexico, and smallholder coffee farming families, whose design was influenced by agroecological principles and participatory action research. The case study demonstrates how agroecology serves to facilitate mutual, horizontal learning among partners, with the capability of reducing vulnerability among farmers dependent on market commodities such as coffee while ensuring food security. It also demonstrates the evolutionary nature of the process of change, rather than one of simple adoption.

In Chapter 9, Cathy Hawes uses indicators of sustainability to assess the impact of management interventions in agroecological and conventional cropping systems. To monitor impact, she proposes a systems approach that accounts for multiple ecosystem functions and the trade-offs between them. This whole-systems approach not only addresses the biodiversity-related indicators upon which agri-environment schemes and agroecological practices have historically focused, but also accounts for more systemic aspects of change in management or intervention that impart wider impacts on all components of the system related to sustainability, including economic and environmental functions as well as ecological processes.

The next three chapters deal with agroecological principles and practices in livestock systems.

In Chapter 10, Bertrand Dumont, Magali Jouven, Thierry Bonaudo, Raphaëlle Botreau, and Rodolphe Sabatier propose a framework for designing agroecological livestock farming systems. They first introduce key ecological processes that can be integrated into the management of livestock farms. They further propose five guiding principles to implement combinations of agroecological practices in livestock farming systems, providing examples from ruminant, pig, poultry, and integrated crop–livestock farms. Further discussed are technical and organisational levers that can be utilised in the implementation of agroecological transitions in livestock farming systems, which extend from the farm scale to the landscape/territory and food-supply chain.

A second chapter addressing livestock systems is provided by Alain Peeters and Alexander Wezel. They analyse the relationship between agroecology and grass-based farming systems, defining different categories of agroecological principles and practices. The chapter focuses on grazing/browsing livestock in temperate climate regions, illustrating a broad range of examples of agroecological practices in herbivory systems from Europe, Japan, New Zealand, and North America.

Chapter 12 presents agroecological management in fish production. Authors Joël Aubin, Joël Robin, Alexander Wezel, and Marielle Thomas focus on traditional fish pond systems, providing examples from France, Brazil, and Indonesia. They evaluate the extent to which fish pond systems comply with the principles of agroecology, the development of pond management practices consistent with these principles, and the synergistic interactions of pond and terrestrial systems that enhance the agroecological character of agriculture. Specific agroecological principles for fish pond systems are defined and different agroecological practices determined.

The last three chapters of the book are devoted to teaching agroecology and, more specifically, agroecological practices.

In Chapter 13, Charles Francis, Mary Wiedenhoeft, Robert Dehaan, and Paul Porter discuss the holistic context for teaching farming and food systems in agroecology education. They provide insight in a U.S. Midwest summer course in agroecosystem analysis, which implements a learning agenda to emphasise whole systems and farm-level evaluation of sustainability. Students are compelled to consider individual practices in the context of holistic farm systems, including farmers' worldviews and production goals. The chapter summarises experiences and current reflections on nearly two decades of the implementation of this course, within which student teams interview farmers, process and discuss observations and new knowledge, and conduct comparisons of sustainability indicators among farms.

Chapter 14 provides examples of teaching agroecological practices in France. Aurélie Ferrer, Florian Celette, Marion Casagrande, Joséphine Peigné, Jean-François Vian, and Alexander Wezel present teaching strategies which integrate instruction on agroecological practices into the educational pathway of a five-year higher education programme. The successive modules in the curriculum, addressing agroecological practices through varying perspectives and levels of depths, are discussed, including their objectives, content and structure, and related pedagogical methods. At the end of the chapter, professional training offered to farmer and other stakeholders is described as another example of teaching agroecological practices.

In the final teaching chapter, Charles Francis, Tor Arvid Breland, Geir Lieblein, Lennart Salomonsson, Suzanne Morse, and Anna Marie Nicolaysen explore examples of hands-on learning practices that provide students direct experience with the broad components of complex farming systems, rather than simply attempting to teach and contextualise practices at a distance. They have found that specific key competencies are needed by students to become agroecologists with deep understanding of agroecological practices as essential and influential components of sustainable farming systems. These competencies include observation, reflection, dialogue, participation, and visioning. They use examples of how each competency contributes to learning more about appropriate future farming practices within a systems perspective. They conclude the chapter with observations and reflections about the learning process by students over 15 years in Norway, and comment on how these skills have been applied throughout the next steps of their thesis research and in future careers.

The concluding chapter by Alexander Wezel and Charles Francis, evaluates the potential of agroecological cropping practices to shape our future agricultural systems and the potential of policy to support their implementation. The capacity for agroecological cropping practice adoption is evaluated in the context of the important issue of scale, including how research and the subsequent application of results are approached when most farmers and other food system professionals tend to address single issues at a time, generally at the lowest levels of spatial and temporal scale. In a second part, the chapter provides descriptions of policies supporting the implementation of agroecological practices. Examples from the E.U., New Zealand, and the U.S. are used to illustrate ways that policies currently influence recommendations that emerge from agroecology, and how policy could be used as a creative tool in the future to guide our agricultural systems forward.

Alexander Wezel

Contents

Chapter 1

Agroecological Principles for the Conversion of Farming Systems

Clara I. Nicholls*, Miguel A. Altieri† and Luis Vazquez‡

*Department of International and Area Studies, University of California, Berkeley, California, USA
†Department of Environmental Science, Policy and Management, University of California, Berkeley, California, USA
‡Latin American Scientific Society of Agroecology (SOCLA), Cuba

1. Introduction

Modern agriculture consists of the replacement of natural plant communities with artificially supported crop communities. Human manipulation and alteration of ecosystems for the purpose of establishing agricultural production has turned modern agroecosystems into highly simplified systems, to the point that they are structurally and functionally very different from natural ecosystems. The inherent self-regulation characteristics of natural communities are lost when humans modify such communities by promoting monocultures (Altieri and Nicholls 2004).

The more intensely such communities are modified, the more abundant and serious the ecological unbalances of simplified cropping systems. For example, Zhao *et al.* (2015) found in a two-year study of 17, 1500 m-radius sites in China, that input of nitrogen fertiliser and cropland expansion compromised the ability of natural enemies to control cereal aphids leading to a disturbance of the interspecific relationships and enhancing reliance on pesticides.

Reliance on homogeneous monoculture production systems is no longer socially, economically and ecologically desirable as these systems compromise biodiversity, utilise resources inefficiently, are highly energy dependent, impose a major ecological footprint, are susceptible to pest outbreaks and are also vulnerable to climatic variability (Thiessen *et al.* 2015). In order to face the trade-off between productivity and resilience, a major task for agriculturalists is to find ways to reinstate more ecological rationale into crop production. This new challenge means finding locally suitable designs inspired by observation of local ecosystems and/or successful diversified farms that exist in a particular region (Malezieux 2012). The design of future farming systems must be based on an understanding of how ecosystems are organised and function, and on the realisation that agroecosystems undergoing ecological conversion operate as complex systems with emergent properties, and therefore management decisions must allow for the special behaviours and properties of complex systems.

Agroecology provides the basic ecological principles for how to evaluate why such systems work and provides elements for the design and management of agroecosystems that are both productive and natural resource conserving, but that are also culturally sensitive, socially just and economically viable (Altieri 1995, Gliessman 1998). In this chapter, we argue that modern agroecosystems require systemic change, but new redesigned farming systems will not emerge from simply implementing a set of practices (rotations, composting, cover cropping, etc.), but rather from the application of already well-defined agroecological principles. These principles can be applied by way of various practices and strategies, and each will have different effects on productivity, stability, and resiliency within the farm system. Agroecological management leads to optimal recycling of nutrients and organic matter turnover, closed energy flows, water and soil conservation and balance pest-natural enemy populations, all key processes in maintaining the agroecosystem's productivity and its self-sustaining capacity (Altieri 2002a).

The agroecological strategy exploits the complementarities and synergisms that result from the various combinations of crops, tree and animals in spatial and temporal arrangements. This is why agroecologists have for some time recognised that intercropping, agroforestry and other diversification methods mimic natural ecological processes, and that therefore the sustainability of complex agroecosystems lies in the ecological models they follow (Altieri 2002b). The challenge to align agricultural systems with ecological principles is immense, especially in the current context of agricultural development where specialisation, short-term productivity and economic efficiency are emphasised. The question of whether the problems of modern agriculture can be ecologically alleviated within the context of the present capital-intensive structure of agriculture is beyond this review.

2. The conversion of farming systems

The reversion of agroecosystems that have already undergone major ecological simplification implies a process of conversion from a high-input monoculture management system to a low-external-input and diversified system (Lamine and Bellon 2009). Most farmers start the conversion process slowly, taking time to gain experience with a more diverse cropping system, to experiment on a small scale, and thus reduce risk, and to learn to be flexible enough to adapt to changing conditions. In general, after 3–4 years of conversion, changes on the soil properties become apparent. The percentage of nitrogen, phosphorus and potassium, pH, organic matter and microbial biomass levels reach values many times significantly higher than at the start of the conversion (Pimentel *et al.* 2005). Many studies of organic agriculture have revealed better performance than conventional systems on various sustainability metrics, including species richness and abundance, soil fertility, nitrogen uptake by crops, water infiltration and holding capacity, and energy use and efficiency (Kremen and Miles 2012).

The conversion to organic management affects the whole farming system, not only single enterprises. Crop rotations are the main management practices that organic farmers overwhelmingly utilise during conversion as these influence forage production, fertility building and are an integral part of weed, pest, and disease management strategies. A major emphasis during conversion is improving overall soil quality by incorporating organic matter into the soil via the application of animal manures or compost, as well as skilful cover cropping and well planned rotations. In most organic systems, cover crops are the source of the vast bulk of organic carbon inputs needed for the desired soil microbial community and adequate nutrient pool (Lotter 2003). Unfortunately pushed by market forces that privilege specialisation, farmers tend to replace practices such as rotations and cover cropping with a set of energy and capital intensive organic "technology packages, e.g. microbial and botanical insecticides, and biofertilisers", and input substitutions, making their operations dependent and intensive.

Many authors have conceptualised conversion as a transitional process with three marked phases (MacRae *et al.* 1990):

1. Increased efficiency of input use through integrated pest management or integrated soil fertility management.
2. Input substitution or substitution of environmentally benign inputs.
3. System redesign: diversification with an optimal crop/animal assemblage, which encourages synergism so that the agroecosystem may provide its own soil fertility, natural pest regulation, and crop productivity.

Many of the practices that are currently being promoted as components of sustainable agriculture fall in categories 1 and 2. Both of these stages offer clear benefits in terms of lower environmental impacts as they decrease agrochemical input use and often can provide economic advantages compared to conventional systems. Incremental changes are likely to be more acceptable to farmers as drastic modifications may be viewed as highly risky. But does the adoption of practices that increase the efficiency of input use or that substitute biologically based inputs for agrochemicals, but leave the monoculture structure intact, really have the potential to lead to the productive redesign of agricultural systems? A true agroecological conversion calls into question monoculture and the dependency on external inputs (Lamine and Bellon 2009).

In general, the fine-tuning of input use through approaches such as Integrated Pest Management (IPM) does little to move farmers towards an alternative to high input systems. In most cases IPM translates to "intelligent pesticide management" as it results in selective use of pesticides according to a predetermined economic threshold, which pests often "surpass" in monoculture situations. Input substitution, used by the large majority of organic farmers follows the same paradigm of conventional farming, overcoming the limiting factor, but this time with biological or organic inputs. Many of these "alternative inputs" have become commodified, therefore farmers continue to be dependent on input suppliers (Rosset and Altieri 1996). In California, many organic farmers cultivating grapes and strawberries apply between 12 and 18 different types of biological inputs per season. In addition to enhancing production costs, many products used for one purpose affect other aspects of the system. For example, Sulphur which is widely used to control foliar diseases of grapes, can also wipe out populations of *Anagrus* parasitic wasps, key regulators of leafhopper pests. Thus, farmers become trapped in an "organic treadmill".

Gliessman (2010), argues that improvements in efficiency of input use and input substitution are not enough to address the challenges facing modern agriculture. Instead, he argues that farming systems must be redesigned based on a new set of ecological relationships. This entails approaching conversion as an ecological transition of agriculture based on notions of agroecology and sustainability.

System redesign arises from the application of agroecological principles that lead to the transformation of agroecosystem function and structure by promoting management guided to ensure ecological processes like nutrient cycling, pest regulation, etc. (Altieri and Nicholls 2012). Ultimately system redesign consists in the establishment of an ecological infrastructure that through plot- to landscape-scale diversification, encourage ecological interactions that generate soil fertility, nutrient cycling and retention, water storage, pest/disease regulation, pollination, and other essential ecosystem services.

3. Principles versus practices

As an applied science, agroecology uses ecological principles for the design and management of diversified agroecosystems where external inputs are replaced by natural processes such as natural soil fertility, allelopathy and biological control (Table 1). Agroecology does not promote technical recipes but rather above principles, which when applied in a particular region take different technological forms depending on the local socio-economic needs of farmers and their biophysical circumstances. Each practice is linked to one or more principle thus contributing to its manifestation in the function of the agroecosystems (Table 2). The applied practices set in motion ecological interactions that drive key processes for agroecosystem function (nutrient cycling, pest regulation, productivity, etc.) (Figure 1).

Agroecology is not about promoting a few magic bullet solutions that are divorced from local contexts and that can be disseminated following top-down approaches. It relies on the quality of complex interactions that

Table 1. Agroecological principles for the design of biodiverse, energy efficient, resource-conserving and resilient farming systems.

1.	Enhance the recycling of biomass, with a view to optimising organic matter decomposition and nutrient cycling over time.
2.	Strengthen the "immune system" of agricultural systems through enhancement of functional biodiversity — natural enemies, antagonists, etc., by creating appropriate habitats.
3.	Provide the most favourable soil conditions for plant growth, particularly by managing organic matter and by enhancing soil biological activity.
4.	Minimise losses of energy, water, nutrients and genetic resources by enhancing conservation and regeneration of soil and water resources and agrobiodiversity.
5.	Diversify species and genetic resources in the agroecosystem over time and space at the field and landscape level.
6.	Enhance beneficial biological interactions and synergies among the components of agrobiodiversity, thereby promoting key ecological processes and services.

Source: Adapted from Altieri (1995) and Reijntjes *et al.* (1992).

Table 2. Relative contribution of several management practices to one or more agroecological principles.

	Principle to which they contribute					
Management practice	Recycling of biomass	Enhancement of functional biodiversity	Enhancement of soil biology	Minimise losses of energy, water, nutrients and genetic resources	Diversification at the field and landscape level	Enhancement of biological interactions and synergies
Compost application	x		x			
Cover crops and/or green manures	x	x	x	x	x	x
Mulching	x		x	x		
Crop rotation	x		x	x	x	
Use microbial/botanical pesticides		x				
Use of insectary flowers to enhance natural enemies		x			x	x
Living fences		x	x		x	x
Intercropping	x	x	x	x	x	x
Agroforestry	x	x	x	x	x	x
Animal integration	x		x	x	x	x

Note: Most of these practices will also appear in the other chapters.

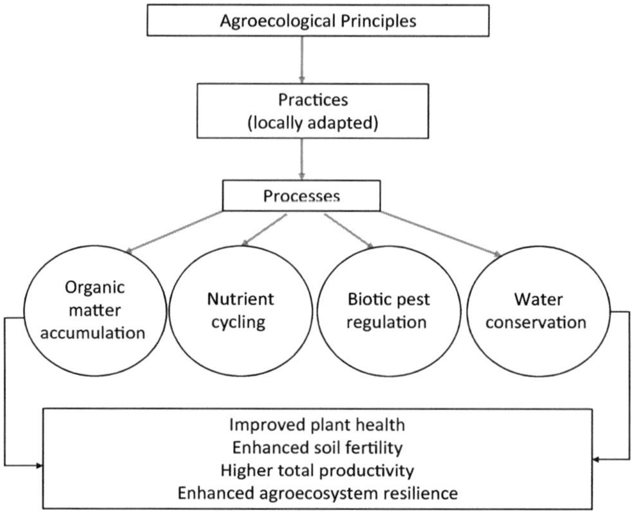

Figure 1. Agroecological principles for the conversion farming systems.

result from the adequate combination of various practices whose operationalisation in particular circumstances will necessarily have to change depending on each context, since each environment has its own characteristics (Malezieux 2012). The array of cultural practices used by each farmer results in functional differences that cannot be accounted for by any single practice. This is what Andow and Hidaka (1989) called "a production syndrome" defined as a set of management practices that are mutually adaptive and lead to high performance. However, subsets of this collection of practices may be substantially less adaptive, that is, the interaction among practices leads to improved system performance that cannot be explained by the additive effects of individual practices. One of the frustrations of research in the organic/conventional yield gap has been the inability of low-input practices to outperform conventional practices in side-by-side experimental comparisons, despite the success of many organic and low-input production systems in practice. A consistent yield gap of 19–25% is reported when comparing organic and conventional agricultural systems, but interestingly the yield gap is reduced substantially when organic farmers adopt multicropping and complex crop rotations (Ponisio *et al.* 2015).

Depending on how it is concretely applied and complemented or not by other practices, one particular practice can sometimes act as an "ecological turntable" by activating key processes such as recycling, biological control, antagonism, and allelopathy, all essential for the health and productivity of a particular farming system. Cover crops, for example, can exhibit several multiple effects simultaneously including suppression of weeds, soil borne diseases and pests, protect the soil from rain and runoff, improve soil aggregate stability, add active organic matter, fix nitrogen and scavenge for nutrients (Altieri 1995).

Clearly, each production system represents a distinct group of management practices and by implication, ecological relations. This re-emphasises the fact that agroecological designs are site-specific, and what may be applicable elsewhere are not the techniques but rather the ecological principles that underlie sustainability. It is of no use to transfer technologies from one site to another, if the set of ecological interactions associated with such techniques cannot be replicated.

4. Agroecological diversification

System redesign is the last stage in the agroecological conversion process and consists in practical steps to break the monoculture nature of many cropping systems and thus reduce their ecological vulnerability, by restoring agricultural biodiversity at the field and landscape level. Promotion of biodiversity within agricultural systems is the cornerstone strategy of system redesign, as increasing diversity within functional groups promotes key processes (pest regulation, nutrient cycling, etc.) fundamental for agroecosystem function (Moonen and Barberi 2008). Higher diversity within the cropping system leads to higher diversity in associated biota. Increased biodiversity also leads to more effective pest control and pollination and to tighter nutrient cycling (Altieri and Nicholls 2012). Diversification should start at the beginning of the conversion process.

Emergent ecological properties develop in diversified agroecosystems that allow the system to function in ways that enhance soil fertility and pest regulation while maintaining crop production in the midst of environmental variability. There is a vast body of literature documenting the benefits of diversified farming systems and the ecological interactions and mechanisms at work that explain why maintaining species diversity in fields acts as a buffer against pests and also against weather variations, thus providing the foundations for plant health, crop productivity and resiliency (Kremen and Miles 2012).

Complementary interactions between the various biotic components can also bc of a multiple nature. Some of these interactions can be used to induce positive and direct effects on the biological control of specific crop pests or on soil fertility regeneration and/or enhancement. The exploitation of these interactions in real situations involves agroecosystem design and management strategies aimed at optimising functional biodiversity. Thus, the optimal behaviour of agroecosystems depends on the level of interactions between the various biotic and abiotic components. By assembling a functional biodiversity, it is possible to initiate synergisms which subsidise agroecosystem processes by providing ecological services such as the activation of soil biology, the recycling of nutrients, the enhancement of beneficial arthropods and antagonists.

A community of organisms in an agroecosystem becomes more complex when, a larger number of different kinds of plants are included, leading to more interactions among arthropods and microorganisms, components of above and below ground food webs. As diversity increases, so do opportunities for coexistence and beneficial interference between species that can enhance agroecosystem sustainability (van Emden and Williams 1974). Diverse systems encourage complex food webs, which entail more potential connections and interactions among members, creating many alternative paths for energy and material flow. For this reason, a more complex community exhibits more stable production and less fluctuations in the number of undesirable organisms (Power and Flecker 1996).

A noticeable impact on ecosystem functioning can be achieved by increasing vegetational richness in species poor crop systems (monocultures). Each newly added species to a simple monoculture has a higher probability of being complementary, and newly added species are more likely to bring in redundant characteristics, enhancing agroecosystem resilience and stability by the presence of redundant species which gain importance following agroecosystem changes or disturbance (Moonen and Barberi 2008). The key is to identify the type of biodiversity that is desirable to maintain and/or enhance in order to carry out ecological services, and then to determine the best practices that will encourage the desired biodiversity components (Altieri and Nicholls 2004). Diversification occurs in many forms at the field (variety mixtures, rotations, polycultures, agroforestry, crop-livestock integration) and at the landscape level (hedgerows, corridors, etc.), giving farmers a wide variety of options and combinations for the implementation of this strategy. A set of combined spatial and temporal diversification strategies described in Table 3, confer farming systems resilience to climate change as greater

Table 3. Temporal and spatial designs of diversified farming systems and their main agroecological effects.

Crop rotations	Temporal diversity in the form of cereal–legume sequences. Nutrients are conserved and provided from one season to the next, and the life cycles of insect pests, diseases, and weeds are interrupted.
Polycultures	Cropping systems in which two or more crop species are planted within certain spatial proximity result in biological complementarities that improve nutrient use efficiency and pest regulation thus enhancing crop yield stability.
Agroforestry systems	Trees grown together with annual crops, in addition to modifying the microclimate, maintain and improve soil fertility as some trees contribute to nitrogen fixation and nutrient uptake from deep soil horizons while their litter helps replenish soil nutrients, maintain organic matter, and support complex soil food webs.
Cover crops and mulching	Flattening cover crop mixtures on the soil surface in conservation farming is a strategy to reduce soil erosion and lower fluctuations in soil moisture and temperature, improve soil quality, and enhance weed suppression resulting in better crop performance. The use of pure or mixed stands of grass and legumes, e.g. under fruit trees, can reduce erosion and provide nutrients to the soil and enhance biological control of pests.
Crop–livestock mixtures	High biomass output and optimal nutrient recycling can be achieved through crop–animal integration. Animal production that integrates fodder shrubs planted at higher densities, intercropped with improved, highly-productive pastures and timber trees all combined in a system that can be directly grazed by livestock enhances total productivity without need of external inputs.

Source: Adapted from Altieri (1995) and Gliessman (1998).

agroecosystem diversity may buffer against shifting rainfall and temperature patterns and possibly reverse downward trends in yields over the long-term as a variety of crops and varieties respond differently to such shocks.

Biodiversity enhances ecosystem function because components that appear redundant at one point in time become important when some environmental change occurs. The redundancies allow for continued ecosystem functioning in the midst of climatic or other changes. On the other hand, a diversity of species acts as a buffer against failure due to environmental fluctuations, by enhancing the compensation capacity of the agro-ecosystem, because if one species fails, others can play their role, thus leading to more predictable aggregate community responses and enhanced ecological resiliency (Altieri *et al.* 2015).

5. Hidden connections

Although crop diversification strategies in the form of multispecies rotations, cover crops, agroforestry, and intercrops are key in the conversion process, when complemented by regular applications of organic amendments (crop residues, animal manures, and composts) surprising effects on plant health, soil quality and productivity can be noticed. These hidden connections have been totally missed by entomologists and other agricultural researchers who have explained pest outbreaks in cropping systems as a consequence of the absence of natural enemies or the effects of pesticide resistance by insect pests or secondary pest outbreaks due to disruptions of biological control (Altieri *et al.* 2012). Western scientists have been largely unaware of the theory of trophobiosis offered by French scientist Francis Chaboussou, who as early as 1967 contended that pest problems were also linked to disturbances in the nutritional balances of crop plants and destruction of life in the soil (Chaboussou 2004). He explained that heavy applications of soluble nitrogen (N) fertilisers (and also certain pesticides) increase the cellular amounts of nitrogen, ammonia and amino acids, faster than the rate at which plants synthesise them for proteins. These reductions in the rate of protein synthesis result in temporary accumulation of free nitrogen, sugars and soluble amino acids in the foliage, substances needed for growth and reproduction by insect herbivores and also plant pathogens. Chaboussou's empirical evidence led him to postulate that insect pests and diseases grow and multiply faster when plants contain more soluble free nutrients due to the inhibition of protein synthesis. He also believed that a healthy soil life is fundamental for a balanced uptake of mineral nutrients by the plant, especially micronutrients. A lack of micronutrients also causes inhibition in protein synthesis and therefore leads to a build-up in nutrients needed by pests and pathogens (Chaboussou 2004).

In the last 20 years, a number of research studies have emerged corroborating Chaboussou's assertions, showing that the ability of a crop plant to resist or tolerate insect pests and diseases is tied to optimal physical, chemical and mainly biological properties of soils. Soils with high organic matter and active soil biological activity generally exhibit good soil fertility as well as complex food webs and beneficial organisms that prevent infection (Altieri and Nicholls 2003). In a series of controlled greenhouse experiments, when given a choice of maize grown on organic versus chemically fertilised soils collected from three nearby farms, European corn borer (*Ostrinia nubilalis*) females significantly laid more eggs in the chemically fertilised plants (Phelan *et al.* 1995). Interestingly, there was significant variation in egg laying among chemical fertiliser treatments within the conventionally managed soil, but in plants under the organic soil management, egg laying was uniformly low. Pooling results across all three farms showed that variance in egg laying was approximately 18 times higher among plants in conventionally managed soil than among plants grown under an organic regimen. Similar studies conducted in China by Hsu *et al.* (2009) indicated that *Pieris rapae crucivora* butterflies preferred to lay eggs on foliage of chemically fertilised cabbage plants and the larvae grew faster on plants fertilised with synthetic fertiliser. The results of this study suggested that proper organic treatment can increase a plant's biomass production and exhibit a lower pest occurrence.

This dampening of plant susceptibility to insects and disease led Phelan *et al.* (1995) to propose the concept of biological buffering, which asserts that a more complex soil community supported by the influx of active organic matter tend to moderate fluctuation in the soil environment and greater ecological stability. During the conversion process additional mechanisms that transfer this stability above ground through greater plant resistance may include: (a) modulation of plant mineral nutrient availability by the soil food web, and/or (b) an enhanced plant systemic defence induced by beneficial microbes interacting with plant roots (Phelan 2009).

When weeds are allowed in the field at certain tolerable levels, nitrogen mediated crop–insect interactions can become even more complex. When examining weedy broccoli fields in Canada, Patriquin *et al.* (1988) found higher numbers of aphid enemies in the diverse systems but they also discovered that the reproductive rate of aphids was proportional to the supply of amino acids in the phloem. Legumes nodulate and fix gaseous nitrogen from air when the supply of mineral nitrogen in the soil is deficient; when it is not, they preferentially take up soil nitrogen. Weeds among the faba beans take up and thereby reduce soil mineral nitrogen to a level below that which suppresses nodulation. This causes the faba beans to nodulate more and to obtain more nitrogen from nitrogen fixation. Under these conditions, there is closer coupling of nitrogen uptake and assimilation than when mineral nitrogen predominates, and consequently accumulation of amino acids in the phloem is reduced. The reproductive rate of the aphids is restricted accordingly. When the weeds are removed, the soil nitrogen supply increases, phloem nitrogen increases, and the plants are more attractive and more nutritious to the aphids. Weedy fava bean plants also had higher yields than the plants in the weed-free plots because the benefits of increased nodulation outweighed any losses due to weeds. This beneficial interaction between weeds and the crop works only if levels of soil nitrate are relatively low (e.g. 10 ppm) to begin with. When plots were fertilised with urea, the interaction does not occur, consequently weeds overgrew the crop and greatly reduced yields.

6. Conclusions

A key agroecological principle applied since the initiation of the conversion process is the diversification of the agroecosystem by adding regenerative components such as combining plants in intercropping arrangements, crops and trees in agroforestry systems, animals and trees in silvopastoral systems, using legumes as cover crops or in rotations, or raising fish in rice paddies. By enhancing functional biodiversity, a major goal of the conversion process is achieved, which is to strengthen the weak ecological functions in the agroecosystem, allowing farmers to gradually eliminate inputs altogether by redesigning the farm system by relying primarily on ecosystem functions (Vazquez *et al.* 2012).

The integrity of an agroecosystem undergoing conversion relies on synergies of plant diversity and the continuing function of the soil microbial community, and its relationship with organic matter. Soils with high organic matter and active soil biological activity generally exhibit good soil fertility as well as complex food webs and beneficial organisms that prevent pathogen infection and pest incidence. It may be argued that agroecosystems whose nutrient cycling is predominantly modulated by the soil food web possess greater ecological stability, with resilience and resistance to perturbation (Altieri *et al.* 2015). Management should therefore be oriented to enhance the ability of crop plants to resist or tolerate insect pests and diseases by manipulating the biological properties of soils complemented by a vegetational infrastructure that harbours natural enemies of pests as well as pollinators. Enhancing below-ground and above-ground positive ecological interactions through integration of soil and pest management practices, a robust and sustainable way for optimising total agroecosystem function can be found (Figure 2).

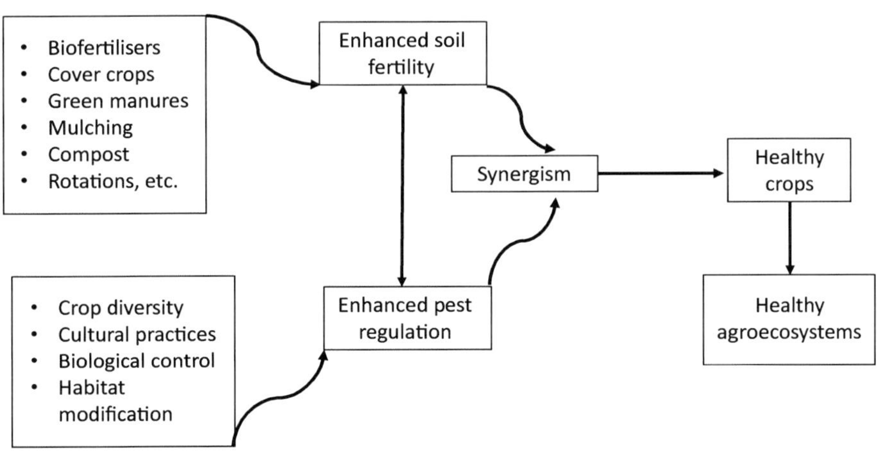

Figure 2. The potential synergism between soil fertility management and pest regulation practices.

Basing the conversion process on particular practices tends to address components in isolation, focusing on the optimisation of one component (soil fertility, plant nutrition, crop growth, etc.) failing to exploit the properties that emerge through the interaction of the farm components. Input substitution thus becomes primarily reactive, shifting efforts to solving problems as they arise, ameliorating symptoms rather than discovering root causes (Phelan 2009). Agroecologists regard pest problems or nutrient deficiencies as the symptom of the failure of an ecological process (biological control or nutrient cycling) and thus endeavour to find out the roots of such unbalance. Instead of focusing on one particular component of the agroecosystem, agroecology emphasises the interrelatedness of all agroecosystem components and the complex dynamics of ecological processes. Thus, agroecology is an alternative approach that goes beyond the use of alternative inputs to develop integrated agroecosystems with minimal dependence on external, off-farm inputs. The emphasis is on the design of complex agricultural systems in which ecological interactions and synergisms between biological components replace inputs to provide the mechanisms for sponsoring soil fertility, productivity, and crop protection (Altieri 1995).

References

Altieri, M.A., 1995. *Agroecology: The Science of Sustainable Agriculture*. Westview Press, Boulder, CO, USA, 411 p.

Altieri, M.A., 2002a. Agroecological principles and strategies for sustainable agriculture. In: Uphoff, N.T. (Ed.), *Agroecological Innovations: Increasing Food Production with Participatory Development*. Earthscan Publication, London, UK, pp. 40–46.

Altieri, M.A., 2002b. Agroecology: The science of natural resource management for poor farmers in marginal environments. *Agriculture, Ecosystems and Environment* **93**, 1–24.

Altieri, M.A., Nicholls, C.I., 2003. Soil fertility management and insect pests: Harmonizing soil and plant health in agroecosystems. *Soil Tillage Research* **72**, 203–211.

Altieri, M.A., Nicholls, C.I., 2004. *Biodiversity and Pest Management in Agroecosystems*. The Harworth Press, Binghamton, New York, USA, 248 p.

Altieri, M.A., Nicholls, C.I., 2012. Agroecology: Scaling up for food sovereignty and resiliency. *Sustainable Agriculture Reviews* **11**, 29–46.

Altieri, M.A., Ponti, L., Nicholls, C.I., 2012. Soil fertility, biodiversity and pest management. In: Gurr, G.M., Wratten, S.D., Snyder, W.E., Read, D.M.Y. (Eds.), *Biodiversity and Insect Pests: Key Issues for Sustainable Management*. John Wiley & Sons, Chichester, UK, pp. 72–84.

Altieri, M.A., Nicholls, C.I., Henao, A., Lana, M.A., 2015. Agroecology and the design of climate change-resilient farming systems. *Agronomy for Sustainable Development* **35**, 869–890.

Andow, D.A., Hidaka, K., 1989. Experimental natural history of sustainable agriculture: syndromes of production. *Agriculture, Ecosystems and Environment* **27**, 447–462.

Chaboussou, F., 2004. *Healthy Crops: A New Agricultural Revolution*. Jon Carpenter Publishing, Oxford, England, 258 p.

Gliessman, S.R., 1998. *Agroecology: Ecological Process in Sustainable Agriculture*. Ann Arbor Press, Michigan, USA, 356 p.

Gliessman, S.R., 2010. *Agroecology: The Ecology of Sustainable Food Systems*, 2nd edn. CRC Press, Boca Raton, FL, USA, 388 p.

Hsu, Y.T., Shen, T.C., Hwang, S.Y., 2009. Soil fertility management and pest responses: A comparison of organic and synthetic fertilization. *Journal of Economic Entomology* **102**, 160–169.

Kremen, C., Miles, A., 2012. Ecosystem services in biologically diversified versus conventional farming systems: benefits, externalities, and trade-offs. *Ecology and Society* **17** (4), 1–40.

Lamine, C., Bellon, S., 2009. Conversion to organic farming: A multidimensional research object at the crossroads of agricultural and social sciences. A review. *Agronomy for Sustainable Development* **29**, 97–112.

Lotter, D.W., 2003. Organic agriculture. *Journal of Sustainable Agriculture* **21**, 37–51.

Malezieux, E., 2012. Designing cropping systems from nature. *Agronomy for Sustainable Development* **32**, 15–29.

MacRae, R.J., Hill, S.B., Mehuys, F.R., Henning, J., 1990. Farm scale agronomic and economic conversion from conventional to sustainable agriculture. *Advances in Agronomy* **43**, 155–198.

Moonen, A.C., Barberi, P., 2008. Functional biodiversity: An agroecosystem approach Agriculture, *Ecosystems and Environment* **127**, 7–21.

Patriquin, D.G., Baines, D., Lewis, J., MacDougall, A., 1988. Aphid infestation of faba beans on an organic farm in relation to weeds, intercrops and added nitrogen. *Agriculture, Ecosystems and Environment* **20**, 279–288.

Phelan, P.L., 2009. Ecology-based agriculture and the next green revolution: Is modern agriculture exempt from the laws of ecology? In: Bohlen, P.J.,

House, G. (Eds.), *Sustainable Agroecosystem Management: Integrating Ecology, Economics, and Society*. CRC Press, Boca Raton, FL, pp. 97–135.

Phelan, P.L., Mason, J.F., Stinner, B.R., 1995. Soil-fertility management and host preference by European corn borer, *Ostrinia nubilalis* (Hübner), on *Zea mays* L.: A comparison of organic and conventional chemical farming. *Agriculture, Ecosystems and Environment* **56**, 1–8.

Pimentel, D., Hepperly, P., Hanson, J., Douds, D., Seidel, R., 2005. Environmental, energetic, and economic comparisons of organic and conventional farming systems. *Bioscience* **55**, 573–582.

Ponisio, L.C., M'Gonigle, L.K., Mace, K.C., Palomino, J., de Valpine, P., Kremen, C., 2015. Diversification practices reduce organic to conventional yield gap. *Proceedings of the Royal Society B: Biological Sciences* **282**, 20141396. http://dx.doi.org/10.1098/rspb.2014.1396

Power, A.G., Flecker, A.S., 1996. The role of biodiversity in tropical managed ecosystems. In: Orians, G.H., Dirzo, R., Cushman, J.H. (Eds.), *Biodiversity and Ecosystem Processes in Tropical Forests*. Springer Verlag, New York, USA, pp. 173–194.

Reijntjes, C., Haverkort, B., Waters-Bayer, A., 1992. *Farming for the Future. An Introduction to Low-external Input and Sustainable Agriculture*. MacMillan Press, London.

Rosset, P.M., Altieri, M.A., 1996. Agroecology versus input substitution: a fundamental contradiction of sustainable agriculture. *Society and Natural Resources* **10**, 283–295.

Thiessen Martens, J.R., Entz, M.H., Wonneck, M.D., 2015. Redesigning Canadian prairie cropping systems for profitability, sustainability, and resilience. *Canadian Journal of Plant Science* **95**, 1049–1072.

van Emden, H.F., Williams, G.F., 1974. Insect stability and diversity in agroecosystems. *Annual Review of Entomology* **19**, 455–475.

Vazquez, L.L., Matienzo Brito, Y., Simonetti, J.A., Veitia Rubio, M., Paredes, E.R., Fernandez, E.G., 2012. Contribucion al disneo agroecologico de sistemas de produccion urbanos y suburbanos para favorecer procesos ecologicos. *Agricultura Oragnica* (Cuba) **18**, 14–19.

Zhao, Z.H., Hui, C., He, D.H., Li, B.L., 2015. Effects of agricultural intensification on ability of natural enemies to control aphids. *Scientific Reports* **5**, 8024.

Chapter 2

Agroecology and Agroecological Cropping Practices

Alexander Wezel* and Erin Silva†

**Department of Agroecology and Environment, ISARA, Lyon, France*

† Department of Plant Pathology, University of Wisconsin, Madison, Wisconsin, USA

1. Introduction

World agriculture imminently faces the major challenge of providing sufficient food for a growing global population, forecasted to increase to 9.1 billion in 2050 (United Nations 2015). Without accounting for overproduction, food waste, and distribution issues symptomatic of supply chain problems throughout industrialised and non-industrialised nations, food production at a global scale must minimally increase 30% over current levels. Moreover, in tandem with the call for higher yields and greater production output, increasing demands for higher quality, nutritionally-dense foods and environmentally, socially, and economically sustainable agricultural practices have also emerged. The effort required to increase food production to meet the needs of an additional 2 billion people is daunting, and the question of how our global agricultural systems can effectively respond to this challenge remains without a clear, uniform answer. Hence, agricultural food production practices must be revisioned and redefined to successfully achieve these goals across diverse agricultural approaches, be it conventional, integrated or organic agriculture.

Opinions and proposed pathways leading to the creation of higher yielding food systems with positive sustainability outcomes can be quite different among stakeholders. Currently, no common agreement exists as to the most appropriate agricultural production practices that should be adopted by the world's farmers. Production options range from practices that are highly technology-based to those based on the principles of ecology. Some argue that precision farming (Srinivasan 2006) or use of genetically modified crops (e.g. Huang *et al.* 2002, Prasifka *et al.* 2009, National Academy of Science 2010, see also discussion in Chapter 6 of this book) could help agricultural production levels match future food demand. Others favour greater integration of ecologically-based practices such as natural biological control of pests and the integration of natural landscape elements into agricultural approaches in order to decrease reliance of pesticides (e.g. Altieri and Nicholls 2004, Gurr *et al.* 2004), or no or reduced tillage practices which increase soil biological activity and nutrient availability (Peigné *et al.* 2007, Vian *et al.* 2009).

The concepts of agroecology, offer great potential to identify and design agricultural practices which contribute to the development and adoption of solutions to the environmental, social, and economic challenges facing our agricultural systems. Strategies including organic crop fertilisation, diversified crop rotations, and biological pest control are well-known agricultural practices that have been historically used by farmers across the globe. However, with the widespread adoption of industrialised agricultural models throughout the last several decades, these once-common practices have been increasingly described as agroecological practices (e.g. Altieri 1995, Arrignon 1987). Although the term "agroecological practices" emerged in the 1980s within the development of the field of agroecology (Wezel *et al.* 2009), we still lack definition as to which specific practices qualify as agroecological practices, and which incorporate characteristics that identify them as such (Wezel *et al.* 2014a).

This chapter summarises the concepts of agroecology and its diverse interpretations. Secondly, it provides definitions and overviews of different agroecological cropping practices. The subsequent chapters of this book will analyse, illustrate and discuss more specifically these practices, including their development and implementation, as well as teaching strategies for the wider dissemination of these practices to students, practitioners, and farmers.

2. Interpretations and concepts of agroecology

Since the first use of the term "agroecology" in the early 20th century (Wezel and Soldat 2009), its meanings, definitions, interpretations, and approaches have changed enormously up to the present. During the 1930–1960s, agroecology experienced slow development as a scientific discipline. From the 1970s onwards, interpretations of agroecology expanded and diversified. Beyond its association as scientific discipline, agroecology as a movement gradually emerged in the 1970s and as a set of practices beginning in the 1980s (Figure 1).

In general, the basis for the agroecological movements was set within the environmental movements of the 1960s and 1970s, which often emerged as a consequence of the unintended impacts of industrialised agriculture after the Green Revolution. Agricultural advances during this period often focused on improving yields and economic returns in the short-term, but rarely accounted for longer-term environmental and social factors. Public policies seldom considered the environmental impacts of agriculture, nor the social consequences of unidimensional rural

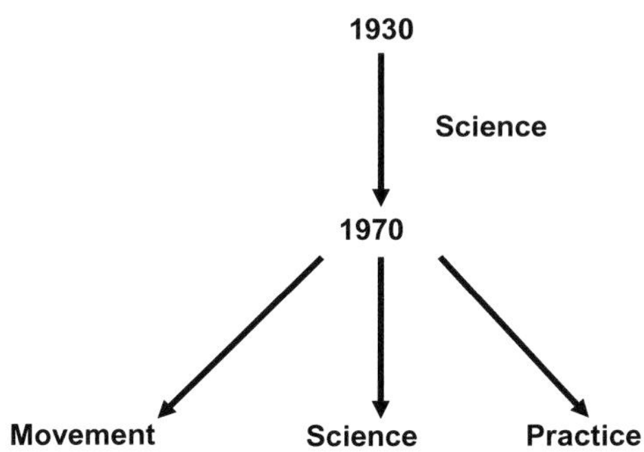

Figure 1. Major interpretations of agroecology from a historical perspective.

development focused solely on yields and economics. However, with increasingly visible and recognisable negative consequences of agricultural chemical use, environmentalists emphasised the adverse impacts of these toxic substances, in particular pesticides, on our natural resources, landscapes, and wildlife. While these nascent efforts to address the negative consequences of an increasingly industrialised agricultural model and adopt more environmentally-sound approaches were pushed forward by environmentalists, the movement throughout the 1960s to the 1980s was not defined or described by the term "agroecology".

In the 1990s, agroecology continued to become more associated with specific agricultural and social movements, especially in Latin America and in the USA, where the term was used to express a new way of considering agriculture and its relationships with society. Also during this time period, a third usage of the term agroecology emerged beyond that of a science and movement — it also began to describe a set of agricultural practices aiming to develop more environment-friendly or sustainable forms of agriculture. The foundations for this shift were established in the middle to late 1970s, as Latin American ecologists, agronomists, and ethnobotanists working in Mexico and Central America used agroecology to provide underlying principles for sustainable agricultural development (Hernandez Xolocotzi 1977). Agroecological concepts assisted local farmers in improving their indigenous farming practices without the adoption of the high input, chemical-intensive agriculture, promoted by international corporations (see Altieri 1989, 1995, Gliessman 2007). Techniques such as adoption of regenerative soil fertility strategies, conservation of natural resources, and enhancement of agrobiodiversity served as practical bases upon which to support burgeoning agroecological movements in Latin America. These efforts included specific examples illustrating more adaptive and technical methods in agriculture, such as water and livestock management or anti-erosion measures as a basis for rural and sustainable development in arid and subhumid areas (Arrignon 1987).

These three main interpretations of agroecology — as a movement, as a science and as a practice — can be further refined in relation to (i) definitions for the scientific discipline, (ii) major objectives of the movements, and (iii) different scales of implementation (Figure 2). For example, the scale of the agroecosystem is pertinent in the determination

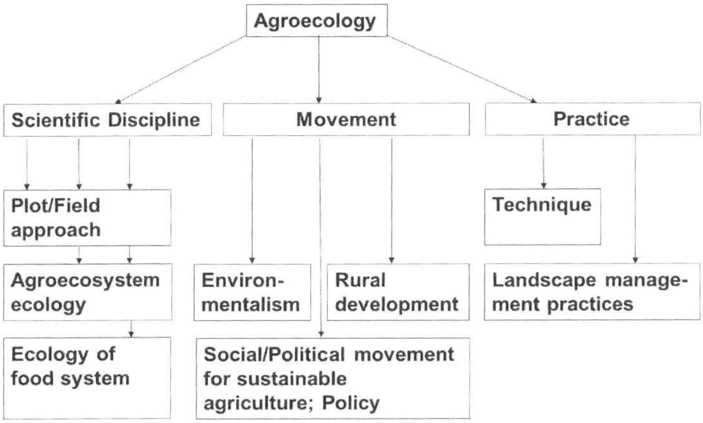

Figure 2. Diversity of current meanings of agroecology and their relation to scale (adapted from Wezel *et al.* 2009). Policy and landscape management practices are new compared to the meanings presented in 2009.

of the approach of the scientific discipline and environmental and rural development movements, as well as for certain practices at landscape scale. In recent years, new aspects of policy development and landscape management practices have also emerged in the expansion of the interpretations of agroecology both as a movement and a practice.

2.1. *Country-specific predominance of agroecology*

Interpretations of agroecology can differ depending on national and regional contexts, often due to the historical evolution of the agroecology framework within them. In many countries, the integration of agroecology into policy and agricultural strategies involves combining aspects of the field as a movement, as a science and as a practice to varying degrees, sometimes very strongly. In Germany, for example, agroecology has a long tradition as a scientific discipline, without association with a particular movement or set of practices (Wezel *et al.* 2009). In contrast, in the USA and Brazil, agroecology is used to describe all three activities, with predominance toward science in the USA and with a stronger movement and/or practice emphasis in Brazil. In France, until recently, agroecology was mainly known as a practice, with increasing recognition over the past several years as a scientific discipline and a movement. In countries where agroecological movements are well established, the concept of practice is strongly connected to, or even incorporated within, these movements. As practice merges with movement, the broader objectives of agroecological movements are further developed, facilitating transitions at both the farm and food system scales to create sustainable agroecosystems (e.g. Gliessman 2007), including through synergies with existing models such as traditional, alternative, or organic farming.

As the effective transition to more agroecological models rests upon farmers adopting alternative practices, agroecology aims to enhance adoption through encouraging farmers and extension personnel to participate in the design of these new systems. With their involvement in system design and practice, their contributions into the related social and political movements are also strengthened. This is particularly the case for Brazil, and to a certain extent for the USA and France. In these situations, links can be often be traced between knowledge generation (the science), technological applications (the practice), and political vision (the movement). More recently, connections between practice and policy are becoming even stronger, as the first cases of agroecological policy have been established by governments. For example, France has become the first nation in Europe to establish an explicit national agroecological strategy (i.e. "Agroecological project for France" in December 2012; *Ministère de l'Agriculture, de l'Agroalimentaire et de la Forêt* 2013).

2.2. *Current variations in definitions and scales in agroecology as a science*

The analysis of agroecology's historical evolution demonstrates a transition from addressing the plot and field scales (1930–1960s) to the farm and agroecosystem scales (1970–2000s), although smaller scale approaches are also still implemented today (Wezel and Soldat 2009). At present, the definitions of agroecology provided by Francis *et al.* (2003) and Gliessman (2007) move beyond the context of the more concrete spatial scale of the field and farm/agroecosystem with relatively defined borders and enter the wider dimensions of the food system.

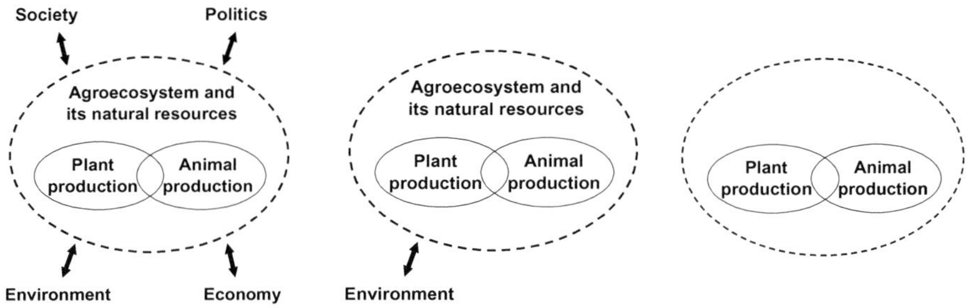

Figure 3. The different definitions and views of agroecology in current research (left: food systems approach, middle: agroecosystem approach, right: plot or field approach) (from Wezel and Soldat 2009).

At the smaller scales, agroecological research at the plot/field/animal level aims to develop new farming practices which provide healthier and more robust systems, e.g. those that foster more efficient use of natural resources, improve nutrient cycling, and increase on-farm diversity, including that of soil organisms, crops, and livestock (Wezel *et al.* 2015). At these more discrete scales, however, research rarely considers the interactions and implications associated with the implementation of these techniques at the agroecosystem or larger landscape scale, including those on the broader food system (Figure 3).

Moving beyond the more discrete scales described above, agroecology shifts into a second agroecosystem approach, widening its lens beyond the fence lines of the farm field and capturing a more extensive view. Here, ongoing research focused on the agroecosystem scale involves exchanges and interactions between the field and farm ecosystem with the surrounding environment. Typical aspects considered within the research and design of agroecological practices at this scale include impacts of farming systems on plant and animal communities, food web interactions for biological control, biodiversity and nature conservation, and drinking water quality and land degradation in agricultural landscapes and agroecosystems. Within this agroecosystem approach, definitions and concepts might vary depending on the delimitation of an agroecosystem. For some researchers, the farm serves as a unique and self-contained agroecosystem, where relationships between farmers' practices and natural resources are analysed. Others view the agroecosystem as a larger expanse, that is, a local or regional landscape where interactions between different types of agriculture and the natural resources of the landscape are investigated. A more recent concept for this agroecosystems scale approach has been developed with "Agroecology-Territories" by Wezel *et al.* (2016).

The food systems approach has emerged more recently as the broadest approach with which to consider agroecological design and study. First described by Francis *et al.* (2003), from this perspective agroecology can be defined as "the integrative study of the ecology of the entire food systems, encompassing ecological, economic, and social dimensions". Gliessman (2007) further defined agroecology as "the science of applying ecological concepts and principles to the design and management of sustainable food systems". Within the food systems approach, concrete spatial scales are abandoned, instead embracing the wider dimensions of the food system. In addition to addressing a range of scales including local, regional, national, and global, this approach also addresses agroecological aspects for which discrete borders may be impossible to delimit, including impacts of food production systems, culture and society, and the economics and politics, all of which are connected and interwoven in different ways. Today, agroecological literature addresses a broad diversity of food system topics, such as issues related to food sovereignty, alternative and local food networks, social agricultural networks, food crisis, food security, right to food, agri-food systems, and food markets (Wezel *et al.* 2015). Illustrations about the food systems approach can be also found in Wezel and David (2012) and Wezel *et al.* (2015).

3. Agroecological cropping practices

Agroecological practices can be characterised as agricultural practices aiming to produce significant amounts of food while valorising ecological processes and ecosystem services by integrating them as fundamental elements in the development of the said practices, as opposed to simply relying on external inputs such as chemical fertiliser and synthetic pesticide applications, or on technological solutions such as genetically modified organisms (Wezel *et al.* 2014a). This assumes that biological processes are able to replace chemical or physical inputs (or, at minimum, to interact favourably with them) while limiting external costs, particularly environmental costs. Based on processes that decrease external inputs and negative environmental consequences, such as nutrient cycling, biological Nitrogen-fixation, natural regulation of pests and diseases, soil and water conservation, biodiversity conservation, and carbon sequestration, agroecological practices contribute to improving the sustainability of agroecosystems. Some of these practices have already been applied in varying degrees in different parts of the world for decades, while others have been more recently developed with a limited rate of adoption.

As described above, the term "agroecological practices" emerged in the 1980s, within the development of agroecology as both a movement and a science (Wezel *et al.* 2009). As a result, throughout the last two decades, different agricultural practices have been increasingly described as "agroecological practices" (e.g. Altieri 1995, Arrignon 1987). Diverse examples of agroecological practices are found in literature, including cover crops, green manures, intercropping, agroforestry, biological control, resource and biodiversity conservation practices, or livestock integration (Arrignon 1987, Altieri 1995, Gliessman 1997, Uphoff 2002). A recent review evaluated different cropping practices under temperate climates, including the specific characteristics that identify them as agroecological practices (Wezel *et al.* 2014a). Whereas agroecological practices for cropping systems have been described in various publications, a marked dearth in literature related to agroecological practices in livestock systems still remains. It is only recently that Dumont *et al.* (2013), Bonaudo *et al.* (2014), and Wezel and Peeters (2014) defined and described agroecological practices for livestock husbandry. Also, Chapters 10–12 in this book will deal with different livestock systems and agroecological practices.

The remainder of the chapter provides an overview of agroecological practices. First, a broader discussion will be presented related to agroecological cropping systems approaches under temperate climates, followed by a short description of the specific practices. Many of these practices, in addition to others not mentioned, are also used in tropical climates, but are not presented here. More detailed information about advantages and constraints of the different practices can be found in Wezel *et al.* (2014a) and in different chapters of this book. Also, illustrated examples for agroecological cropping practices from France and other countries in Europe are provided in Wezel *et al.* (2014b).

In Table 1, different agroecological cropping practices are grouped into four categories, depending on their impact on the system: (1) efficiency increase, substitution practices (E and S), and (2) redesign and diversification practices (R and D). Efficiency increase refers to practices that reduce input consumption (e.g. water, pesticides, and fertilisers) and improve crop productivity. Substitution practices refer to the substitution of an input or a practice (e.g. replacing chemical pesticides by natural pesticides). Finally, redesign refers to the change of the whole cropping, or even farming system, and diversification refers to practices which integrate a higher diversity of cultivars, crops, or production systems.

Diversification plays a critical role in agroecology, as primary goals of agroecosystem design include increased resilience (under environmental, social, and economic variability), increased conservation of (agro) biodiversity, and increased capacity for farmers to vary their production strategies and approaches, each of which is profoundly affected by the level of diversification in the system. Many agroecological practices promote, to varying degrees, system diversification (Figure 4). This ranges from integrating more diversity at the field and farm levels (e.g. cultivar selection, crop association, crop rotation strategy within and across years, and production system), or even to integrating biodiversity in the surrounding area (semi-natural landscape elements). The overview provided in Figure 4, also distinguishes the scale of application.

3.1. *Cultivar mixing, and use of site-adapted cultivars and crops*

Increasing plant diversity throughout agricultural fields can improve productivity through the reduction of pest abundance and damage (Tooker and Frank 2012). In most cropped areas characteristic of industrialised agriculture, hectares of land are not only managed as monocultures, but also planted in crop varieties characterised by a narrow germplasm base, resulting in genetically similar neighbours (Goodman 2005, Bonato *et al.* 2006). Greater genetic uniformity in crops can increase pest and disease vulnerability and the risk of crop loss (National

Table 1. Different agricultural practice categories and their respective agroecological practices of agricultural systems under temperate climates (adapted from Wezel *et al.* 2014b).

Agricultural practice category	Practice impact	Description of practice
Efficiency increase and substitution practices		
Cultivar and crop choice	E,S	Use of site-adapted cultivars and crops
	E,D	Cultivar mixing
Crop fertilisation management	E	Split fertilisation
	S	Biofertiliser
	E,S,R	Organic fertilisation
Weed, pest, and disease management	S	Natural pesticides
	S	Biological pest control
Water management	E	Drip irrigation
	R,D	Soil erosion control with integration of landscape elements
Redesign and diversification: practices of cropping systems		
Tillage management	E,S,R	Direct seeding into living cover crops or mulch
	E,S,R	Reduced tillage
Soil covering	S,R,D	Cover crops
Crop rotations	S,R,D	Diversified crop rotations
Crop associations	E,S,R,D	Intercropping
	E,S,R,D	Relay intercropping
Weed, pest, and disease management	S,R,D	Allelopathic plants
	S,R,D	Push and pull systems
Redesign and diversification: integration of production systems		
Agroforestry	E,S,R,D	Integration of timber, fruit or nut trees on cropped fields
Management of landscape elements at field, farm, or landscape scale	S,R,D	Integration of semi-natural landscape elements for biological control
	S,R,D	Integration of semi-natural landscape elements for pollination
	R,D	Integration of semi-natural landscape elements for erosion control

Note: E = efficiency increase, S = substitution, R = redesign, D = diversification. Note that one practice can correspond to one or more categories of the ESRD framework.

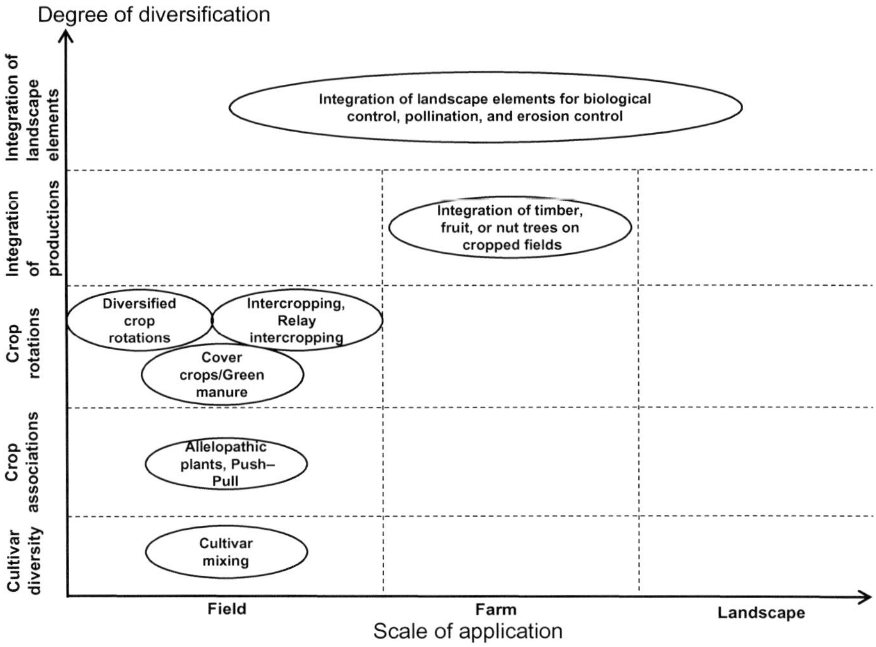

Figure 4. Agroecological practices which integrate a higher diversity of cultivars, crops, production systems, and landscape elements. Practices are placed along a diversification gradient from intraspecies diversity via crop and production systems diversity to integration of landscape elements.

Research Council 1993). To increase plant genotypic diversity at field scale, cultivar or crop mixtures can be planted. Moreover, selecting cultivars which are adapted to specific local cropping systems and regions can help to improve crop resistance to abiotic stresses (N and water deficiency), pathogens, and diseases (Tilman *et al.* 2002). Additionally, crop species or cultivars can be selected which favour symbiotic interactions with beneficial soil microorganisms, such as arbuscular mycorrhizal fungi (AMF) and plant growth promoting rhizobacteria (PGPR), which stimulate plant growth via different mechanisms such as increased nutrient acquisition, protection against pathogens, and modulation of phytohormone synthesis (Garcia de Salamone *et al.* 1996, Kaeppler *et al.* 2000, An *et al.* 2010).

3.2. *Split fertilisation, organic fertilisation management, and biofertiliser*

Split fertiliser applications can improve nutrient use efficiency and minimise nutrient losses in agricultural crop production (Randall and Sawyer 2008). By synchronising the supply of nutrients at specific times to meet the crop demand, nutrient leaching and run-off can be minimised, reducing ground and surface water contamination by fertilisers (Campbell *et al.* 1995, Robertson 1997). However, this technique requires increased labour, technology, and modelling capabilities. While the accurate estimation of crop nitrogen demand and its subsequent delivery can be difficult, precision farming, including the use of sensor and GPS technologies, could improve the application of fertilisers by varying rates and mixtures as needed with reference to inter and intrafield variability in crops.

Organic manures and compost have been shown to increase soil biological activity (Birkhofer *et al.* 2008, Steenwerth and Belina 2008), enhancing the mineralisation of organic carbon and nitrogen. However, the application of these bulky amendments may lead to higher labour and transport costs. Additionally, as nutrient availability from organic fertility sources is highly dependent on soil biology and environmental conditions, the optimisation of plant-available nitrogen to synchronise with crop demand can be difficult to achieve (Sanchez *et al.* 2004).

Biofertiliser products consist of free-living microorganisms that are applied to seed, plant surfaces, or soils. The microorganisms colonise the rhizosphere or the interior of the plant, and thus increase the supply or availability of nutrients to the host plant (Vessey 2003). Three major groups of microorganisms that serve as active agents in these products include: AMF, PGPR, and nitrogen fixing rhizobia (Malusá *et al.* 2012). This technology still needs further improvement, including research to more clearly understand the functioning of the soil-plant-microorganism system to allow for greater efficacy (Malusá *et al.* 2012).

3.3. *Biological pest control and natural pesticides*

Agroecological practices, including biological pest control, natural pesticides, allelopathic plants, and push-pull approaches, can positively contribute to the management of weeds, pests, and diseases in agricultural systems. Integration of these practices into farming strategies can be limited by their effectiveness on target pest species, the costs associated with implementation and management, and the availability of technical knowledge and guidance. Biological pest control includes both the release of natural enemies of pests (i.e. predators, parasitoids or pathogens) into the agroecosystems and augmentation of native populations of natural enemies. Furthermore, biological control includes use of pheromones to disrupt mating cycles and reduce sexual reproduction of targeted insect pests.

Natural pesticides, often also called botanical pesticides, botanicals, or biopesticides, have been developed to substitute for synthetic pesticides. Botanical pesticides have arisen from several naturally-occurring sources, including the seeds of trees, plant essential oils, floral extracts (e.g. pyrethrum), and extracts of herbaceous and woody plants (Coulibaly *et al.* 2002, Charleston *et al.* 2005, Isman 2006, 2008, Batish *et al.* 2008, Regnault-Roger and Philogène 2008). Biopesticides also include symbiotic and beneficial bacteria, AMF inoculants, and other fungi that can inhibit the growth of deleterious organisms (Whipps 2001, Vessey 2003). These types of pesticides impact pests by antibiosis, competition, induction of plant resistance mechanisms, inactivation of pathogen germination, and/or degradation of the toxicity of the pathogens (Whipps 2001). Greater integration of natural pesticides into integrated pest management programmes remains relatively limited, in part due to the lack of efficacy trialling for these compounds, including their effectiveness in larger-scale applications (Marrone 2007). Natural pesticide options also remain limited, in part due to variable efficacy of pest control, availability, national regulations and registration processes, and costs (Isman 2008).

3.4. *Drip irrigation*

While not specifically valorising, ecological processes and ecosystem services, drip irrigation can also be considered an agroecological practice, as it greatly decreases the reliance on increasingly scarce water resources in agricultural production. Especially in horticultural systems, drip irrigation has a high potential to reduce water inputs, improve water use efficiency, and more effectively meet crop water demand in time and space. With precision application reducing the overall amount of water applied, drip irrigation also reduces risk of soil salinisation. The major constraints to wider adoption of this practice include high initial capital investment and ongoing management costs. The agroecological benefits of drip irrigation can be further enhanced with the integration of cover crops; rows of cover crops are planted between the irrigated cash crop rows, providing further benefits of reducing evaporation of soil moisture, decreasing soil erosion, increasing soil organic matter, and contributing to soil nitrogen when legumes are included as a cover crop species (Lopes *et al.* 2011).

3.5. *Reduced tillage and direct sowing into living cover crops or mulch*

Reduced tillage is characterised by minimal soil disturbance without soil inversion. Typically, using these techniques, soil is worked to a shallow depth of 5–15 cm prior to seeding; the reduction in soil disturbance contributes to the preservation of soil organic matter at the surface and within the first few centimetres of the soil, enhancing soil biodiversity and promoting biological activity (Ball *et al.* 1998, Peigné *et al.* 2009, Vian *et al.* 2009). Minimal and reduced tillage techniques are often achieved through strategic use of field equipment, implements, and planters, including chisels, sweeps, field cultivators, and drills. In reduced tillage systems using greater integration of ecological weed management practices and decreased reliance on chemical methods, effective weed control remains a major constraint (Brainard *et al.* 2013). In organic farming, the adoption of reduced tillage strategies often result in increased need for in-season cultivation for weed control, thus increasing machine traffic, labour time and energy costs (Peigné *et al.* 2007). Additionally, under cold-temperate climates and certain soil conditions, the lack of deeper tillage can result in increased soil compaction (Soane *et al.* 2012).

Increasingly, reduced tillage strategies have been developed that involve the direct seeding of cash crops into living or killed cover crop mulches (Mirsky *et al.* 2012). Although tillage is used prior to the planting of the cover crop, with the adoption of these systems, tillage and cultivation passes are greatly reduced or eliminated during cash crop production. Reduction of tillage is achieved, in large part, due to the high degree of weed suppression achieved by cover crop, which provides a physical barrier preventing weed germination and emergence. In organic production, rolled-crimped cover crop systems have been developed to reduce reliance on cultivation for weed management and limit soil disturbance during the cash crop season. Typically, these systems involve the establishment of a fall-sown cover crop that is terminated in the spring with a roller-crimper, a ground-driven implement which creates a thick cover crop mulch (Figure 5(a)). Just prior to or shortly after termination, the cash crop is directly sown into the cover crop biomass, with no further weed management practices required during the production season (Silva 2014) (Figure 5(b)).

3.6. *Cover crops*

Cover cropping is a widely applied agroecological practice, enhancing the accumulation of soil C, reducing risks of soil erosion and nitrate leaching, and improving the retention and availability of both water and nutrients (Sanchez *et al.* 2004, Scholberg *et al.* 2010). The integration of cover cropping into farming system approaches also improves soil health indicators, such as soil biological activity and soil structure (Blanco-Canqui *et al.* 2011). If legumes are included as a cover crop, either singly or within a cover crop mixture, nitrogen can be made available through plant associations with symbiotic bacteria capable of fixing atmospheric nitrogen and supplied to the subsequent crop, limiting dependence on external fertiliser inputs (Birkhofer *et al.* 2008, Steenwerth and Belina 2008, Fustec *et al.* 2010). Integration of cover crops into crop rotations can serve as a tool to enhance diversity, capturing the associated benefits of more complex crop rotations. Increasingly, farmers are using cover crop mixtures, or "cocktails", allowing for the provision of multiple ecosystem services within a given cover crop phase (Tribouillois *et al.* 2016) (Figure 6). Despite demonstrated benefits, cover crop adoption remains hindered by production constraints, including higher farm labour demand, risk of negative insect and disease interactions (e.g. slugs under cover crops), and uncertain economic return (Moore *et al.* 2016). Research is

(a) (b)

Figure 5. (a) Roller-crimper implement used to terminate a fall-sown cereal rye (*Secale cereale*) cover crop during the subsequent spring, immediately followed by direct sowing of soybean (*Glycine max*). (b) The soybean crop will emerge through the cover crop residue, which provides season-long weed suppression, allowing for the elimination of tillage and cultivation during the soybean production season. (Photos E. Silva.)

Figure 6. A summer cover crop mixture of forage radish (*Raphanus sativus*), sunnhemp (*Crotalaria juncea*), and sorghum–sudangrass (*Sorghum* × *drummondii*). The mixture provides multiple ecosystem services, including "biodrilling" from the radish, nitrogen fixation from the sunnhemp, and weed suppression from the sorghum–sudangrass. (Photo E. Silva.)

continuing to more clearly document positive benefits of cover crops, including yield increases for the subsequent cash crops, reduction in weed populations, and suppression of soil diseases, insects, and nematodes (Blanco-Canqui *et al.* 2015).

3.7. *Diversified crop rotations*

The diversification of crop rotations provides multiple agroecosystem benefits. Improvement of soil condition remains a major service provided by crop rotational diversity, including positive impacts on soil fertility, optimisation of nutrient availability, increase in organic matter content, and improvements in soil structure and aggregate stability (Watson *et al.* 2002, Dogliotti *et al.* 2004). If leguminous plants are integrated into the rotation, either as a cash crop or cover crop, atmospheric nitrogen can be fixed, further benefitting soil fertility and providing an important source of nitrogen for subsequent crops. With improvements in soil fertility and nutrient holding capacity through the increase in soil organic matter, as well as the integration of cash crops varying in the intensity of their nitrogen requirements, the need for external fertiliser is reduced. Crop diversification also serves to fulfil objectives of soil protection and conservation; through the integration of winter cash crops or cover crops planted in fallow periods during the fall or spring, year-long ground cover is provided. Not only do these practices stabilise soil and minimise erosion, they also contribute to retention and sequestration of otherwise leached nutrients, preventing contamination of surface and ground water and improving nutrient availability. Diversified crop rotations also serve a third objective of reducing pest and disease occurrence through disruption of pest life cycles, through avoiding successive host crops in the case of insects and diseases and altering the timing of soil disturbance, planting, and crop growth cycles in the case of weed infestations.

Illustrated below are examples of crop rotations observed under conventional and organic agriculture in south-eastern France. Generally, crop rotations under non-irrigated conventional agriculture are characterised by short rotations of cereals, oilseed crops, and spring crops (Figure 7) (Agreste 2010, 2012, 2013). If irrigation is available, conventional systems are primarily based on maize monoculture, as maize is considered the most profitable crop under the given climate and soil conditions. This monoculture system contributes to greater weed issues (e.g. *Sorghum halepense*), pest populations (e.g. corn rootworm), and soil degradation (soil crusting), all impairing crop performance. To mitigate these phenomena, the monoculture is interrupted every 3–6 years with 1 year of winter cereal.

Figure 8 illustrates examples of diversified crop rotation under organic farming located in south-eastern France. When irrigation is not available, a typical rotation of 12–14 years in pre-mountainous areas is lucerne–winter wheat–secondary cereals–grass clover mixture–winter wheat. A second type of non-irrigated rotation of 8–10 years is rather found in plain areas and includes lucerne followed by a combination of maize–soybean–winter wheat. Winter cereal crops generally dominate within the different rotations, particularly winter wheat. This system is generally less dependent on inputs and valorises forage legumes as green manures and cover crops during intercrop periods to control weed infestations and to reduce mechanical weeding. It also integrates, in some cases, temporary grasslands, particularly in zones where livestock is still present and where forage could be used to feed animals.

If access to irrigation is provided, typical cropping strategies are based on shorter rotations of 3–4 years, with soybean–winter wheat–maize (which is sometimes repeated a second year) (Figure 6). In such a rotation, winter wheat benefits from the nitrogen contribution of the previous leguminous crop. Spring and winter crops, and irrigated and non-irrigated crops, are grown alternately. However, these more intensive crop rotations are very dependent on inputs. While the availability of irrigation provides a high degree of efficiency, the intensity of the rotation that irrigation allows also creates a high demand in nitrogen for wheat and maize, along with challenges in managing weed infestations. The latter is partly managed with soil tillage and mechanical weeding. Such practices may maintain crop productivity, but can also degrade soil condition. Consequently, organic farmers have generally decided to sow a temporary grassland with lucerne or clover after 9–12 years to regenerate soil fertility

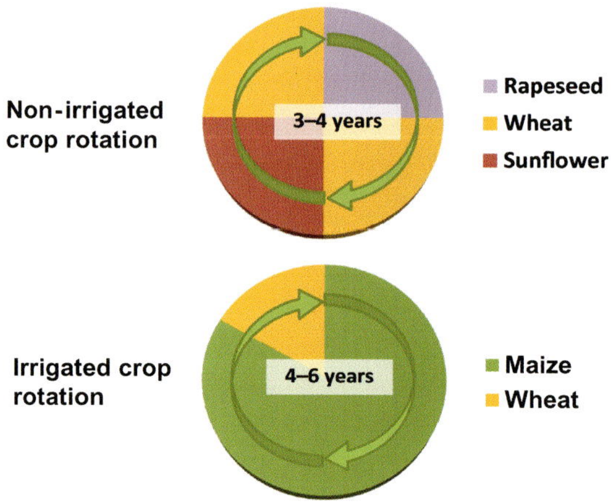

Figure 7. Typical crop rotations observed in conventional cereal farms in south-eastern France. With access to irrigation, the spring crop maize dominates the rotation. Without irrigation, crop rotations are shorter but more diversified.

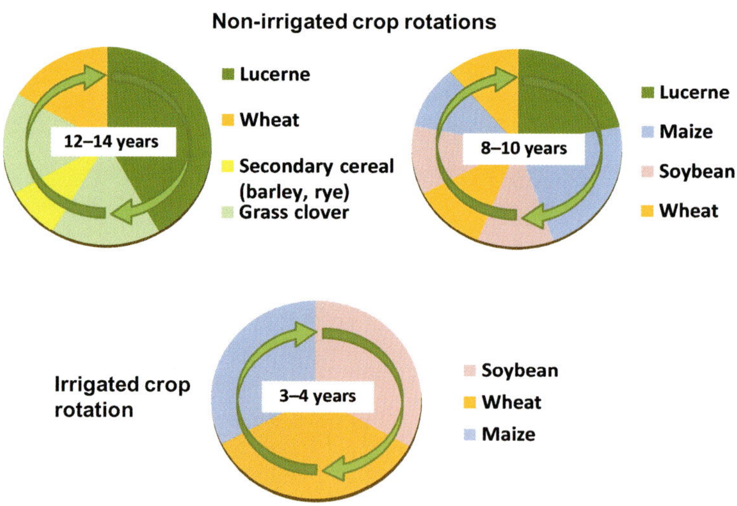

Figure 8. Typical crop rotations observed in organic cereal farms in south-eastern France. With irrigation, rotations remain short and are only based on three crops. Longer and more diversified rotations are found when irrigation is not available.

and control weeds. Some farmers diversify the basic crop rotation with sunflower, rye or durum wheat, but these crops are considered less profitable or more difficult to produce because of climatic conditions or limited nitrogen fertilisation.

3.8. *Intercropping and relay intercropping*

Intercropping systems are characterised by the simultaneous association and cultivation of two or more crops, which can be arranged spatially in different ways. The intensity and type of interactions will depend on the chosen arrangement and associated species (Malézieux *et al.* 2009). A basic row system involves intercropping two or more crops in rows or strips (Figure 9). Sowing of crops does not necessarily occur at the same time. Harvest times may also differ, but crops are

Figure 9. An example of an intercropping system in North America. Winter wheat (*Triticum aestivum*) is sown in late fall, with soybeans (*Glycine max*) planted directly into the established wheat crop during the following spring. The winter wheat is harvested over the growing soybean crop, allowing the soybean to continue to mature. (Photo E. Silva.)

usually grown simultaneously for a significant part of cropping season. Relay intercropping differs from the basic row system, as two or more crops are grown together only for part of their life cycles, thus limiting interactions between species (Vandermeer 1989). Other categories of intercropping may include agroforestry (see following text), with plant associations partially comprised of perennial species. Intercropping systems hold many potential advantages, including greater land productivity, resource use efficiency, mitigation of weed infestations, improvement of soil physical structure and soil fertility, stability of outputs, and resilience to disturbance (Vandermeer 1989, Hauggaard-Nielsen *et al.* 2006, Carof *et al.* 2007). However, as with many agroecological practices, they can be more difficult to manage, as competition for light, water and nutrient resources can occur between the associated crops (Willey 1990, Van Noordwijk *et al.* 1996).

3.9. *Allelopathic plants, and push and pull systems*

Allelopathic plants produce chemical compounds which can negatively impact the growth and development of weeds, pests, or diseases (Weston 1996, Tabaglio *et al.* 2008). Pest and weed suppressive effects can be achieved through their use either as a cash crop (e.g. rapeseed to suppress soil disease organisms) or as a cover crop or intercrop (e.g. cereal rye's allelopathic effects on germinating weed seeds). Allelopathic plants can also have repellent or attractive effects against certain agricultural insect pests. This phenomenon is a key element in the development of "push–pull" management strategies, which are based on repelling or deterring insect pests from cash crops (push), and then attracting them to trap plants around or embedded within fields to "pull" them away from crops (e.g. Khan and Pickett 2004). In one example of these systems, crops (or cultivars within crop species) which contain high levels of allelopathic chemicals with attraction properties are planted in specific areas of the field to act as a "trap crop" and draw in target pests, allowing for more precise and limited spray applications of chemical or biological insecticides (Hokkanen 1991, Shelton and Badenes-Peres 2006). Presently, the use of allelopathic plants has limited applications, despite the wide range of possible benefits expected from such plants.

3.10. *Integration of timber, fruit, or nut trees on cropped fields*

The integration of different types of trees into cropping systems are known as agroforestry practices. The use of perennial woody crops as a component of annual cropping strategies can reduce nutrient leaching and soil erosion, increase land productivity, increase carbon sequestration, enhance biodiversity, and increase diversity of the production system with complementary production of wood, fruits and nuts for various uses (e.g. Buck *et al.* 1998, Eichhorn *et al.* 2006, Mosquera-Losada *et al.* 2009). Although most agroforestry systems integrate timber trees, more specialised systems exist that incorporate fruit, nut, and forage trees. In some cases, these fruit or nut tree systems are coupled with extensive grazing of grassland below or between the trees. Constraints observed with agroforestry systems include higher management and labour needs, loss of cropped land for the main crop, and competition of trees with adjacent crops for water, nutrients, and light.

3.11. *Integration of semi-natural landscape elements at field, farm, or landscape scale*

Landscape elements integrated into farm system plans can offer habitats and over-wintering sites for beneficial insects or other pest predators, including the provision of resources such as nectar, pollen, and alternative prey (Figure 10). With these food and habitat resources available, populations of these insects are enhanced, reducing the need for pesticide applications. While a diversity of habitat and food resource elements are critical to obtain positive effects on pest control, the percentage of land covered at the field, farm or landscape scale and their spatial arrangement are also important factors in determining their overall impact on the system (Gardiner *et al.* 2009). Integrating a high degree of plant species and floral diversity ensures constant provision of resources such as nectar and pollen, important for the support of different beneficial insects throughout the entire year and during certain periods of their life cycles.

Heterogeneity across landscape elements is also critical for adequate crop pollination. With higher natural plant and floral diversity,

Figure 10. Landscape elements surrounding a cereal field (left), and conservation biological control via a ladybird beetle, a natural predator of aphids, on organic wheat in south-eastern France (right). (Photos A. Wezel.)

varied landscape plantings attract pollinators to adjacent croplands, hosting them beyond the crop flowering period and serving as nesting sites for reproduction and over-wintering. These elements are becoming increasingly crucial on our landscapes, as pollinator populations and diversity are at risk in many countries worldwide due to various factors, among them decline of habitats and intensive use of pesticides being a few.

Beyond serving as habitat to insects and pollinators that benefit farm management, landscape elements can provide services that benefit the broader ecosystem beyond the farm. In-field and around-field landscape elements can serve to protect soil against wind and water erosion and surface water against contamination, protecting shared community resources. In addition, they can also support biodiversity conservation in agricultural regions.

In general, the major constraints to the expanded incorporation of landscape elements into farming system strategies include reduction of cropped area, thus decreasing the potential volume of food produced, and need for increased management by farmers if the elements are included on their immediate property. Moreover, negative interactions can occur if these areas become habitats for pest species, and their contributions to natural pest control may vary considerably.

4. Agroecological practices in European policies

The European Union and its members thus far have not established clear policies for agroecological practices and sustainable agriculture. In addition, national action plans and programmes on this topic still remain marginal. France is the sole country among the 28 Member States to have instituted an explicit strategy, "Agroecological Project for France" in December 2012. However, the newly defined Common Agricultural Policy (CAP) for 2014–2020 includes additional elements which align with agroecological practices, beyond already existing agri-environment measures.

A primary change in the new CAP is an additional financial subheading named "Green payment", representing 30% of direct aids. This green component, based on compulsory practices to be followed by farmers addressing both climate and environment policy goals, is established in Pillar 1 (direct payments); prior to this change, policy solely reinforced environmental measures within Pillar 2 (rural development). Greening practices take the form of simple, generalised, non-contractual and annual actions that go beyond the common requirements and the regulatory cross-compliance which is the EU directive for good agricultural and environmental practices. The three proposed compulsory practices of greening include: (i) crop diversification, (ii) maintenance of permanent grasslands, and (iii) establishment or maintenance of ecological focus areas (European Commission 2016a). More details are presented in the last chapter of the book. Under the second pillar of the CAP, different agri-environment measures are also proposed by the different member countries (European Commission 2016b). Examples which are covered by national/regional schemes are also presented in the last chapter of the book.

5. Conclusions

Agricultural production should provide sufficient food for the world's population while being environmentally friendly, socially acceptable, and economically beneficial for farmers. In addition, the basic food commodities should also be available at affordable prices for low-income populations without negatively impacting nutritional quality. The foundations of the agricultural model needed to achieve these goals lay within the different practices which farmers apply to crop and livestock production. Some of these practices can be considered as agroecological practices, if they effectively valorise ecological processes and ecosystem services through their integration as fundamental elements in the development of agricultural strategies, moving away from reliance on synthetic inputs such as chemical fertiliser and synthetic pesticide application, or technological solutions such as genetically modified organisms.

Many agroecological practices already exist around the world and are applied in varying degrees in different regions and under different climatic conditions. Most of these practices include increasing diversity, either (i) at the field level by mixing cultivars, using crop associations, and diversified crop rotations, (ii) at the farm level by increasing diversity of production types (annual crops, perennial crops, forestry, and livestock), and integrating biodiversity via semi-natural landscape elements, and (iii) at the landscape level by integrating natural and semi-natural rangelands into mixed crop-livestock systems, or by integrating a broad range of landscape elements to improve biological control, pollination, and erosion control. However, integrating higher diversity also implies, for farmers, managing more complex systems, and in many cases also taking higher economic and technological risks.

Applying some of the agroecological practices presented here may readily and immediately result in positive effects; what is more challenging, however, is not the application of any single practice, but the development a systems approach where these agroecological practices are used to support economic and social sustainability in addition to environmental sustainability. Therefore, optimum management of agroecosystems requires the ongoing integration of multiple practices into agricultural strategies that are both feasibly implemented by farmers and adapted to local farming conditions.

Acknowledgements

Some parts of the text are based on publications on agroecological practices which have been written together with M. Casagrande, F. Celette,

C. David, A. Ferrer, A. Letort, J. Peigné, and J.F. Vian. I hereby thankfully acknowledge their contributions.

References

Agreste, 2010. Grandes cultures — Alternance. pp. 9–13. http://www.agreste.agriculture.gouv.fr/IMG/file/dossier8_cultures_alternance.pdf (accessed May 2016).

Agreste, 2012. L'irrigation en Rhône-Alpes. 4 p. http://www.agreste.agriculture.gouv.fr/IMG/pdf/r8212a27.pdf (accessed May 2016).

Agreste, 2013. Inosys, une typologie des exploitations, 6 p. http://www.agreste.agriculture.gouv.fr/IMG/pdf/R8213A22.pdf (accessed May 2016).

Altieri, M.A., 1989. Agroecology: A new research and development paradigm for world agriculture. *Agriculture, Ecosystems and Environment* **27**, 37–46.

Altieri, M.A., 1995. *Agroecology: The Science of Sustainable Agriculture*. Westview Press, Boulder, CO, USA, 433 p.

Altieri, M.A., Nicholls, C.I., 2004. *Biodiversity and Pest Management in Agroecosystems*. Food Product Press, New York, USA, 336 p.

An, G.H., Kobayashi, S., Enoki, H., Sonobe, K., Muraki, M., Karasawa, T., Ezawa, T., 2010. How does arbuscular mycorrhizal colonization vary with host plant genotype? An example based on maize (*Zea mays*) germplasms. *Plant Soil* **327**, 441–453.

Arrignon, J., 1987. *Agro-écologie des zones arides et sub-humides*. Editions G.-P. Masonneuve & Larose et ACCT, Paris, France, 283 p.

Ball, B.C., Tebrügge, F., Sartori, L., Giráldez, J.V., González, P., 1998. Influence of no-tillage on physical, chemical and biological soil properties. In: Tebrügge F., Böhrnsen A. (Eds.), *Experiences with the Applicability of No-tillage Crop Production in the West-European Countries, Review Papers, Summaries and Conclusions of the Concerted Action*. Justus-Liebig University, Giessen, Germany, pp. 7–27.

Batish, D.R., Sing, H.P., Kohli, R.K., Kaur, S. 2008. Eucalyptus essential oil as a natural pesticide. *Forest Ecology and Management* **256**, 2166–2174.

Birkhofer, K., Bezemer, T.M., Bloem, J., Bonkowski, M., Christensen, S., Dubois, D., Ekelund, F., Fließbach, A., Gunst, L., Hedlund, K., Mäder, P., Mikola, J., Robin, C., Setälä, H., Tatin-Froux, F., Van der Putten, W.H., Scheu, S., 2008. Long-term organic farming fosters below and aboveground biota: implications for soil quality, biological control and productivity. *Soil Biology and Biochemistry* **40**, 2297–2308.

Blanco-Canqui, H., Mikha, M.M., Presley, D.R., Claasen, M.M., 2011. Addition of cover crops enhances no-till potential for improving soil physical properties. *Soil Science Society of America Journal* **75**, 1471–1482.

Blanco-Canqui, H., Shaver, T.M., Lindquist, J.L., Shapiro, C.A., Elmore, R.W., Francis, C.A., Hergert, G.W., 2015. Cover crops and ecosystem services: Insights from studies in temperate soils. *Agronomy Journal* **107**, 2449–2474.

Bonato, A.L.V., Calvo, E.S., Geraldi, I.O., Arias, C.A.A., 2006. Genetic similarity among soybean (Glycine max(L) Merrill) cultivars released in Brazil using AFLP markers. *Genetics and Molecular Biology* **29**, 692–704.

Bonaudo, T., Burlamaqui Bendahan, A., Sabatier, R., Ryschawy, J., Bellon, S., Leger, F., Magda, D., Tichit, M., 2014. Agroecology principles for the redesign of integrated crop-livestock systems. *European Journal of Agronomy* **57**, 43–51.

Brainard, D.C., Haramoto, E., Williams II, M.M., Mirsky, S., 2013. Towards a no-till no-spray future? Introduction to a symposium on nonchemical weed management for reduced-tillage cropping systems. *Weed Technology* **27**, 190–192.

Buck, L.E., Lassoie, J.P., Fernandes, E., 1998. *Agroforestry in Sustainable Agricultural Systems*. CRC Press, Boca Raton, FL, USA, 411 p.

Campbell, C.A., Myers, R.J.K., Curtin, D., 1995. Managing nitrogen for sustainable crop production. *Fertilizer Research* **42**, 277–296.

Carof, M., de Tourdonnet, S., Saulas, P., Le Floch, D., Roger-Estrade, J., 2007 Undersowing wheat with different living mulches in a no-till system. I. Yield analysis. *Agronomy for Sustainable Development* **27**, 347–356.

Charleston, D.S., Kfir, R., Dicke, M., Vet, L.E.M., 2005. Impact of botanical pesticides derived from Melia azedarach and Azadirachta indica on the biology of two parasitoid species of the diamond back moth. *Biological Control* **33**, 131–142.

Coulibaly, O., Mbila, D., Sonwa, J.J., Adesina, A., Bakala, J. 2002. Responding to economic crisis in sub-Saharan Africa: New farmer-developed pest management strategies in cocoa-based plantations in Southern Cameroon. *Integrated Pest Management Review* **7**, 165–172.

Dogliotti, S., Rossing, W.A.H., van Ittersum, M.K., 2004. Systematic design and evaluation of crop rotations enhancing soil conservation, soil fertility and farm income: A case study for vegetable farms in South Uruguay. *Agricultural Systems* **80**, 277–302.

Dumont, B., Fortun-Lamothe, L., Jouven, M., Thomas, M., Tichit, M., 2013. Prospects for agroecology and industrial ecology for animal production in the 21st century. *Animal* **7**, 1028–1043.

Eichhorn, M.P., Paris, P., Herzog, F., Incoll, L.D., Liagre, F., Mantzanas, K., Mayus, M., Moreno, G., Papanastasis, V.P., Pilbeam, D.J., Pisanelli, A., Dupraz, C., 2006. Silvoarable systems in Europe — past, present and future prospects. *Agroforestry Systems* **67**, 29–50.

European Commission, 2016a. Greening. http://ec.europa.eu/agriculture/direct-support/greening/index_en.htm (accessed May 2016).

European Commission, 2016b. Agri-environment measures. http://ec.europa.eu/agriculture/envir/measures/index_en.htm (accessed May 2016).

Francis, C., Lieblein, G., Gliessman, S., Breland, T.A., Creamer, N., Harwood, R., Salomonsson, L., Helenius, J., Rickerl, D., Salvador, R., Wiedenhoeft, M., Simmons, S., Allen, P., Altieri, M., Flora, C., Poincelot, R., 2003. Agroecology: The ecology of food systems. *Journal of Sustainable Agriculture* **22**, 99–118.

Fustec, J., Lesuffleur, F., Mahieu, S., Cliquet, J.-B., 2010. Nitrogen rhizodeposition of legumes. A review. *Agronomy for Sustainable Development* **30**, 57–66.

Garcia de Salamone, I.E., Döbereiner, J., Urquiaga, S., Boddey, R.M., 1996. Biological nitrogen fixation in Azospirillum strain-maize genotype associations as evaluated by the 15N isotope dilution technique. *Biology and Fertility of Soils* **23**, 249–256.

Gardiner, M.M., Fiedler, A.K., Costamagna, A.C., Landis, D.A., 2009. Integrating conservation biological control into IPM systems. In: Radcliffe, E.B., Hutchison, W.D., Cancelado, R.E. (Eds.), *Integrated Pest Management. Concepts, Tactics, Strategies and Case Studies*. Cambridge University Press, Cambridge, UK, pp. 151–162.

Gliessman, S.R., 1997. *Agroecology: Ecological Processes in Sustainable Agriculture*. CRC Press, Boca Raton, FL, USA.

Gliessman, S.R., 2007, *Agroecology: The Ecology of Sustainable Food Systems*. CRC Press, Taylor & Francis, New York, USA, 384 p.

Goodman, M.M., 2005. Broadening the U.S. maize germplasm base. *Maydica* **50**, 203–214.

Gurr, G.M., Wratten, S.D., Altieri, M.A., 2004. *Ecological Engineering for Pest Management. Advances in Habitat Manipulation for Arthropods*. CSIRO Publishing, Australia, 244 p.

Hauggaard-Nielsen, H., Andersen, M.K., Jornsgaard, B., Jensen, E.S., 2006. Density and relative frequency effects on competitive interactions and resource use in pea-barley intercrops. *Field Crops Research* **95**, 256–267.

Hernandez Xolocotzi, E. (Ed.), 1977. *Agroecosistemas de Mexico: Contribuciones a la Ensenanza, Investigacion, y Divulgacion Agricola*. Colegio de Postgraduados, Chapingo, Mexico.

Hokkanen, H.M.T., 1991. Trap cropping in pest management. *Annual Review of Entomology* **36**, 119–138.

Huang, J., Pray, C., Rozelle, S., 2002. Enhancing the crops to feed the poor. *Nature* **418**, 678–684.

Isman, M.B., 2006. Botanical insecticides, deterrents, and repellents in modern agriculture and increasingly regulated world. *Annual Review of Entomology* **51**, 45–66.

Isman, M.B., 2008. Botanical insecticides: for richer, for poorer. *Pest Management Science* **64**, 8–11.

Kaeppler, S.M., Parke, J.L., Mueller, S.M., Senior, L., Stuber, C., Tracy, W.F., 2000. Variation among maize inbred lines and detection of quantitative trait loci for growth at low phosphorus and responsiveness to arbuscular mycorrhizal fungi. *Crop Science* **40**, 358–364.

Khan, Z.R., Pickett, J.A., 2004. The ‘push-pull’ strategy for stemborer management: A case study in exploiting biodiversity and chemical ecology. In: Gurr, G.M., Wratten, S.D., Altieri, M.A. (Eds.), *Ecological Engineering for Pest Management. Advances in Habitat Manipulation for Arthropods.* CSIRO Publishing, Australia, pp. 155–164.

Lopes, C.M., Santos, T.P., Monteiro, A., Rodrígues, M.L., Costa, J.M., Chaves, M.M., 2011. Combining cover cropping with deficit irrigation in a Mediterranean low vigor vineyard. *Scientia Horticulturae* **129**, 603–612.

Malézieux, E., Crozat, Y., Dupraz, C., Laurans, M., Makowski, D., Ozier-Lafontaine, H., Rapidel, B., de Tourdonnet, S., Valantin-Morison, M., 2009. Mixing plant species in cropping systems: concepts, tools and models. A review. *Agronomy for Sustainable Development* **29**, 43–62.

Malusá, E., Sas-Paszt, L., Ciesielska, J., 2012. Technologies for beneficial microorganisms inocula used as biofertilizers. *The Scientific World Journal* **2012**, 491206.

Marrone, P.G. 2007. Barriers to adoption of biological control agents and biological pesticides. CAB reviews: Perspectives in agriculture, veterinary science, *Nutrition and Natural Resources* **2**, 1–12.

Ministère de l’Agriculture, de l’Agroalimentaire et de la Forêt, 2013. Projet agroécologique pour la France. http://agriculture.gouv.fr/Projet-agro-ecologique-pour-la (accessed April 2015).

Mirsky, S.B., Ryan, M.R., Curran, W.S., Teasdale, J.R., Maul, J., Spargo, J.T., Moyer, J., Grantham, A.M., Weber, D., Way, T.R., Camargo G.G., 2012. Conservation tillage issues: Cover crop-based organic rotational no-till grain production in the mid-Atlantic region, USA. *Renewable Agriculture and Food Systems* **27**(1), 31–40.

Moore, V.M., Mitchell, P.D., Silva, E.M., Barham, B.L., 2016. Cover crop adoption and intensity on Wisconsin's organic vegetable farms. *Agroecology and Sustainable Food Systems* **40**, 693–713.

Mosquera-Losada, M.R., McAdam, J., Romero-Franco, R., Santiago-Freijanes, J.J., Rigueiro-Rodrígues, A., 2009. Definitions and components of agroforestry practices in Europe. In: Rigueiro-Rodrígues, A., McAdam, J., Mosquera-Losada, M.R. (Eds.), *Agroforestry in Europe. Current Status and Future Prospects*. Springer, Germany, pp. 3–19.

National Academy of Science, 2010. *The Impact of genetically engineered crops on farm sustainability in the United States*. http://www.nationalacademies.org/includes/genengcrops.pdf (accessed May 2016).

National Research Council, 1993. *Agricultural Crop Issues and Policies*. National Academy Press, Washington, DC, USA.

Prasifka, J.R., Hellmich, R.L., Weiss, M.J., 2009. Role of biotechnology in sustainable agriculture. In: Radcliffe, E.B., Hutchison, W.D., Cancelado, R.E. (Eds.), *Integrated Pest Management. Concepts, Tactics, Strategies and Case Studies*. Cambridge University Press, Cambridge, UK, pp. 260–272.

Peigné, J., Ball, B., Roger-Estrade, J., David, C., 2007. Is conservation tillage suitable for organic farming? A review. *Soil Use Management* **23**, 129–144.

Peigné, J., Cannavaciuolo, M., Gautronneau, Y., Aveline, A., Giteau, J.L., Cluzeau, D., 2009. Earthworm populations under different tillage systems in organic farming. *Soil and Tillage Research* **104**, 207–214.

Randall, G.W., Sawyer, J.E., 2008. Nitrogen application timing, forms, and additives. In: UMRSHNC (Upper Mississippi River Sub-basin Hypoxia Nutrient Committee), Final Report: *Gulf Hypoxia and Local Water Quality Concerns Workshop*, American Society of Agricultural and Biological Engineers, St. Joseph, Michigan, USA, pp. 73–85.

Regnault-Roger, C., Philogène, B.J.R., 2008. Past and current prospects for the use of botanicals and plant allelochemicals in integrated pest management. *Pharmaceutical Biology* **46**, 41–52.

Robertson, G.P., 1997. Nitrogen use efficiency in row-crop agriculture: Crop nitrogen use and soil nitrogen loss, In: Jackson, L. (Ed.), *Ecology in Agriculture*, Academic Press, New York, USA, pp. 347–365.

Sanchez, J.E., Harwood, R.R., Willson, T.C., Kizilkaya, K., Smeenk, J., Parker, E., Paul, E.A., Knezek, B.D., Robertson, G.P., 2004. Managing soil carbon and nitrogen for productivity and environmental quality. *Agronomy Journal* **96**, 769–775.

Scholberg, J.M.S., Dogliotti, S., Leoni, C., Cherr, C.M., Zotarelli, L., Rossing, W.A.H., 2010. Cover crops for sustainable agrosystems in the Americas.

In: Lichtfouse, E. (Ed.), *Genetic Engineering, Biofertilisation, Soil Quality and Organic Farming*. Springer, Dordrecht, The Netherlands, pp. 23–58.

Shelton, A.M., Badenes-Perez, F.R., 2006. Concepts and applications of trap cropping in pest management. *Annual Review of Entomology* **51**, 285–308.

Silva, E.M., 2014. Management of five fall-sown cover crops for organic no-till production in the Upper Midwest. *Agroecology and Sustainable Food Systems* **38**, 748–763.

Soane, B.D., Ball, B.C., Arvidsson, J., Basch, G., Moreno, F., Roger-Estrade, J., 2012. No-till in northern, western and south-western Europe: A review of problems and opportunities for crop production and the environment. *Soil and Tillage Research* **118**, 66–87.

Srinivasan, A., 2006. *Handbook of Precision Agriculture: Principles and Applications*. Haworth Press, USA.

Steenwerth, K., Belina, K.M., 2008. Cover crops and cultivation: Impacts on soil N dynamics and microbiological function in a Mediterranean vineyard agroecosystem. *Applied Soil Ecology* **40**, 370–380.

Tabaglio, V., Gavazzi, C., Schulz, M., Marocco, A., 2008. Alternative weed control using the allelopathic effect of natural benzoxazinoids from rye mulch. *Agronomy for Sustainable Development* **28**, 397–401.

Tilman, D., Cassman, K.G., Matson, P.A., Naylor, R., Polasky, S., 2002. Agricultural sustainability and intensive production practices. *Nature* **418**(8), 671–677.

Tribouillois, H., Cohan, J.-P., Justes, E., 2016. Cover crop mixtures including legume produce ecosystem services of nitrate capture and green manuring: assessment combining experimentation and modelling. *Plant Soil* **40**, 347–364.

Tooker, J.F., Frank, S.D., 2012. Genotypically diverse cultivar mixtures for insect pest management and increased crop yields. *Journal of Applied Ecology* **49**, 974–985.

United Nations (2015). The 2015 Revision of World Population Prospects. Population Division of the Department of Economic and Social Affairs of the United Nations Secretariat. https://esa.un.org/unpd/wpp/ (accessed June 2016).

Uphoff, N., 2002. *Agroecological Innovations. Increasing Food Production with Participatory Development*. Earthsan, London, UK, 306 p.

Vandermeer, J. 1989. *The Ecology of Intercropping*. Cambridge University Press, New York, USA.

Van Noordwijk, M., Lawson, G.J., Soumare, A., Groot, J.J.R., Hairiah, K., 1996. Root distribution of tree and crops: competition and/or complementarity. In: Ong, C.K., Huxley, P. (Eds.), *Tree-Crop Interactions: A Physiological Approach*. CAB International, Wallingford, UK, pp. 319–365.

Vessey, J.K., 2003. Plant growth promoting rhizobacteria as biofertilizers. *Plant Soil* **255**, 571–586.

Vian, J.F., Peigne, J., Chaussod, R., Roger-Estrade, J., 2009. Effects of four tillage systems on soil structure and soil microbial biomass in organic farming. *Soil Use Management* **25**, 1–10.

Watson, C.A., Atkinson, D., Gosling, P., Jackson, L.R., Rayns, F.W., 2002. Managing soil fertility in organic farming systems. *Soil Use Management* **18**, 239–247.

Weston, L.A., 1996. Utilization of allelopathy for weed management in agroecosystems. *Agronomy Journal* **88**, 860–866.

Wezel, A., Bellon, S., Doré, T., Francis, C., Vallod, D., David, C., 2009. Agroecology as a science, a movement or a practice. A review. *Agronomy for Sustainable Development* **29**, 503–515.

Wezel, A., Brives, H., Casagrande, M., Clément, C., Dufour, A., Vandenbroucke, P., 2016. Agroecology Territories: Places for sustainable agricultural and food systems and biodiversity conservation. *Agroecology and Sustainable Food Systems* **40**(2), 132–144.

Wezel, A., Casagrande, M., Celette, F., Vian, J.F., Ferrer, A., Peigné, J., 2014a. Agroecological practices for sustainable agriculture. A review. *Agronomy for Sustainable Development* **34**(1), 1–20.

Wezel, A., David, C., 2012. Agroecology and the food system. In: Lichtfouse, E. (Ed.), *Agroecology and Strategies for Climate Change. Sustainable Agriculture Reviews* Vol. 8, Springer, Dordrecht, The Netherlands, pp. 17–34.

Wezel, A., David, C., Ferrer, A., Letort, A., Féret, S., Peigné, J., Vian, J.V., Celette, F., 2014b. *Agroecological Practices Supporting Provision of Goods and Services in Agriculture: Examples from France and Europe*. ISARA Lyon, France, 67 p.

Wezel, A., Fleury, Ph, David, C., Mundler, P., 2015. The food system approach in agroecology supported by natural and social sciences: Topics, concepts, applications. In: Benkeblia, N. (Ed.), *Agroecology, Ecosystems and Sustainability*. CRC Press, Boca Raton, FL, USA, pp. 181–199.

Wezel, A., Peeters, A., 2014. Agroecology and herbivore farming systems — principles and practices. *Options Meditérranéennes A*, **109**, 753–767.

Wezel, A., Soldat, V., 2009. A quantitative and qualitative historical analysis of the scientific discipline agroecology. *International Journal of Agricultural Sustainability* **7**(1), 3–18.

Whipps, J.M., 2001. Microbial interactions and biocontrol in the rhizosphere. *Journal of Experimental Botany* **52**, 487–511.

Willey, R.W., 1990. Resource use in intercropping systems. *Agriculture Water Management* **17**, 215–231.

Chapter 3

Applying Agroecological Principles for Regenerating Soils

Matthew R. Ryan* and Joséphine Peigné†

*Department of Crop and Soil Sciences,
Cornell University, Ithaca, New York, USA
†Department of Agroecology and Environment,
ISARA, Lyon, France

Agroecology is the integrative study of the ecology of the entire food system, encompassing ecological, economic, and social dimensions. It focuses on working with and understanding the interactions between plants, animals, humans, and the environment within agricultural systems. By bringing ecological principles to bear in agroecosystems, through ecological intensification, novel management approaches can be identified, building on key interactions and strengthening virtuous cycles in agricultural production that would not otherwise be considered. (Food and Agriculture Organization of the United Nations 2014)

1. Introduction

A transformative change is needed in how farmers think about and manage their soil in order to meet increasing needs for greater productivity and environmental stewardship. Instead of minimising harm from degrading practices, emphasis should be placed on maximising improvements in soil health. Increasing productivity and food security with regenerative agriculture is possible through ecological intensification. This approach aims to boost productivity by using practices that increase supporting and regulating ecosystem services (e.g. nutrient cycling, biological nitrogen fixation, biological control of pests, etc.). Ecological intensification is based on: (1) *ecological enhancement*, which involves increasing the level of an underpinning service (e.g. pest regulation, pollination, soil nutrient, water supply, etc.), and (2) *ecological replacement*, which involves replacing exogenous inputs such as synthetic nitrogen fertiliser with ecological processes such as biological nitrogen fixation (Bommarco *et al.* 2013). As the need to reduce dependency on fossil fuels and adapt cropping systems to climate change becomes clearer and urgent, ecological intensification will become increasingly important (Tittonell 2014).

Here, we review relevant agroecological principles and the use of various management practices for regenerating soil health and improving crop production. As soil management is so intertwined with other aspects of crop and livestock production, we take a systems approach and aim to provide a broad overview of applied agroecological soil management, while considering profitability, resilience, and environmental impact over the short- and long-term. Starting with a review of ecological processes, we explore applied management and introduce a new theoretical framework for agroecological soil nitrogen management.

2. Soil health

Soil health is the capacity of a soil to function, sustain productivity, maintain environmental quality, and promote plant and animal health (Doran and Parkin 1994, Gugino *et al.* 2009). In the past, agriculturists have focused on soil fertility or the chemical status of soils. Although the chemical and physical properties of soils are important, soil scientists now recognise the critical role of the biological properties of soils (Scholes and Scholes 2013). Together, the chemical, physical, and biological properties of a soil interact and regulate ecological processes. Below we describe the three components of soil health, how they affect ecological processes that are essential for proper soil functioning, and ecosystem services that soils provide.

Agroecological management practices strive to use ecological processes to manage soil health instead of exogenous inputs. In many cases, exogenous inputs can create problems for farmers that require additional inputs to correct. For example, synthetic fertiliser can lower soil pH and introduce salts that reduce soil health and limit crop performance. Cropping systems that are based on natural ecosystems and managed with ecological processes are self-reinforcing and tend to improve over time.

2.1. *Soil biodiversity*

Biodiversity of soil organisms includes living organisms that perform all or part of their development cycle in the soil. These organisms are responsible for ecological processes in agricultural soils, and thus are the source of many ecosystem services. Among the different ecological processes, the following are particularly important for the preservation of soil fertility:

- **Formation and stabilisation of soil structure:** Organisms living in soil affect its structure. For example, earthworms create pores and enhance soil structure by their burrowing activities, and the network of hyphae from soil fungi binds organo-mineral soil particles together, thereby increasing resistance to slaking (i.e. the disintegration of soil aggregates from exposure to water).
- **Mineralisation, humification, and organisation of soil organic matter:** Soil meso and macrofauna contribute to soil organic matter cycling through the burial and degradation of crop residues. Microorganisms also participate in the cycling of soil organic matter by performing a wide variety of biogeochemical transformations including the cycling of carbon, nitrogen, and sulfur.
- **Mutualistic symbiosis and resource acquisition:** Biological nitrogen fixation is the conversion of atmospheric nitrogen (N_2) to ammonia (NH_3) in the presence of nitrogenase, from bacteria living in symbiosis with leguminous plants and free-living bacteria in the soil.

Most plants can establish symbiotic relationships with mycorrhizal fungi, which increase the ability of plant roots to acquire phosphorus and water (Gianinazzi *et al.* 2010).

- **Crop protection against pathogens and soil pests:** Several studies suggest that the presence of soil organisms in the rhizosphere (i.e. area surrounding plant roots) stimulates plant defences against pests (e.g. induced systemic resistance). Blouin *et al.* (2005) showed that the presence of earthworms improves the resistance of rainfed rice to nematode attacks. Similarly, mycorrhizal fungi and some bacteria in the rhizosphere can protect plants from soilborne diseases.

Soil organisms continuously interact with each other and their environment, and these complex interrelationships create the soil food web. It is important to note that although many organisms living in the soil are the source of ecosystem services, other organisms are responsible for crop damage and disservices. However, soil food webs with high biodiversity are typically less susceptible to pest and pathogen outbreaks and tend to increase resistance to human disturbance and resilience (i.e. ability to return to the original state or a state of equilibrium close to the initial state prior to disturbance).

2.2. *Soil structure*

Soil structure affects most physical, chemical, and biological processes occurring in the soil. For example, soil structure of the first several centimetres of soil often determine the difference between water infiltration and runoff. Soil structure also affects environmental factors that regulate biological processes in the soil such as temperature, humidity, and oxygenation. Playing a major role in habitat quality, soil structure affects the activity of soil organisms and indirectly influences the dynamics of carbon, nitrogen, and other mineral elements. Soil aggregation is also involved in the physical protection of soil organic matter against microbial degradation. Furthermore, soil structure influences seed germination, seedling emergence, root growth, and capacity of roots to capture resources in the soil.

2.3. *Soil organic matter*

Soil organic matter is a critically important attribute of soil health and thus the content and quality of soil organic matter are considered key indicators for assessing soil health and the sustainability of agricultural systems. The main roles of soil organic matter are:

- **Provide metabolisable energy to soil organisms:** Most soil organisms are heterotrophic and therefore use the soil organic matter for growth. This degradation activity of soil organic matter by soil organisms in turn provides nutrients for crops (Haynes 2005).
- **Soil cation exchange capacity:** The macromolecules that make up the stable soil organic matter are negatively charged and contribute to the retention of cations, and thus can improve soil nutrient supply.
- **Soil structure stabilisation:** The stabilised organic materials involved, as well as clay, help to bind soil particles together and form soil aggregates (Balesdent *et al.* 2000), which reduces soil erosion and improves aeration and drainage. This is because soil aggregates create micro and macropores in the soil.
- **Water retention:** The amount of water retained in the soil depends primarily on porosity and depth, mineralogy, and the amount of organic matter in the soil. Increasing the content of soil organic matter improves water retention capacity at low tension. Soil organic matter thus has a direct effect on the amount of water available to plants and an indirect effect through a change in the distribution and size of soil pores (Loveland and Webb 2003).

3. Soil management

Farmers play an important role in society, not only because they provide food and other agricultural products, but also because they are stewards of the land and are responsible for the protection of ecosystems. Their management practices can have a tremendous impact on environmental quality. Having a long-term perspective and considering the future impact of management practices when making decisions can reduce the potential for negative outcomes. Investing in soil health can pay off over time, especially when severe deficiencies are addressed.

Regenerative soil management requires that any major problem be addressed to increase soil functioning and crop performance. Even when relying on agroecological management practices, it is important that farmers know their soil and conduct tests to determine if there are deficiencies. Soil health tests can be performed at an increasing number of commercial laboratories using indicators such as soil organic matter, respiration, and aggregate stability (Davidson 2014, Gugino *et al.* 2009, Haney 2013, Idowu *et al.* 2009). Certain deficiencies might require off-farm inputs such as lime to raise soil pH. Other issues, such as excessively high weed populations in the soil weed seedbank, should be managed carefully to correct the problem while minimising any potential negative impacts of the corrective measures, e.g. using cover crops to offset negative impact of using tillage to reduce weed populations. In addition to soil nutrient and pest considerations, soil hydrologic properties should be monitored and corrected if necessary. Installing tile drainage, swales, and other structural water management features can increase soil functioning and increase overall cropping system performance.

As weather can be variable, and is projected to become more variable with climate change, it is important that farmers anticipate and account for extreme events such as excessive precipitation and short-term drought. As described below, farmers can invest in "ecological insurance" by using agroecological management practices, which increase the absorptive capacity of their operations (Lemanceau *et al.* 2015). However, farmers also need to embrace adaptive management and increase their capacity to respond to perturbations. Adaptive management can also increase efficiency and improve cropping system performance by applying management tactics only when necessary. Adjusting fertiliser rates based on soil nutrient levels or cover crop growth, switching crops based on pest pressure or nutrient legacies, and choosing to use or avoid tillage based on weed populations are all examples of how adaptive management can be used to optimise cropping system performance.

4. Agroecological principles

Several key principles can guide farmers as they adopt an agroecological approach to soil management. These principles are based on observation and research of natural ecosystems and indigenous agricultural systems that have sustained populations over long periods of time. Several previous reviews have identified similar agroecological principles (Altieri 2000, Magdoff 2007, Gliessman and Rosemeyer 2009, Snapp 2011, Vandermeer and Perfecto 2013). Most agroecological principles are based on the notion that farmers should manage their soils and crops in a way that mimics natural ecosystems. Natural ecosystems are efficient and do not require exogenous inputs to be productive nor do they create waste. Unlike conventional monoculture agroecosystems, natural ecosystems tend to be more diverse and this diversity drives the ecological processes that help regulate and sustain productivity. Natural ecosystems are often resilient and can more quickly rebound to high levels of productivity after a disturbance.

- **Employ biology:** Use biology and ecological processes instead of synthetic inputs when possible. In this way, farmers are working with, rather than against, nature. Managing ecological processes can also serve to reinforce systems and make them stronger, whereas over-reliance on synthetic inputs tends to degrade soils and make them more susceptible to pest outbreaks and negative impacts of disturbance.
- **Foster diversity:** Integrate diversity across temporal and spatial scales. Crop diversity and diversity of management practices creates soil biodiversity and regulates the ecological processes that drive productivity in agroecosystems. Crop and management diversity prevents pests from building up and ensures production stability. Although functionality can plateau with increasing diversity over the short-term, redundancy both at the species and functional group (i.e. organisms with similar behaviour) level is important because it imparts resiliency over the long-term.
- **Cycle nutrients:** Reduce waste and maximise system level nutrient use efficiency. Match inputs with outputs, minimise unnecessary

nutrient removal, and return nutrient to the soil with organic amendments. Crop residues and straw should only be removed if it is subsequently returned with animal manure or compost.
- **Integrate perennials:** Incorporate perennial plants into agroecosystems. Perennial plants feed soil microorganisms throughout the year, use resources more efficiently, and stabilise the agroecosystem. Perennial plants within and adjacent to crop fields can enhance ecological processes such as biological control and help protect annual crops (e.g. reduce wind and surface water flow) which can be vulnerable to various disturbances.
- **Combine production:** Integrate crop and livestock production to take advantage of complementary interactions. Rotating crops and livestock within the same area can regenerate soil health. In addition to depositing valuable nutrients and reducing pests, integrating livestock can facilitate root turnover and feed soil microorganisms that are responsible for the ecological processes that sustain production.
- **Nurture soil:** Manage soil as a living entity. Healthy soil requires food to eat, water to drink, and air to breathe. Pesticides and salts are toxins that negatively impact soil biology. Soil biology is complex with many trophic levels that all affect the ecological processes that support and sustain production.

Together, these principles coupled with knowledge of crops, soils, pest ecology, achieved through scouting, on-farm research, soil testing, and learning from other successful farmers, can guide farmers towards more robust and productive farming systems. Although optimal management practices will depend on site-specific conditions, farmer goals, and resources that are available, these principles can help famers use ecological processes as levers to protect soil, regenerate soil health, and increase crop production.

5. Regenerative practices

Crop rotation, cover cropping, and intercropping are common agroecological soil management practices that can regenerate soil health.

By applying knowledge of crop and soil ecology, crop and non-crop vegetation can be strategically arranged spatially and temporally to protect soil and achieve specific goals including increasing nutrient availability, suppressing pests, and improving soil structure and overall function.

5.1. *Crop rotation*

Crop rotation is one of the oldest methods of regenerating soil health, a practice that can increase soil fertility, reduce pest populations, and increase crop yields. Crop rotation has been used for thousands of years, first by early farmers who alternated between crop production and fallow periods, which then evolved to alternating cereals with pulses and fodder crops. Rotating crops with different nutrient acquisition traits (e.g. legume versus non-legume, shallow versus deep root system) can increase soil nutrient availability. Legume fodder crops such as red clover (*Trifolium pratense* L.) or alfalfa (*Medicago sativa* L.) can often provide all of the required nitrogen to subsequent maize (*Zea mays* L.) (Lawrence *et al.* 2008). Some crop root exudates facilitate greater nutrient acquisition. For example, buckwheat (*Fagopyrum esculentum* Moench) has been shown to increase phosphorus availability to subsequent crops through acidification of rhizosphere, which solubilises phosphorus from labile pools (Teboh and Franzen 2011).

Crop rotation also integrates diversity, which helps to stabilise cropping systems and protects crops from pest outbreaks. Smith *et al.* (2008) quantified the effects of crop rotation diversity under the extreme condition of no external fertiliser or pesticide inputs and reported a strong positive relationship between maize yield and the number of crops in rotation. In this case, maize grown in the most diverse crop rotation, which included two other cash crops and three cover crops, yielded twice as much as maize grown in a continuous monoculture (Smith *et al.* 2008). Diverse crop rotations are often more efficient and have been shown to increase crop yield and profitability while reducing agrichemical inputs and environmental pollution compared to simplified crop rotations (Davis *et al.* 2012).

5.2. *Cover crops*

Although simply rotating crops can be very helpful, integrating cover crops into crop rotations during fallow periods when soil is otherwise left bare and then returning all of the biomass that is produced to the soil, can expedite the positive impact on soil health. Cover crops are crops that are seeded and grown, but not harvested or grazed (Figure 1). However, a loosening of this definition has recently occurred and now many famers occasionally harvest their cover crop. This is quite normal in smallholder agriculture where obtaining a harvestable product is essential, even if the crop is primarily grown for other purposes such as regenerating soil health or suppressing pests (Bunch 2012). Traditionally, the primary purpose of growing cover crops was to protect soil from erosion. However, cover crops are recognised as being multifunctional and providing numerous benefits or ecosystem services. Ecosystem services provided by cover crops are often directly linked to their growth and thus biomass is often correlated with the magnitude of beneficial effects. Because cover crops are typically grown between cash crops when the soil would otherwise be bare, they contribute organic matter and can increase soil carbon and

Figure 1. Field day participants evaluate over 180 plots of cover crop and cover crop mixtures at the United States Department of Agriculture Natural Resource Conservation Service Big Flats Plant Materials Center in New York, USA (credit: Matthew Ryan).

nitrogen over time. Cover crops reduce rill and interrill erosion through a number of mechanisms including intercepting raindrops and reducing rainfall impact force and subsequent soil detachment, reducing flow velocity and increasing sediment deposition, and increasing soil structure and water infiltration. A meta-analysis of 122 studies on the effects of crop diversity showed that, compared to monocultures, diversified crop rotations that included cover crops increased total soil carbon by 8.5% and total soil nitrogen by 12.8%, whereas diversified crop rotations without cover crops did not result in increased soil carbon or nitrogen (McDaniel *et al.* 2014). In a separate meta-analysis using data from 37 sites, introducing cover crops into crop rotations was linearly correlated with a mean annual soil carbon sequestration rate of 0.32 ± 0.08 Mg ha^{-1} yr^{-1} (Poeplau and Don 2015). These results are congruent with research highlighting the critical, but historically underappreciated, role of soil microorganisms and labile organic matter that quickly decomposes for building soil organic matter (Kallenbach *et al.* 2015, Lehmann and Kleber 2015).

Different cover crops perform different functions and thus cover crop selection should be used based on desired outcomes. Catch crops are typically grass cover crops such as cereal rye and oats. These cover crops are typically seeded in the fall after maize and other crops that received nitrogen fertiliser. Catch crops help cycle nutrients and prevent them from leaching through the soil profile during fallow periods. Green manures are typically legume cover crops such as hairy vetch and red clover that are grown primarily to provide nitrogen to a subsequent cash crop. These cover crops can fix atmospheric nitrogen via symbiotic relationship with bacteria, and thus provide nitrogen to subsequent crops after they decompose. Smother crops are cover crops that grow fast, such as buckwheat and that are grown primarily to compete with weeds through intense shading. Biofumigant crops are cover crops such as brassicas (e.g. white mustard and forage radish) and grasses that suppress soilborne disease and insect pests through chemicals that are exuded through their roots and during decomposition.

Previous research has demonstrated that cover crops can also provide a rotation effect by enhancing subsequent crop performance (Andraski and Bundy 2005). For example, in an experiment comparing maize grain yield response to nitrogen fertiliser, maize yields were consistently greater in plots with a preceding red clover cover crop compared to plots without the red clover, even at high nitrogen fertiliser rates (Figure 2) (Henry *et al.* 2010). The rotation effect is one example of how cover crops can enhance overall productivity (Gaudin *et al.* 2013).

Cover crops can help farmers "perennialise" their annual cropping systems. By growing during fallow periods, they provide many of the same benefits that are found in perennial ecosystems. They occupy empty niche space, cycle nutrients that would otherwise be lost, prevent unwanted plants from growing, and help support biodiversity in cropping systems. They help feed beneficial soil microorganisms that are important for healthy crops through exudates from their roots and as they decompose. Cover crops also represent an excellent way to integrate crop diversity into cropping systems. Planting mixtures of species with different functional traits can magnify this effect and enable cover crop to provide multiple functions simultaneously. Cover crop mixtures should be designed to maximise synergy between species, which is typically achieved by combining species with different functional traits such as deep and shallow rooted species, grasses and legumes (Bybee-Finley *et al.* 2016).

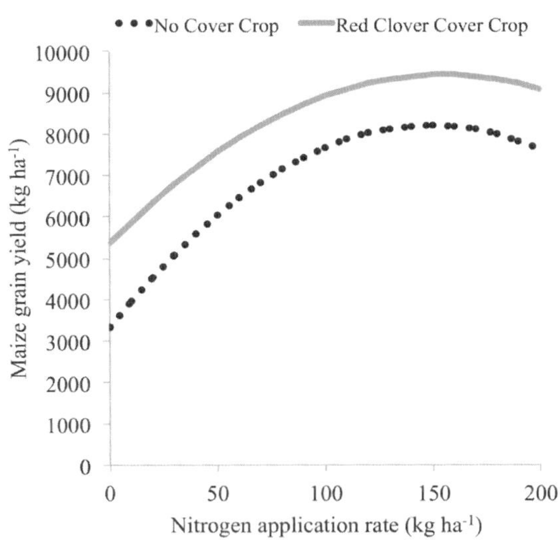

Figure 2. Maize yield in plots with and without a preceding red clover cover crop across a gradient of supplemental nitrogen fertiliser. Modified from Henry *et al.* (2010).

Although cover crops provide many benefits, there are also several potential drawbacks that should be considered before integrating them into crop rotations. Cover crops can provide habitat for pests such as slugs and other invertebrates that feed on seeds and seedling. They can also serve as a reservoir for disease causing pathogens. In some cases, cover crops can become weeds and compete with crops for resources or interfere with mechanical operations. However, many of these problems can be avoided with careful cover crop species selection and proper management.

5.3. *Intercropping*

Intercropping is the simultaneous cultivation of multiple crops in the same area (Figure 3), and is commonly used throughout the world to increase production efficiency, especially in smallholder farming. However, intercropping is also used in industrialised cropping systems. Crops can be

Figure 3. Intercropping red clover into winter wheat (*Triticum aestivum* L.) before (a) and after wheat harvest (b) and intercropping cover crops into maize silage before (c) and after maize silage harvest (d) (credits: a and b, Joséphine Peigné; c, Michael Ha; and d, Chris Pelzer).

grown together in a number of different ways in intercropping systems including relay intercropping (i.e. crops overlap for a short period of time), strip intercropping (i.e. crops are arranged in separate adjacent rows), and mixed intercropping (i.e. crops are sown together). Growing multiple crops together mimics natural ecosystems and can provide several advantages when there is complementarity between crop species.

In most cases, complementarity arises from soil resource partitioning where crops can access different pools of nutrients thus producing more than either crop grown in monoculture. Facilitation, where one crop enhances the growth of another crop, is also a form of complementarity. This type of complementarity is typically limited to interactions between grass and legumes, where legumes fix atmospheric nitrogen and increase nitrogen availability to the grass crop. This process is generally slow and is more commonly observed in perennial crop mixtures where the legume can increase soil nitrogen over time. However, intercropping has also been shown to increase phosphorus availability in a study comparing wheat and chickpea grown in monoculture and intercropping in high and low phosphorus soils. A recent analysis of strip intercropping showed a 4% increase in total soil carbon in plots that were intercropped compared to plots where crops were grown in monoculture (Cong *et al.* 2015). In this case, increased nitrogen fixation from intercropping is thought to allow greater accumulation of soil carbon over time leading to long-term improvements in soil health. As with cover crops, intercropping can also introduce potential challenges, mainly competition between crops. In some cases, the main crop can be outcompeted by the secondary crop. However, competitive interactions can be managed with cultural practices (e.g. seeding rates, spacing, and timing) and interventions to reduce asymmetric competition.

6. Applying organic materials

The application of organic materials can increase soil organic matter levels and affect the chemical, physical, and biological properties of soils. Exogenous organic materials that are imported into the cropping system contain both a mineral and organic fraction. The mineral fraction provides mineral elements that are directly available for the crop. The organic fraction is transformed through the biological activity of soil organisms (i.e. humification and mineralisation) and contributes to the improvement of soil structure (Diacono and Montemurro 2010).

The quality of exogenous organic materials can vary substantially and are typically classified into two categories according to their effects on soil health and stability (i.e. decomposition rate), which depends on the origin of the material (e.g. manure, compost, wood chips, grass clippings) and its composition.

Exogenous organic materials:

- Organic amendments are exogenous organic materials, which act primarily on soil properties (e.g. structural stability, water retention, and stimulation of biological activity) and should be transformed by soil organisms to release nutrients to the crop. Organic amendments also facilitate increases in soil organic matter stocks (Gattinger *et al.* 2012).
- Organic fertilisers mineralise quickly and therefore have little long-term influence on soil fertility. In contrast to amendment, they bring nutrients (nitrogen, phosphorus, and potassium) that are readily available to crops.

Generally, composted products contain organic matter that is more stable than fresh manures, liquid manures, crop residues, and green manures. Compost made from urban waste (e.g. leaves and yard clippings) is typically more stable than manure based compost. Although most compost does not provide large amounts of readily available nitrogen, compost applications can increase soil nitrogen availability over time. Compost can contain large amounts of phosphorus and potassium and improve soil fertility by enriching the pool of stable soil organic matter. This pool provides a supply of nutrients, stabilises soil structure, and increases cation exchange capacity and water holding capacity of the soil. High temperatures achieved during the thermophilic composting process can decrease pathogen and weed seed viability. Compared to manure, compost can also reduce the risk of nitrogen leaching through the soil profile.

Non-composted organic materials decompose in the soil faster than composted material. Non-composted materials provide organic nitrogen during the year, with a significant proportion of very quickly mineralisable nitrogen depending on the carbon nitrogen ratio of the material. Such materials improve soil fertility and increase the labile soil organic matter pool (i.e. soil organic matter that is easily used by the microorganisms) and thus stimulate soil life. The composition of non-composted manure is highly variable depending on the origin and the amount of bedding material (e.g. sawdust). Optimal application rates of organic materials will depend on the purpose of the application (e.g. increasing the content of organic matter in soil), crop nutrient needs (mainly nitrogen, phosphorus, and potassium), the type of organic material (Diacono and Montemurro 2010). Calculating appropriate application rates of organic inputs requires knowledge of crop requirements, the quality of material, its behaviour in soil, and the soil management practices used (Diacono and Montemurro 2010). Soils that are prone to crusting might require frequent inputs of organic materials for improving the stability of soil aggregates, whereas nutrient-limited soils might require mineral fertilisers to optimise production.

7. Reducing tillage

Soil tillage affects all components of soil health (i.e. physical, chemical, and biological). Although soil tillage can suppress pests, incorporate amendments, warm the soil, and create a fine seedbed for plants, there are also drawbacks. In addition to making soil more susceptible to erosion, tillage can disrupt soil organisms, reduce soil moisture, and increase input costs associated with fuel and labour. In organic cropping systems, some tillage is essential for weed management and thus organic farmers need to offset the negative impacts of tillage with agroecological management practices such as cover cropping and applying organic amendments. This requires constant balancing and thus organic, and also conventional, farmers should consider the effect of tillage on both soil regeneration and pest suppression (e.g. weeds, plant pathogens, and herbivorous insects).

Tillage alters soil structure. However, this is not the only factor that influences the evolution of soil structure; climates, soil organisms, and agricultural equipment also have an effect. Tillage systems are often classified based on the amount of residue from the previous crop that is retained on the soil surface (Köller and El Titi 2003). Using this classification, tillage that retains less than 15% residue cover on the soil surface is considered conventional tillage, whereas tillage that retains greater than 30% residue cover on the surface is considered conservation tillage. Conservation tillage can be achieved with a wide range of tillage tools that do not invert the soil. Although most conservation tillage tools are designed for shallow disturbance, soil can be worked deeper as long as sufficient residue is maintained on the soil surface. The most extreme conservation tillage systems involve direct seeding, which also called "direct drilling", "zero tillage", or "no-tillage". In this case, the soil is only worked in a small area where the seed is deposited by discs or coulters attached to the planter. Strip tillage is another approach that involves tilling a small strip of soil with a separate operation prior to planting the crop.

Compared to annual mouldboard plough tillage, conservation tillage reduces mechanical work and associated inputs (e.g. diesel fuel), reduces soil erosion and improves soil structure and aggregate stability (Capowiez *et al.* 2009), and decreases plough pans and soil compaction due to fewer passes with equipment. Such improvements in soil structure can enhance biological activity, increase soil fertility (Holland 2004), and facilitate soil health regeneration (Kladivko 2001, Säle *et al.* 2015). It is important to note that although conservation tillage can provide a number of benefits compared to conventional tillage, the effects are influenced by soil type, climate, and crop management practices (Palm *et al.* 2014). For example, depending on soil texture, conventional tillage might be more effective than conservation tillage in terms of addressing constraints such as soil compaction, and increasing crop performance (Peigné *et al.* 2013).

Tillage can also be reduced in different ways that are not classified based on the amount of residue retained on the soil surface, including reducing tillage frequency and reducing tillage intensity. Approaches to these tillage reductions strategies are described below:

7.1. *Reduce frequency with integrating perennial crops in the crop rotation*

Incorporating perennial legume forage crops into crop rotations can build soil health with no tillage and permanent soil cover by legumes (Watson *et al.* 2008). Perennial crops reduces the frequency of tillage in the crop rotation and in the case of association of legumes improves nitrogen content in the soil and also soil structure because of their extensive root systems that are present for longer periods than annuals crops. However, perennial forage crops are often destroyed by ploughing or deep tillage. The use of deep tillage to terminate perennial forage crops can degrade soil health that has been built up during no tillage phase. However, few studies exist on the effect of one ploughing after several years of no tillage on soil and there is no clear conclusion, even if soil fertility effectively tends to be reduced (Dang *et al.* 2015). Perennial grain crops such as intermediate wheatgrass can also improve soil structure, increase nutrient cycling, and serve to enhance soil health (Culman *et al.* 2013). Although grain yields of perennial grain crops are currently substantially lower than annual grain crops, research is underway to identify strategic ways to integrate these crops into agroecosystems. For example, harvesting both grain and forage from perennial grain crops can increase their profitability (Bell *et al.* 2014).

7.2. *Reduce intensity with chisel ploughing and other shallow disturbance tools*

Another strategy to regenerate soil health is to replace mouldboard ploughing by non-soil inversion tillage techniques such as chisel ploughing or disc harrowing. The reduction of tillage intensity, concentrates soil organic matter at soil surface or upper soil layers compared to conventional tillage, which stimulates soil microbial activity (Vian *et al.* 2009). The effects of reducing tillage intensity on soil macrofauna are less clear. In a French experiment, comparing tillage techniques in organic farming, chisel ploughing did not systematically increase earthworm biomass compared to mouldboard ploughing (Peigné *et al.* 2009). However, chisel ploughing had the positive effects on earthworm populations in a very similar Swiss trial with different soil, climate, and crops conditions (Kuntz *et al.* 2013). In addition to the type of tillage, the number of passes, depth of soil disturbance, and the timing of tillage in relation to soil conditions also affect tillage intensity.

Although reducing tillage intensity can improve soil health, it often leads to increased weed populations and weed competition with crops, especially in organic cropping systems. In conventional cropping systems, the absence of soil inversion is often associated with greater herbicide use (Lin *et al.* 1995), which can negatively impact soil organisms, non-target vegetation, and water quality. In organic cropping systems, soil tillage and inter-row cultivation is often used to control weeds, which can facilitate erosion, soil compaction, and degradation of soil health (Peigné *et al.* 2015). Smith *et al.* (2011) described this phenomenon as, the weed management–soil quality paradox, which states "although effective weed management and enhancement of soil quality are necessary components of sustainable crop production, practices focused on enhancing one component could inhibit the other". Below we introduce an alternative approach, one that aims to overcome this apparent trade-off between weed suppression and soil health.

7.3. *Reduce frequency with rotational no-till*

Alternating between crops that are established using no-till planting practices and crops that are established in tilled seedbeds is another way to reduce soil tillage frequency. Although soil structure benefits can be reduced by tillage after periods of no-tillage, research has shown soil organic matter and yield benefits from biennial tillage (Venterea *et al.* 2006). When coupled with other regenerative practices, rotational no-till can provide important benefits and allow farmers to address management challenges such as compaction and incorporation of amendments.

Rotational no-till coupled with the use of cover crops has also been promoted for use in organic cropping systems where herbicides are prohibited. Cover crop-based, organic rotational no-till involves the use of tillage to establish cover crops that are then mechanically terminated prior to no-till planting organic cash crops such as soybean and maize (Figure 4) (Mirsky *et al.* 2012, 2013, Nord *et al.* 2012). This practice of using high biomass cover crops to suppress annual weeds was originally developed in South America and has since been promoted across the. Tillage is used to control perennial weeds prior to establishing cover crops. Although this system is still being developed, cover crop-based, organic rotational

(a) (b)

Figure 4. (a) Terminating a cereal rye cover crop with a front-mounted roller-crimper prior to no-till planting soybean at the Cornell Musgrave Research Farm (credit: Jeff Liebert). (b) Organic soybean growing in mulch from a rolled cereal rye cover crop (credit: Matthew Ryan).

no-till is a viable way organic farmers can reduce tillage and associated labour and fuel costs while maintaining high yields of maize and soybean. Research is needed to understand the impacts of using high-biomass cover crops as mulch on soil health, but preliminary research shows potential for greater water infiltration and soil moisture holding capacity compared to traditional tillage based organic management practices (Ryan *et al.* 2015). No-till farmers that are not organic have also found that high-biomass cover crops can provide a number of soil health benefits while helping to increase weed suppression (Davis 2010, Price *et al.* 2016).

8. Agroecological theory and application

Using ecological theory to guide management agricultural systems can provide several benefits. First, knowledge of soil, crop, and pest ecology can improve management and increase productivity. Second, applying ecological principles can reduce negative impacts of agriculture and enhance ecosystem services that are provided by agricultural landscapes. Theory also helps to focus our understanding of agroecosystems and contributes to the continuous evolution of knowledge about agriculture and the natural world. Here we synthesise information related to agroecological soil management and introduce a new theoretical framework for managing ecological processes in the soil that are associated with carbon and nitrogen. Our goal is to examine ecological theory from natural ecosystems and apply this theory to soil management practices.

8.1. *Nitrogen pulse-performance hypothesis*

Mineralisation of soil organic matter, nitrogen immobilisation, and biological nitrogen fixation are key ecological processes that often occur in pulses over time in agroecosystems. These interrelated processes can either enhance or inhibit system performance depending on which crops are being grown. Although increasing soil organic matter is promoted, and rightfully so, utilising organic matter and the nitrogen that is released during the process of mineralisation for crop production is an opposing goal (Janzen 2006). Certainly increasing soil organic matter and providing nitrogen to crops are both important for crop production, but there is no clear framework for balancing trade-offs between these goals. Here, we posit that regenerative agriculture should strive for balance between mineralisation, immobilisation, and fixation over an extended period by working with natural cycles and enhancing processes with agroecological management practices.

Benefits of crop rotation, which were outlined earlier, stem from sequencing crops in an order that takes advantage of ecological processes. Having an understanding of the temporal dynamics of these processes, particularly mineralisation, nitrogen immobilisation, and biological nitrogen fixation, can facilitate management and improve productivity. From an applied perspective, avoiding crop sequences and management practices that reduce crop available nitrogen just prior to growing heavy nitrogen using crops might be obvious. Certainly, many maize farmers choose to plant maize after legume crops such as alfalfa or red clover, which can reduce fertiliser requirements and increase yields as described earlier. However, intentionally managing soil nitrogen to reduce availability and increase carbon storage in soils is less common. Depending on the crops being grown, this practice can reduce pests and increase crop performance. For example, organic soybeans will fix more atmospheric nitrogen and can perform better after a grass cover crop with a high Carbon Nitrogen ratio compared to a legume cover crop with a low Carbon Nitrogen (Schipanski *et al.* 2010). Lowering soil nitrogen prior to growing legumes reduces competition from weeds and reduces potential for Nitrogen losses.

Here we explore ecological theory and introduce the nitrogen pulse-performance hypothesis which is loosely based on ideas presented in *Nature's Pulsing Paradigm* of Odum *et al.* (1995). The nitrogen pulse-performance hypothesis states that: (1) soil function, crop productivity, and overall performance can be optimised by arranging legume and high non-legume crops and management practices in a system that promotes oscillations in mineralisation, immobilisation, and biological nitrogen fixation, and (2) reinforcing internal pulses with management practices and low levels of external inputs can create resonance and increase performance. In this sense, we consider resonance to be the result of external inputs (e.g. tillage and fertiliser) driving the system to oscillate with greater amplitude.

> *An attractive hypothesis is that by pulsing and appropriate coupling of external and internal pulsing, a higher performance is obtained, which generates more work, thus reinforcing the system's characteristics. If pulsing is general, then what is sustainable in ecosystems is a repeating oscillation* (Odum *et al.* 1995).

8.2. *Nitrogen pulse-performance application*

Modern maize and soybean crop rotations create an oscillation of soil nitrogen over time associated with nitrogen fertiliser application in maize but not soybean. This is beneficial because low soil nitrogen in maize decreases performance, whereas high soil nitrogen in soybean makes the crop more susceptible to competition from weeds and other pests and reduces biological nitrogen fixation. Farmers can use management practices and inputs to magnify these nitrogen dynamics. For example, in maize production, starter fertiliser and tillage prior to planting can supplement fertiliser applications and the internal pulse of mineralisation in the spring from soil organisms responding to warmer soil temperatures. Similarly a catch crop, such as cereal rye, seeded after maize harvest can provide a high Carbon Nitrogen residue in the spring, which can immobilise nitrogen before soybean.

In organic cropping systems, two-year maize–soybean crop rotations are extremely rare as such simplified systems often fail without inputs that are commonly used in conventional production (Cavigelli *et al.* 2008). The three-year, maize–soybean–wheat rotation is more common and several long-term cropping systems experiments have used this crop rotation, and thus it is fairly well characterised. Although there have been variations on the theme, it is common for cereal rye to be seeded after maize harvest and for red clover to be interseeded in winter wheat (Schipanski and Drinkwater 2011). This crop rotation has gained popularity with organic grain crop farmers because of the beneficial role that wheat can play in weed suppression and because of the nitrogen that the red clover supplies for maize. A recent examination of ecosystem services from cover crops using this rotation illustrates how soil nitrogen levels oscillate over time (Schipanski *et al.* 2014). The three-year rotation is arranged to take advantage of these pulses of nitrogen. In this case, the red clover cover crop results in greater mineralisation just prior to main planting, which can both decrease fertiliser requirements and increase yield potential (Figure 5). Other research shows how supplementing internal pulses with strategic use of inputs, such as

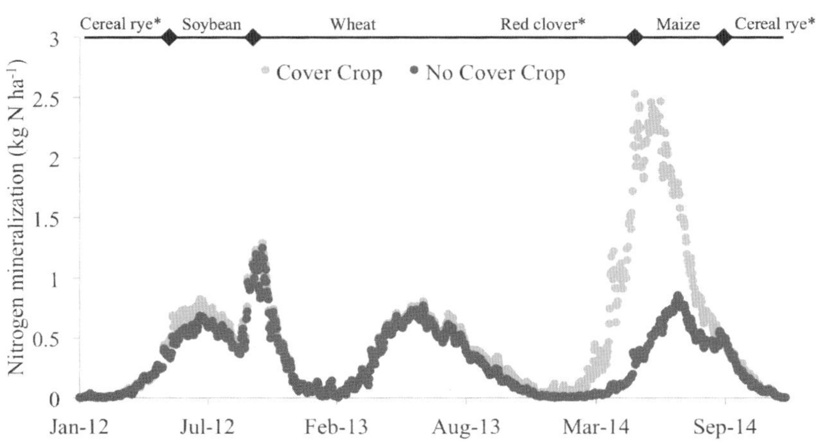

Figure 5. Nitrogen mineralisation dynamics over time in a 3-year soybean–wheat–maize rotation with and without cereal rye and red clover cover crops (modified from Schipanski *et al.* 2014). Tillage is represented by black diamonds separating crops (*cover crops were not included in the No Cover Crop system). Nitrogen mineralisation was modelled using the Cycles cropping systems model, which is a multiyear, multicrop, daily time step cropping systems simulation model (Kemanian and Stöckle 2010).

intercropping, cover crop mixtures, and banded poultry litter, can increase soil functioning and overall performance (Poffenbarger *et al.* 2015, Regehr *et al.* 2015).

In the long-term, Cornell Organic Cropping Systems Experiment, which used a maize–soybean–spelt rotation and compared four management systems that differed in nutrient and weed management, poultry litter additions prior to spelt increased yield and profitability whereas poultry litter before maize increased weeds and reduced profitability (Caldwell *et al.* 2014). In this case, the red clover prior to maize provided adequate nitrogen and the supplemental pulse of nitrogen from poultry litter was only beneficial in the spelt crop. Small grains are known to respond to nitrogen applications. Interestingly, small grains have among the greatest yield gaps between organic and conventional production (Seufert *et al.* 2012). This is likely due to inadequate nitrogen availability to the small grain. However, the yield gap between organic and conventional maize is much lower, suggesting that nitrogen supply is not limiting during the warmer months when nitrogen demand in maize is high. Thus, supplemental pulses of nitrogen for cool season crops could be more important than for warm season crops depending on how temperature affects mineralisation. Supplementing organic small grains with additional nitrogen will also likely increase crop yields and reduce the yield gaps with conventional crops.

Although we have not yet explicitly tested the nitrogen pulse productivity hypothesis, we believe it provides a useful framework for studying soil nutrient management and using ecological knowledge to improve cropping system performance. The hypothesis is based on the observation that a large amount of energy is used in agriculture to provide nutrients, manage weeds, and maintain a perpetual state of secondary succession with abundant resources and no competition (Smith 2015). This energy use is proportional to the environmental degradation from agriculture and has largely replaced ecological processes that farmers have relied on for thousands of years. Combined with adaptive management, understanding the patterns of ecological processes and working to reinforce rather than resist pulses of energy (e.g. nutrients) can increase efficiency and contribute to ecological intensification.

9. Conclusion

Soil health can be regenerated using a variety of management practices based on agroecological principles (Table 1). These practices facilitate ecological processes and reinforce system integrity because of their positive effects on the abundance and diversity of soil organisms. They are helpful both for crop production and long-term soil health. However, soil health has to be managed at cropping system and farm scale as the combination of regenerative practices is more powerful than single practice to preserve soil. For instance, reduced tillage practices need to be coupled with crop rotation, cover crops, and soil organic matter management to maximise their beneficial impact on soil health (Palm *et al.* 2014). Crop production also benefits greatly from close integration with livestock production systems as manure applications, grazing, and perennial forage production all enhance soil health. Although, rotational no-till and intercropping with high crop diversity might be challenging for farmers in terms of knowledge and technical skill, diversifying crop rotations, using cover crops, applying organic amendments, and reducing tillage are all sound agroecological practices that can be easily integrated into existing cropping systems (Wezel *et al.* 2014). Future research should focus on identifying synergies between agroecological practices and the development of systems that take advantage of positive interactions between ecological processes.

Table 1. Agroecological principles and how they relate to soil and crop management practices.

Agroecological principles	Soil and crop management practices to regenerate soil health		
	Crop rotation, cover crops, intercropping	Organic amendments	Reduced-tillage practices
Employ biology	Greater root density and diversity in the soil increases nutrient supply and crop diversity increases ecosystem services (biological control).	Carbon feeds soil organisms and increases fertility through mineralisation and biological nitrogen fixation.	Reducing tillage preserves soil organisms, which create biopores, and improves soil structure and habitat for soil organisms.
Foster diversity	Create greater crop diversity at spatial (intercrop, crop rotation) and temporal (cover crop, crop rotation) scales.	Organic amendments can increase soil biodiversity.	Increase soil biodiversity in certain conditions and in addition to crops diversity introduction.
Cycle nutrients	Catch crops can prevent leaching and intercropping systems can be more nutrient efficient.	Organic amendments can replace nutrients lost by crop removal.	Decreased tillage can prevent erosion and reduce mineralisation thus minimising nutrient losses.
Integrate perennials	Perennial crops in the crop rotation increase soil nutrient availability and soil organic matter.	Perennial crops increase carbon inputs through root exudates and crop residues.	Perennial crops in the crop rotation decrease tillage frequency at the crop rotation scale.
Combine production	Animals can use crop residues and cover crops directly through grazing and indirectly as forage.	Livestock provide organic inputs in the form of manure.	Growing perennial forage crops and grazing reduces tillage frequency.
Nurture soil	Diversity and permanence of roots and crop residues in the soil increase soil organisms.	Fresh organic matter provides energy to support the soil food web.	Reduced tillage intensity and frequency preserves soil organisms.

References

Altieri, M.A., 2000. Agroecology: Principles and Strategies for Designing Sustainable Farming Systems. In: Uphoff, N. (Ed.), *Agroecological Innovations: Increasing Food Production with Participatory Development.* Earthscan Publications, London, UK, pp. 40–46.

Andraski, T.W., Bundy, L.G., 2005. Cover crop effects on corn yield response to nitrogen on an irrigated sandy soil. *Agronomy Journal* **97**, 1239–1244.

Balesdent, J., Chenu, C., Balabane, M., 2000. Relationship of soil organic matter dynamics to physical protection and tillage. *Soil and Tillage Research* **53**, 215–230.

Bell, L.W., Moore, A.D., Kirkegaard, J.A., 2014. Evolution in crop–livestock integration systems that improve farm productivity and environmental performance in Australia. *European Journal of Agronomy* **57**, 10–20.

Blouin, M., Zuily-Fodil, Y., Pham-Thi, A., Laffray, D., Reversat, G., Pando, A., Tondoh, J., Lavelle, P., 2005. Belowground organism activities affect plant aboveground phenotype, inducing plant tolerance to parasites. *Ecology Letters* **8**, 202–208.

Bommarco, R., Kleijn, D., Potts, S.G., 2013. Ecological intensification: Harnessing ecosystem services for food security. *Trends Ecology and Evolution* **28**, 230–238.

Bunch, R., 2012. Restoring the Soil: *A Guide for Using Green Manure/cover Crops to Improve the Food Security for Smallholder Farmers.* Canadian Foodgrains Bank.

Bybee-Finley, K.A., Mirsky, S.B., Ryan, M.R., 2016. Functional diversity in summer annual grass and legume intercrops in the northeastern United States. *Crop Science* **56**, 1–16.

Caldwell, B., Mohler, C.L., Ketterings, Q.M., DiTommaso, A., 2014. Yields and profitability during and after transition in organic grain cropping systems. *Agronomy Journal* **106**, 871–880.

Capowiez, Y., Cadoux, S., Bouchant, P., Ruy, S., Roger-Estrade, J., Richard, G., Boizard, H., 2009. The effect of tillage type and cropping system on earthworm communities, macroporosity and water infiltration. *Soil and Tillage Research* **105**, 209–216.

Cavigelli, M.A., Teasdale, J.R., Conklin, A.E., 2008. Long-term agronomic performance of organic and conventional field crops in the mid-Atlantic region. *Agronomy Journal* **100**, 785–794.

Cong, W., Hoffland, E., Li, L., Six, J., Sun, J., Bao, X., Zhang, F., Van Der Werf, W., 2015. Intercropping enhances soil carbon and nitrogen. *Global Change Biology* **21**, 1715–1726.

Culman, S.W., Snapp, S.S., Ollenburger, M., Basso, B., DeHaan, L.R., 2013. Soil and water quality rapidly responds to the perennial grain Kernza wheatgrass. *Agronomy Journal* **105**, 735–744.

Dang, Y., Seymour, N., Walker, S., Bell, M., Freebairn, D., 2015. Strategic tillage in no-till farming systems in Australia's northern grains-growing regions: I. Drivers and implementation. *Soil and Tillage Research* **152**, 104–114.

Davidson, D., 2014. Evaluating the quality of your soil. *Crops Soils* **47**, 4–13.

Davis, A.S., 2010. Cover-crop roller-crimper contributes to weed management in no-till soybean. *Weed Science* **58**, 300–309.

Davis, A.S., Hill, J.D., Chase, C.A., Johanns, A.M., Liebman, M., 2012. Increasing cropping system diversity balances productivity, profitability and environmental health. *PLoS ONE* **7**, e47149.

Diacono, M., Montemurro, F., 2010. Long-term effects of organic amendments on soil fertility. A review. *Agronomy for Sustainable Devlopment* **30**, 401–422.

Doran, J.W., Parkin, T.B., 1994. Defining and assessing soil quality. *SSSA Special Publication, Defining Soil Quality for a Sustainable Environment*, **35**, 1–21.

Food and Agriculture Organization of the United Nations, 2014. Final Report for the *International Symposium on Agroecology for Food Security and Nutrition*. Rome, Italy.

Gattinger, A., Muller, A., Haeni, M., Skinner, C., Fliessbach, A., Buchmann, N., Mäder, P., Stolze, M., Smith, P., Scialabba, N.E.-H., 2012. Enhanced top soil carbon stocks under organic farming. *Proceeding of the National Academy of Sciences* **109**, 18226–18231.

Gaudin, A., Westra, S., Loucks, C.E., Janovicek, K., Martin, R.C., Deen, W., 2013. Improving resilience of Northern field crop systems using inter-seeded red clover: A Review. *Agronomy* **3**, 148–180.

Gianinazzi, S., Gollotte, A., Binet, M.-N., van Tuinen, D., Redecker, D., Wipf, D., 2010. Agroecology: the key role of arbuscular mycorrhizas in ecosystem services. *Mycorrhiza* **20**, 519–530.

Gliessman, S.R., Rosemeyer, M., 2009. *The Conversion to Sustainable Agriculture: Principles, Processes, and Practices*. CRC Press/Taylor & Francis Group, Boca Raton, Florida, USA.

Gugino, B.K., Abawi, G.S., Idowu, O.J., Schindelbeck, R.R., Smith, L.L., Thies, J.E., Wolfe, D.W., van Es, H.M., 2009. *Cornell Soil Health Assessment Training Manual*. Cornell University College of Agriculture and Life Sciences, Ithaca, New York, USA.

Haney, R., 2013. *Soil Testing for Soil Health* (Power Point). United States Department of Agriculture.

Haynes, R., 2005. Labile organic matter fractions as central components of the quality of agricultural soils: an overview. *Advances in Agronomy* **85**, 221–268.

Henry, D.C., Mullen, R.W., Dygert, C.E., Diedrick, K.A., Sundermeier, A., 2010. Nitrogen contribution from red clover for corn following wheat in western Ohio. *Agronomy Journal* **102**, 210–215.

Holland, J., 2004. The environmental consequences of adopting conservation tillage in Europe: reviewing the evidence. *Agricultural Ecosystems and Environment* **103**, 1–25.

Idowu, O., Van Es, H., Abawi, G., Wolfe, D., Schindelbeck, R., Moebius-Clune, B., Gugino, B., 2009. Use of an integrative soil health test for evaluation of soil management impacts. *Renewable Agriculture and Food Systems* **24**, 214–224.

Janzen, H., 2006. The soil carbon dilemma: shall we hoard it or use it? *Soil Biology and Biochemistry* **38**, 419–424.

Kallenbach, C., Grandy, A., Frey, S., Diefendorf, A., 2015. Microbial physiology and necromass regulate agricultural soil carbon accumulation. *Soil Biology and Biochemistry* **91**, 279–290.

Kemanian, A.R., Stöckle, C.O., 2010. C-Farm: A simple model to evaluate the carbon balance of soil profiles. *Eur J Agron* **32**, 22–29.

Kladivko, E.J., 2001. Tillage systems and soil ecology. *Soil and Tillage Research* **61**, 61–76.

Köller, K., El Titi, A., 2003. Techniques of soil tillage. In: A. El Titi (ed.), *Soil Tillage Agroecosystems*. CRC Press. Boca Raton, Florida, USA. pp. 1–25.

Kuntz, M., Berner, A., Gattinger, A., Scholberg, J., Mäder, P., Pfiffner, L., 2013. Influence of reduced tillage on earthworm and microbial communities under organic arable farming. *Pedobiologia* **56**, 251–260.

Lawrence, J.R., Ketterings, Q.M., Cherney, J.H., 2008. Effect of nitrogen application on yield and quality of silage corn after forage legume-grass. *Agronomy Journal* **100**, 73–79.

Lehmann, J., Kleber, M., 2015. The contentious nature of soil organic matter. *Nature* **528**, 60–68.

Lemanceau, P., Maron, P.-A., Mazurier, S., Mougel, C., Pivato, B., Plassart, P., Ranjard, L., Revellin, C., Tardy, V., Wipf, D., 2015. Understanding and managing soil biodiversity: a major challenge in agroecology. *Agronomy for Sustainable Devlopment* **35**, 67–81.

Lin, B.-H., Taylor, H., Delvo, H., Bull, L., 1995. Factors influencing herbicide use in corn production in the North Central Region. *Review of Agricultural Economics* **17**, 159–169.

Loveland, P., Webb, J., 2003. Is there a critical level of organic matter in the agricultural soils of temperate regions: a review. *Soil and Tillage Research* **70**, 1–18.

Magdoff, F., 2007. Ecological agriculture: Principles, practices, and constraints. *Renewable Agriculture and Food Systems* **22**, 109–117.

McDaniel, M.D., Tiemann, L.K., Grandy, A.S., 2014. Does agricultural crop diversity enhance soil microbial biomass and organic matter dynamics? A meta-analysis. *Ecological Applications* **24**, 560–570.

Mirsky, S.B., Ryan, M.R., Curran, W.S., Teasdale, J.R., Maul, J., Spargo, J.T., Moyer, J., Grantham, A.M., Weber, D., Way, T.R., Camargo, G.G., 2012. Conservation tillage issues: Cover crop-based organic rotational no-till grain production in the mid-Atlantic region, USA. *Renewable Agriculture and Food Systems* **27**, 31–40.

Mirsky, S.B., Ryan, M.R., Teasdale, J.R., Curran, W.S., Reberg-Horton, C.S., Spargo, J.T., Wells, M.S., Keene, C.L., Moyer, J.W., 2013. Overcoming weed management challenges in cover crop–based organic rotational no-till soybean production in the Eastern United States. *Weed Technology* **27**, 193–203.

Nord, E.A., Ryan, M.R., Curran, W.S., Mortensen, D.A., Mirsky, S.B., 2012. Effects of management type and timing on weed suppression in soybean no-till planted into rolled-crimped cereal rye. *Weed Sciences* **60**, 624–633.

Odum, W.E., Odum, E.P., Odum, H.T., 1995. Nature's pulsing paradigm. *Estuaries* **18**, 547–555.

Palm, C., Blanco-Canqui, H., DeClerck, F., Gatere, L., Grace, P., 2014. Conservation agriculture and ecosystem services: an overview. *Agric Ecosyst Environ* **187**, 87–105.

Peigné, J., Cannavaciuolo, M., Gautronneau, Y., Aveline, A., Giteau, J.-L., Cluzeau, D., 2009. Earthworm populations under different tillage systems in organic farming. *Soil and Tillage Research* **104**, 207–214.

Peigné, J., Casagrande, M., Payet, V., David, C., Sans, F.X., Blanco-Moreno, J.M., Cooper, J., Gascoyne, K., Antichi, D., Bàrberi, P. and Bigongiali, F., 2016. How organic farmers practice conservation agriculture in Europe. *Renewable Agriculture and Food Systems* **31**, 72–85.

Peigné, J., Vian, J.-F., Cannavacciuolo, M., Lefevre, V., Gautronneau, Y., Boizard, H., 2013. Assessment of soil structure in the transition layer between topsoil and subsoil using the profil cultural method. *Soil and Tillage Research* **127**, 13–25.

Poeplau, C., Don, A., 2015. Carbon sequestration in agricultural soils via cultivation of cover crops–A meta-analysis. *Agriculture, Ecosystems and Environment* **200**, 33–41.

Poffenbarger, H.J., Mirsky, S.B., Weil, R.R., Kramer, M., Spargo, J.T., Cavigelli, M.A., 2015. Legume proportion, poultry litter, and tillage effects on cover crop decomposition. *Agronomy Journal* **107**, 2083–2096.

Price, A., Monks, C., Culpepper, A., Duzy, L., Kelton, J., Marshall, M., Steckel, L., Sosnoskie, L., Nichols, R., 2016. High-residue cover crops alone or with strategic tillage to manage glyphosate-resistant Palmer amaranth (*Amaranthus palmeri*) in southeastern cotton (*Gossypium hirsutum*). *Journal of Soil and Water Conservation* **71**, 1–11.

Regehr, A., Oelbermann, M., Videla, C., Echarte, L., 2015. Gross nitrogen mineralization and immobilization in temperate maize-soybean intercrops. *Plant Soil* **391**, 353–365.

Ryan, M., Wayman, S., Crowly, K., Youngerman, C., Curran, W., Mirsky, S., 2015. *Agroecological Strategies for Balancing Tradeoffs in Organic Corn and Soybean Production*. USDA, Organic Project Director Meeting, Washington, D.C., USA.

Säle, V., Aguilera, P., Laczko, E., Mäder, P., Berner, A., Zihlmann, U., van der Heijden, M.G., Oehl, F., 2015. Impact of conservation tillage and organic farming on the diversity of arbuscular mycorrhizal fungi. *Soil Biology and Biochemistry* **84**, 38–52.

Schipanski, M., Drinkwater, L., 2011. Nitrogen fixation of red clover interseeded with winter cereals across a management-induced fertility gradient. Nutr *Cycl Agroecosystems* **90**, 105–119.

Schipanski, M., Drinkwater, L., Russelle, M., 2010. Understanding the variability in soybean nitrogen fixation across agroecosystems. *Plant Soil* **329**, 379–397.

Schipanski, M.E., Barbercheck, M., Douglas, M.R., Finney, D.M., Haider, K., Kaye, J.P., Kemanian, A.R., Mortensen, D.A., Ryan, M.R., Tooker, J., White, C., 2014. A framework for evaluating ecosystem services provided by cover crops in agroecosystems. *Agricultural Systems* **125**, 12–22.

Scholes, M.C., Scholes, R.J., 2013. Dust unto dust. *Science* **342**, 565–566.

Seufert, V., Ramankutty, N., Foley, J.A., 2012. Comparing the yields of organic and conventional agriculture. *Nature* **485**, 229–232.

Smith, R.G., 2015. A succession-energy framework for reducing non-target impacts of annual crop production. *Agricultural Systems* **133**, 14–21.

Smith, R.G., Gross, K.L., Robertson, G.P., 2008. Effects of crop diversity on agroecosystem function: Crop yield response. *Ecosystems* **11**, 355–366.

Smith, R.G., Ryan, M.R., Menalled, F.D., 2011. Direct and indirect impacts of weed management practices on soil quality. In: F.D., Hatfield, J.L. and Sauer, T.J. (Eds.), *Soil Management: Building a Stable Base for Agriculture*.

American Society of Agronomy and Soil Science Society of America, Madison, Wisconsin, USA, pp. 275–286.

Snapp, S.S., 2011. Agroecology: principles and practice. In: *Agricultural Systems: Agroecology and Rural Innovation for Development*. S.S. Snapp and B. Pound (Eds.), Academic Press, Burlington, MA, USA, pp. 53–88.

Teboh, J.M., Franzen, D.W., 2011. Buckwheat (*Fagopyrum esculentum* Moench) potential to contribute solubilized soil phosphorus to subsequent crops. *Communications in Soil Science and Plant Analysis* **42**, 1544–1550.

Tittonell, P., 2014. Ecological intensification of agriculture — sustainable by nature. *Current Opinion in Environmental Sustainability* **8**, 53–61.

Vandermeer, J., Perfecto, I., 2013. Complex traditions: intersecting theoretical frameworks in agroecological research. *Agroecology and Sustainable Food Systems* **37**, 76–89.

Venterea, R.T., Baker, J.M., Dolan, M.S., Spokas, K.A., 2006. Carbon and nitrogen storage are greater under biennial tillage in a Minnesota corn–soybean rotation. *Soil Science Society American Journal* **70**, 1752–1762.

Vian, J.F., Peigné, J., Chaussod, R., Roger-Estrade, J., 2009. Effects of four tillage systems on soil structure and soil microbial biomass in organic farming. *Soil Usage and Management* **25**, 1–10.

Watson, C., Stockdale, E., Rees, R., 2008. Assessment and maintenance of soil fertility in temperate organic agriculture. *CAB Reviews Perspectives in Agriculture Veterinary Science Nutrition and Natural Resources* **3**, 1–11.

Wezel, A., Casagrande, M., Celette, F., Vian, J.-F., Ferrer, A., Peigné, J., 2014. Agroecological practices for sustainable agriculture. A review. *Agronomy for Sustainable Development* **34**, 1–20.

Chapter 4

Transitions to Agroecological Nutrient Management Practices in the USA Corn Belt

Jennifer Blesh* and Ryan E. Galt†

**School of Natural Resources and Environment, University of Michigan, Ann Arbor, Michigan, USA*
†Department of Human Ecology, University of California, Davis, California, USA

1. Introduction

With concurrent food, energy, and climate crises, the question of how to produce food more sustainably is at the centre of many public debates (IAASTD 2009, Foley *et al.* 2011, Kremen and Miles 2012). The global environmental consequences of industrial agriculture are well-documented and include contributions to the greenhouse gas emissions that cause climate change (up to 35% of the global total (Vermeulen *et al.* 2012, IPCC 2013)), losses of soil organic matter and biodiversity, and pollution of surface and ground waters (Matson *et al.* 1997, Diaz and Rosenberg 2008). Further, despite national and international farm policies that emphasise production (yield) over other goals, globally, 795 million people are chronically undernourished (FAO *et al.* 2014) since vastly unequal social systems mediate access to food and productive resources (Lappé *et al.* 1998).

Given the urgency of the linked food and ecological crises, and the vulnerability of the agricultural sector to global change, it is critical that we understand how to manage agroecosystems to support ecological resilience, rural communities, and food security. Standard economic definitions of agricultural efficiency based on productionism (yield per unit area or per worker) are too narrow, and neglect many important dynamics (Buttel 2005). More accurate definitions of agricultural efficiency that take a systems perspective expand the dimensions that matter by considering the type and quality of foods produced, equity of food distribution and access, and the ecological impacts of production (Farley *et al.* 2011).

Ecological knowledge, in particular, can inform management of biotic processes and interactions (i.e. biodiversity) to sustain ecosystem functions such as productivity and nutrient cycling, while reducing fossil fuel-based inputs and associated pollution (Lowrance *et al.* 1984, Shennan 2008, Kremen and Miles 2012). Analysing different agricultural management systems through an ecological perspective provides a more complete understanding of their variable impacts along multiple dimensions such as external input use, emissions to air or water, yield, soil erosion, and tillage intensity. Improved understanding of these interactions can inform transitions to multifunctional agroecosystems that simultaneously meet production and conservation goals, and which should be recognised under multifunctional policy and better compensated (Clark 2005, McCarthy 2005).

In this chapter, we examine the dynamic, interconnected social and ecological processes that support or constrain farmers' transitions to agroecology. In particular, we draw upon an interdisciplinary study integrating approaches from agroecology and political ecology to better understand grain and livestock farmers in the USA Corn Belt who have transitioned to soil fertility management practices based on ecological principles. We begin with a brief overview of agroecology and its application to soil fertility management in temperate climate cropping systems. We then introduce the USA Corn Belt, which is the region where our case studies were located and describe two contrasting frameworks for nutrient management present in this region. We turn to our research findings on the relationship between farm management practices and nitrogen retention that highlight how ecological knowledge can contribute to enhancing environmental sustainability. Building on the ecological results, we then examine social processes and knowledge networks at multiple scales involved in transitioning to practices that reduce nitrogen pollution from farms. We conclude by discussing future research and activities that can facilitate broader transitions to ecological management and multifunctional agroecosystems.

1.1. *Agroecology: Definitions and application to soil fertility management*

Agroecology applies ecological principles and systems thinking to the design, analysis, and management of farms, linking theory with farmer practice (Gliessman 2007). For example, the agroecosystem concept emphasises flows of matter and energy, and interactions between organisms and their environment across scales. Agroecology blends theories and methods from ecology and agronomy, which is its own interdisciplinary field, e.g. including weed science and plant pathology. In addition to spanning several natural science disciplines, agroecology also bridges the natural and social sciences, encompassing social dimensions of agroecosystems (Tomich *et al.* 2011). From the early development of the field, agroecologists recognised the inevitably coupled human–natural linkages that occur in managed ecosystems in which humans make management decisions within particular socio-political and economic contexts (e.g. Odum 1984).

Farmers can manipulate ecological relationships and processes such as predation, competition, or decomposition to manage for particular ecosystem functions or services — the goods or benefits that people derive from ecosystems (MEA 2003). By intentionally managing biotic diversity to increase positive ecological interactions, farmers can reduce or eliminate the external (chemical) inputs that sustain production in industrial agroecosystems (Shennan 2008). It is important to note that all farms are agroecosystems and can be assessed along many dimensions through an ecological perspective. Additionally, agroecological practices are management practices that apply ecological principles to support particular ecosystem functions such as productivity, soil fertility, or nutrient cycling (Wezel *et al.* 2014). This framing of agroecosystems and agroecological practices is broader than allowed for under some definitions of agroecology which place strong emphasis on traditional farmer knowledge and social movements promoting major transformations of agriculture (Altieri and Toledo 2011). However, focusing on management practices informed by ecological principles is useful for exploring how system change might occur within highly industrialised regions where the links to intergenerational knowledge systems about non-industrial practices have been mostly lost, and in which there are few to no radical social movements advocating for an agroecological vision (Qazi and Selfa 2005).

In this chapter, our specific focus under the broad umbrella of agroecology is soil fertility management, and its impacts on carbon and nitrogen cycles. Nitrogen is the nutrient that most frequently limits crop productivity in temperate ecosystems. Beginning with the development of the Haber–Bosch process for synthesising ammonia in 1909, and quickly increasing during the Green Revolution, synthetic nitrogen fertilisers are now widely applied to agricultural landscapes to increase crop yields (Smil 1999, Galloway *et al.* 2008). Nutrient management is a particularly salient concern in Midwestern United States, a region that contributes the majority of the nitrogen and phosphorus pollution causing a hypoxic zone to form in the Gulf of Mexico each summer (Rabalais *et al.* 2002). More broadly, human alteration of the global nitrogen cycle through fertiliser inputs contributes to greenhouse gas emissions, and to over 400 hypoxic zones worldwide (Diaz and Rosenberg 2008).

Increased fertiliser use is tightly linked to the rapid spatial expansion of agroecosystems with low crop diversity that were created to conform to political economic realities within agriculture and the food system (Lyson 2004). As we discuss in the section below, titled Ecological Nitrogen Management, managing soil nutrients more sustainably requires systemic changes to industrial cropping systems. These changes would also impact other aspects of agroecosystem management such as controlling weeds, insect pests, and diseases. Fertility management is thus a core dimension of an agroecological farming approach with broad implications for global ecological sustainability.

1.1.1. Political ecology meets agroecology

An important, complementary lens to employ with agroecological concepts and methods is that of political ecology (Galt 2013, Gonzalez de Molina 2013). A multidisciplinary field that explores society–environment relationships, one long-standing emphasis in political ecology is understanding land users' decision-making within the context of local communities, nation states, and global markets and institutions, and the unequal access to resources often created through these social institutions (Blaikie 1985, Stonich 1993, Zimmerer 1996, Galt 2014). These meso- and macro-level actors and conditions establish structural relations that support or constrain farm-scale transitions to more ecologically-sensitive management (Galt 2014), including ecological nutrient management. Political ecology also draws upon the philosophy of critical realism, thereby sidestepping many problems of positivist ways of knowing by acknowledging a multiplicity of perspectives, including around the important concept of sustainability, and the ways in which power shapes knowledge production (Forsyth 2003, Taylor 2005, Lewontin and Levins 2007). In this chapter we use a political ecological perspective to better understand the connections between land management and farmers' knowledge networks, competencies on and off the farm, and perceptions of civic and professional communities. We also examine farmers' perceptions of sustainability and water quality, as these discourses and goals are important context for understanding land management practices.

1.2. *The United States Corn Belt: Highly productive and highly polluting*

The Mississippi River Basin is an enormous watershed that drains 40% of the contiguous United States. The region known as the Corn Belt spreads across the upper Mississippi River Basin, and is where corn has become the dominant crop on the landscape, replacing mostly native prairie vegetation starting with Euro-American settlement and often violent displacement of Native Americans in the mid-1800s. The Corn Belt has been an extremely important agricultural area for the USA since this settlement. While the area occupies only 8% of the USA territory, as early as in 1919, its farmlands accounted for 40% of the value of all agricultural land in the USA due to its fertility and potential for production (Baker 1927). The high economic value of the land means that it has been farmed more intensively than most other regions, as farmers need to produce higher levels of returns to pay for expensive land (either through mortgages or leases). Farming in the area currently relies on very high levels of external inputs in the form of seeds, machinery, and synthetic pesticides and fertilisers.

The large inputs of nitrogen fertilisers are a core component of a technical package that substitutes chemical inputs for ecological complexity and biological processes, and increases dependence on an industrial approach to crop production. The extraordinary intensification of agriculture in the Corn Belt region has contributed to significant reductions in enterprise diversity and biodiversity (Buttel *et al.* 1990). Farms consolidated, farm size increased, and material inputs and infrastructure became relatively homogeneous, as did the knowledge systems that determine how they are managed and used (Lighthall and Roberts 1995). This transformation was enabled in part by the region's flat topography and very fertile, former prairie soils (Cronon 2009). The combination of biophysical characteristics with public and private sector investment in industrial agriculture, globalised markets for agricultural commodities, supportive national policy (various iterations of the Farm Bill), concentration of livestock in confinement operations, and absence of regulation of the pollution that results from this agricultural model have produced the current landscapes of corn and soybeans that dominate the region's arable land. In addition, at larger scales, agricultural intensification is driven by changes in institutions, governance, discourses, and corporate power (McMichael 2009). These processes push agricultural management in the region along trajectories that decrease agrobiodiversity and discourage agroecological practices that support long-term fertility and minimise negative off-farm impacts.

As a whole, the Corn Belt is the dominant source of the nitrogen pollution — primarily lost as riverine nitrate — that causes the Gulf of Mexico hypoxic zone. Both basin- and county-scale models identify several factors driving nitrogen losses from farms in this region: increased presence of corn monocultures that receive large inputs of nitrogen fertiliser; tile drainage, which speeds drainage of water from fields and channels soluble nitrogen more quickly to surface waters, particularly in the winter and spring; and losses of soil organic matter, which reduces the capacity for soils to store nitrogen (Howarth *et al.* 2002, David *et al.* 2010).

Globally, grains, oilseeds, and legumes are planted on nearly 70% of cropland (Glover *et al.* 2010), and their management directly impacts both human and ecosystem health. In the Corn Belt — where many counties have up to 90% of total land area in either corn or soybeans — less than 25% of total crop production ends up in foods consumed directly by people. Instead, the majority goes to uses such as animal feed, fibre, or biofuels (Foley *et al.* 2011). Humans may eventually eat the animal feed as meat; however, allocating grains from the most productive agricultural lands to animal feed is widely recognised as ecologically and energetically inefficient and thus potentially reduces global food supplies (Foley *et al.* 2011). There is a critical need, then, to explore leverage points for transitions to multifunctional grain and livestock agroecosystems that are resource conserving and produce more diverse, nutritious foods for a range of market types.

The Corn Belt is home to a growing number of community organisations promoting local, organic, and sustainable foods, but many of these groups primarily focus on restructuring production and consumption relations around fruits and vegetables through direct marketing such as farmers' markets and community supported agriculture (Galt 2011). In the USA, such groups tend to focus much less on non-perishable staples like grains, which are typically processed before consumption. However, there are some farmer organisations, such as Practical Farmers of Iowa, which have a substantial number of grain farmers interested in alternative practices (Bell 2004).

Iowa and Ohio occupy western and eastern positions in the Corn Belt, respectively, and are the focus of our case study. At state and county scales, both states contribute significantly to nitrogen pollution (David *et al.* 2010) and are dominated by heavy use of industrial inputs. Yet, at smaller spatial scales in each state there are a handful of farms engaged in a broad range of management types, including organic grain farms, diversified farms with mixed crop-livestock production, and grain farms with winter small grains or annual cover crops. It is important to note that these farms are also present, even if such practices are not well-represented at larger scales on the landscape.

1.3. *Industrial and ecological nutrient management: Two contrasting frameworks*

In contrast to natural ecosystems, agricultural systems have large, regular exports of nitrogen and other nutrients in harvested crops. For continued production, these nutrients must be replenished to sustain soil fertility over time. Industrial and ecological nutrient management have been identified as two contrasting frameworks informing soil fertility management in agroecosystems (Drinkwater and Snapp 2007). Both approaches are present on farms in the Corn Belt, although industrial management is by far the dominant approach. Below we briefly summarise and compare these frameworks with respect to nitrogen management.

1.3.1. Industrial nitrogen management

An industrial approach to nitrogen management, which rapidly gained support in the 1940s and 1950s, aims to increase crop yields by applying inorganic nitrogen fertilisers, since nitrate and ammonium are the nitrogen forms that plants assimilate. This approach has become dominant in developed countries due to reinforcing feedback loops within industrially managed agroecosystems. For example, most farms in the Corn Belt now have low planned and associated biodiversity — with reduced cultivation of crops like winter grains and perennial forages that used to provide key ecological functions — and a corresponding increase in the use of chemical fertilisers and pesticides.

Corn productivity responds strongly to nitrogen inputs, national policies have long promoted high corn yields through commodity payments, and farmer and extension discourses in the region tend to focus on overall production (e.g. aiming for "200 bushel [per acre] corn") rather than economic returns or other indicators of success (e.g. Caldwell 2016). Farmers therefore often apply excess or "insurance" nitrogen to guarantee high productivity. As a result, most Corn Belt farm fields are prone to losses because they are nitrogen saturated (Fenn *et al.* 1998, Drinkwater and

Snapp 2007), consistently receiving nitrogen inputs in excess of harvested exports such that nitrogen availability exceeds the capacity for the agroecosystem to cycle or store the nitrogen internally.

Nitrogen readily changes chemical form through multiple biogeochemical processes, and a given nitrogen atom can therefore follow numerous pathways and impact different environmental reservoirs throughout the whole ecosystem (Galloway *et al.* 2003). The inorganic nitrogen ions, nitrate and ammonium, are easily lost from agroecosystems through multiple processes of the nitrogen cycle. Industrial farm fields have reduced capacity to retain inorganic nitrogen because of low stocks of soil organic matter, increased frequency and intensity of disturbance (e.g. tillage, planting, harvesting), and low crop diversity (e.g. often just corn and soybeans). They therefore have limited timeframes of plant growth, which are dominated by annual rather than perennial species, and extended bare fallow periods. Any applied inorganic nitrogen that is not taken up by corn plants — which is 50–60% of what is applied, on average (Smil 1999, Gardner and Drinkwater 2009) — can be potentially lost in water by leaching or runoff, or in gaseous forms to the atmosphere. The actual amount of nitrogen lost to surrounding ecosystems is governed by interactions between management practices and environmental conditions such as soil texture, temperature, and precipitation.

1.3.2. Ecological nitrogen management

A growing number of researchers and farmers have been developing a systems framework for nutrient management based on principles of ecology (Drinkwater *et al.* 2008). In contrast to industrial management, an ecological nitrogen management framework attempts to balance nitrogen inputs with exports in harvested crops, to reduce chemical inputs, and increase agroecosystem nitrogen cycling efficiency. Several lines of evidence support the potential for ecological practices to reduce nitrogen losses with minimal reductions in crop yield, including quantitative syntheses of the literature (Tonitto *et al.* 2006, Gardner and Drinkwater 2009), greenhouse studies (Nissen and Wander 2003), long-term cropping systems experiments (Clark *et al.* 1998, Drinkwater *et al.* 1998), and on-farm research (Blesh and Drinkwater 2013).

An ecological nitrogen management approach better mimics nitrogen cycling in natural ecosystems where biological nitrogen fixation represents the primary input of "new" nitrogen, and microbial decomposition of organic matter, and interactions between plants and microorganisms, drive the internal nitrogen cycle. Soil organic matter is approximately 50% carbon, and is the main energy source for microbial processes that can conserve nitrogen in soil. By increasing organic matter amendments and agroecosystem plant diversity, the cycling of nitrogen and carbon is better "coupled", meaning these elements flow through ecosystem reservoirs together (Drinkwater and Snapp 2007).

Farmers can build soil organic matter through a number of practices. They can use organic nitrogen sources such as legumes, animal manure, and compost, all of which must decompose via microbial processes to convert the nitrogen to plant-available forms. Farmers can also increase the proportion of the year that a field has living plant cover. A longer period of plant growth increases carbon inputs through photosynthesis. It also increases the window for plants to assimilate soil nitrogen into their biomass. Agroecosystems typically have much lower biodiversity than natural ecosystems; therefore, small increases in biodiversity with more complex crop rotations can lead to large benefits, particularly if plant functional group diversity is increased (Jackson *et al.* 2007, Shennan 2008). Strategies for adding diversity to annual cropping systems include planting a greater number of plant functional types, such as perennials (e.g. forages or pastures) or legumes, as well as non-harvested plants such as cover crops.

Changes in nitrogen source and crop rotation complexity often co-occur (e.g. by planting leguminous cover crops), and their respective effects on nitrogen cycling are difficult to disentangle. In this chapter, we analyse nutrient management practices along these two management dimensions: rotational complexity and relative reliance on legume nitrogen sources (Figure 1). Increased crop rotation diversity often leads to greater and/or more diverse plant residue and root inputs that build or maintain soil organic matter as they decompose (Drinkwater *et al.* 1998, Gregorich *et al.* 2001). More diverse organic matter inputs can support soil microbial processes and increased soil food web complexity, which enhances soil nutrient turnover (Clarholm 1985, Culman *et al.* 2010).

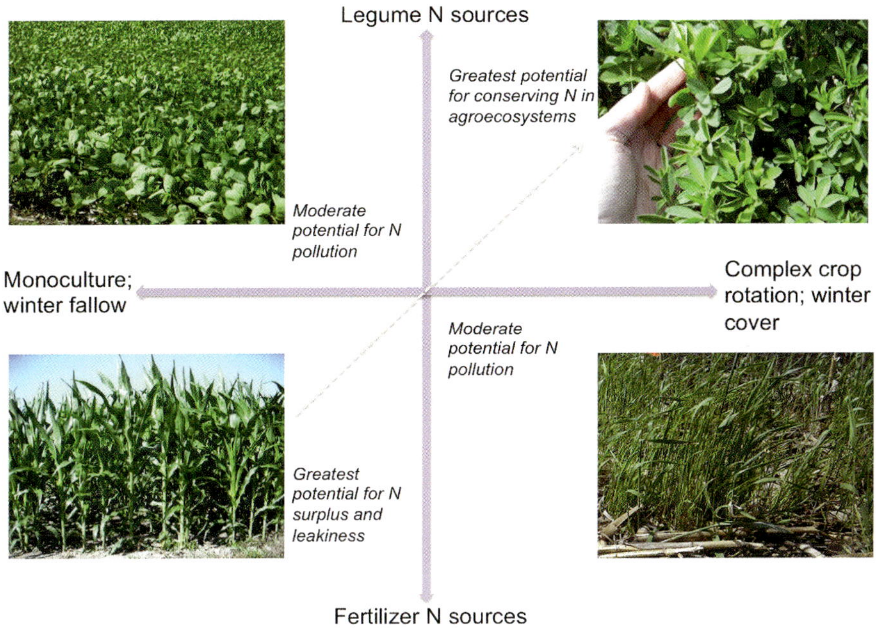

Figure 1. An ecological framework for assessing nitrogen management, placing management systems along two key dimensions that affect nitrogen losses. Crop rotation complexity (e.g. increasing plant functional group diversity such as inclusion of perennial or legume species, or adding non-harvested crops) is on the *x*-axis, and a gradient of nitrogen sources is along the *y*-axis, from sole use of chemical nitrogen fertilisers to sole reliance on legume nitrogen fixation. In this study, management systems that were legume-based had smaller average nitrogen surpluses than systems that were fertiliser-based; however the capacity of nitrogen from any source to be retained in soil depends in part on other management practices (e.g. tillage, irrigation, timing of application) as well as on environmental conditions (e.g. soil type, precipitation).

Increasing crop rotation diversity along with use of legume nitrogen sources ultimately enables reducing or eliminating synthetic fertiliser inputs as the capacity increases for biotic processes to supply nitrogen to crops (Drinkwater and Snapp 2007).

Legumes can be more efficient sources of nitrogen than chemical fertilisers, or potentially even manure or compost (Nelson and Janke 2007) due to several mechanisms. First, they simultaneously add both carbon and nitrogen to the agroecosystem through photosynthesis (carbon fixation) and biological nitrogen fixation in association with Rhizobia bacteria. Although some fixed nitrogen is added to the soil through legume root exudation and turnover, the majority of the newly fixed nitrogen is added to the system when legumes are incorporated by tillage. The organic nitrogen is mineralised to inorganic forms as the legume residues decompose, whereas chemical fertilisation represents a pulse addition of the most soluble nitrogen forms. Second, legume cover crops or forages increase the timeframe of actively growing plant roots that are absorbing nitrogen, which decreases the fallow periods when fields are most vulnerable to nitrogen losses. Finally, adding legumes into agroecosystems increases functional group diversity, which may, in turn, increase microbial diversity or activity and associated processes that build soil organic matter and stabilise/cycle both carbon and nitrogen in soil (Gregorich *et al.* 2001, Liang and Balser 2010, Kallenbach *et al.* 2015).

Increasing crop rotation complexity by including perennial plant species in addition to annuals can also increase nitrogen retention, in part by reducing the frequency of tillage and planting events. Reducing these soil disturbances reduces pulses of rapid decomposition of soil organic matter. Perennial species can also increase the spatial and temporal diversity of roots in an agroecosystem, which increases carbon inputs to soil via root exudation, death, and turnover (Schmidt *et al.* 2011), as well as nitrogen assimilation by plants (Crews and Peoples 2004). The broader window for nitrogen uptake by plants, together with increasing the carbon (energy) supply for microbial capture of nitrogen, can dramatically reduce nitrogen losses.

Adding perennials such as hay or forages back into Corn Belt rotations is necessary for a larger redesign of the region's agriculture; for example, reintegrating crop and livestock management (Russelle *et al.* 2007). Currently, meat production is highly specialised and spatially segregated from crop production, even though much of the crop production goes to animal feed. Despite the dominant forms of production, other agroecosystems that include perennials are present in the region, including management intensive rotational grazing. These farms often have a combination of land uses including mixed grass–legume pastures, hay or forages, in addition to some annual grain crops. Integrated crop-livestock production, and other ecological nitrogen management practices such as use of legume nitrogen sources, can be more knowledge-intensive but have the potential to reduce external inputs and emissions to the environment. For this reason, we examine the ecological and social dynamics of these more diversified systems in relation to the common industrial practices of the region.

2. Methods

We conducted coordinated fieldwork in the summer of 2007 to create a combined (ecological and social) dataset for 13 farmers in the Columbus region of Ohio (Darby Creek watershed) and 21 farmers from the Des Moines lobe region of Iowa. This dataset represents a subset of the farmer sample in a companion ecological study (Blesh and Drinkwater 2013) relating management practices to nitrogen use efficiency by constructing five-year nitrogen balances for two fields on each of 95 farms in the Corn Belt. A nitrogen mass balance is a useful indicator of nitrogen use efficiency (Robertson and Vitousek 2009) equal to the sum of all nitrogen inputs to a field (i.e. from fertiliser, manure, legume nitrogen fixation, and atmospheric deposition) minus the amount of nitrogen removed from the field in harvested crops. A positive mass balance indicates surplus nitrogen additions that are available for loss to surrounding ecosystems either in gaseous forms or via leaching or runoff. Surplus nitrogen, then, is an indicator of the potential for nitrogen pollution, which we used to compare the relative ecological sustainability of different management systems. Some surplus nitrogen may also be stored in soil organic matter if it is accumulating, which is likely not the case for most farm fields in the Corn Belt, based on current and historic trends in common practices, aside from ones that have been recently transitioned to more diverse rotations or to reduced tillage intensity. In the companion study, Blesh calculated nitrogen mass balances using data from farmer interviews, nutrient analysis of corn grain samples and animal manure, and measurements of legume nitrogen fixation inputs for fields with legume cover crops or forages (see Blesh and Drinkwater 2013 for detailed methods). A subset of these data, from farms included in the socio-ecological study, is used here.

Galt interviewed the 34 farmers included in the socio-ecological study using a semistandardised protocol to collect data to complement Blesh's agroecological data. This included questions about farmer demographics, farm details, information networks, assessments of management competencies of the farmer and off-farm service providers, characteristics of civic and professional communities, and perspectives on hypoxia and sustainable agriculture. Our farmer sample was purposeful rather than representative and generalisable to all farmers in Corn Belt states. Our goal was to collect detailed socio-ecological data on a broad range of management approaches, particularly capturing agroecological practices that have the potential to improve environmental sustainability, but which are currently extremely rare on farms in this region.

In addition, in 2009, Blesh used a purposeful sampling strategy to conduct semistructured interviews with nine farmers on the Des Moines lobe whose management practices led to the greatest nitrogen use efficiency as measured in the larger study. These interviews were also part of a related study comparing farmers in central and northeastern Iowa (Blesh and Wolf 2014). The goal of the interviews was to collect in-depth information about how a subset of farmers had transitioned to agroecological management in this highly agro-industrialised state. Farmers were invited to tell and make meaning of their transition stories, and questions focused on "how" they had transitioned, to inform a multilevel analysis of key resources and factors at various scales, from household to national. In the following sections we present an integrated analysis of 28 farms, the final sample size for which we had complete social and ecological data, to explore transitions to agroecological nutrient management in Iowa and Ohio. Our discussion of transition processes, however, also builds upon our broader understanding of agriculture in the Corn Belt as informed by all three components of the interdisciplinary research conducted between 2007 and 2010.

3. Results

3.1. *Nitrogen use efficiency of different management practices in the Corn Belt*

Results from the companion ecological research on 95 farms in four Corn Belt states (Iowa, Ohio, Minnesota and Wisconsin; Blesh and Drinkwater,

2013) revealed several management practices that had nitrogen mass balances close to zero (Figure 2). In particular, farm fields with legume nitrogen fixation as the primary nitrogen source (Figure 2(a)) or fields with perennial species in rotation (Figure 2(b)) had the lowest mean surplus nitrogen (3–8 kg of nitrogen per hectare per year). These practices fall within an ecological nutrient management framework, and demonstrate reduced potential for nitrogen losses. In contrast, three common inorganic fertiliser practices (surface broadcast, injection, and sidedress[1] application) had mean nitrogen surpluses ranging from 30–38 kg of nitrogen per hectare per year. The results for management of manure nitrogen sources varied greatly depending on the application timing and rate. Fall manure application led to a large potential for nitrogen pollution, whereas applying manure in the spring, closer to the timing of crop growth, allowed for reduced manure nitrogen inputs and surpluses. The nitrogen balance for spring-applied manure was not significantly different from the balance for legume-based management.

Building on the ecological results, we divided the farmers in Iowa and Ohio, for which we have joint social and ecological information, into two groups based on their nutrient management practices: (i) farmers who depend on leguminous crops for more than half of their nitrogen inputs (LEG-N; $n = 7$), and, (ii) farmers who rely on non-leguminous nitrogen inputs, primarily synthetic fertiliser, for more than half of their nitrogen inputs (NONLEG-N; $n = 21$). Despite concerted efforts to increase the sample size of legume-based farmers in the Iowa and Ohio study sites, these agroecological practices are extremely rare on the most productive soil types of the Corn Belt. Although only seven farmers are in the LEG-N group, this is a large number, relative to the total population of farmers using these systems in the study areas.

As in the larger ecological dataset (Blesh and Drinkwater 2013), the percentage of total nitrogen inputs from leguminous crops was significantly correlated with reduced nitrogen surplus ($r = -0.41$, $p = 0.02$) for this subset of farmers. Indeed, the nitrogen mass balances between the two groups were significantly different (t-test, $p = 0.01$), indicating that the

---

[1] Nitrogen fertiliser applied in a split application, with at least one application post-emergence.

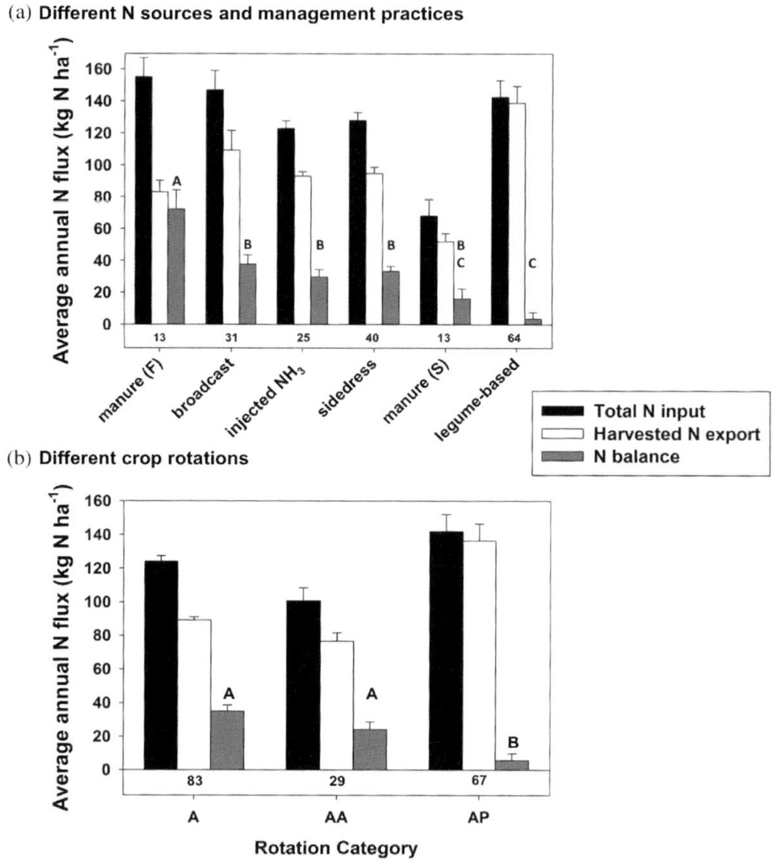

Figure 2. Observed means with standard errors for annual nitrogen inputs, harvested nitrogen exports and nitrogen balance (the difference between inputs and exports) based on five years of data for: (a) different nitrogen sources and nitrogen management practices — fall or spring manure applications; surface broadcast, sidedressed, or injected commercial fertiliser; or greater than 50% of nitrogen from legume nitrogen sources (legume-based), and (b) different crop rotations (on the *x*-axis: summer annuals only (A); summer annuals and winter annuals (AA); or annuals and perennials (AP)). Most of the perennial crop species were mixed legume–grass forages such as alfalfa–orchardgrass. Statistics are for the nitrogen balance. The nitrogen balances with different letters are significantly different ($p < 0.05$). Numbers listed below the *x*-axis indicate the number of observations per practice (after Blesh and Drinkwater 2013, with permission of Ecological Society of America (ESA)).

LEG-N systems have greater nitrogen use efficiency. These two groups show considerable differences in their farm management practices as well (Table 1). In particular, a greater proportion of the LEG-N farmers were certified organic, they grew corn less frequently in their crop rotations, and they had more complex crop rotations with a greater mean number of crop species at the farm-scale compared to NONLEG-N farmers.

3.2. *Socio-economic contexts of different nutrient management practices in the Corn Belt*

Below we explore the differences in competencies and knowledge networks for the two groups of farmers for which we have joint social and ecological data. We used t-tests to compare the means of the variables within the different sections, reporting *p*-values below 0.10 as significant since the sample size for LEG-N farmers is small ($n = 7$).

3.2.1. Farmer and farm characteristics

There were no significant differences between the two groups of farmers in age, days worked off-farm, gross farm sales, or gross household income (Table 2). The one significant difference in farmers' background and demographics was that of formal education ("some college or vocational training"). LEG-N farmers' average was 2.7 and NONLEG-N farmers' average was 3.3. Part of this difference is likely due to types of information prioritised in formal education settings, as the industrial nitrogen management model has been dominant for decades. As one LEG-N farmer noted, "I went to the land-grant university for the latest and greatest, but now they're far behind" in relation to agroecological management.

3.2.2. Information sources

Interview results indicated a number of significant differences in how the two groups valued different information sources (Table 3). Compared to NONLEG-N farmers, LEG-N farmers were more likely to value "other

Table 1. Farm management and crop diversity of farms surveyed in Ohio and Iowa, USA.

	LEG-N farmers ($n = 7$)					NONLEG-N farmers ($n = 21$)					*t*-test
	Mean	S.D.	Min	Max	*n*	Mean	S.D.	Min	Max	*n*	*p*-value
Nitrogen balance, most productive field (kg N ha^{-1} yr^{-1})	10.2	7.6	3.4	25.02	7	31.0	23.1	−2.9	76.3	21	**0.01**
Farm management											
proportion certified organic	0.7	0.5	0	1	7	0.2	0.4	0	1	21	**0.01**
certified organic hectares	192	107	24	293	7	41	122	0	554	21	**0.00**
Field management											
hectares in over-wintering plants	97	134	14	393	7	30	64	0	286	20	**0.04**
frequency of corn in 5-year rotation	2.1	0.4	2	3	7	2.9	0.4	2	4	21	**0.00**
Farm crop diversity											
total number of crop species	3.9	1.1	2	5	7	2.4	0.9	2	5	21	**0.00**
hectares in corn	136	167	32	502	7	269	220	32	1029	20	**0.08**
hectares in soybean	93	121	6	357	7	244	210	10	938	20	**0.04**
hectares in spelt	14	37	0	98	7	0	0	0	0	20	**0.04**
hectares in buckwheat	3	8	0	20	7	0	0	0	0	20	**0.05**
hectares in hay (not alfalfa)	48	92	0	253	7	2	6	0	20	20	**0.01**
hectares in pulses (not soybeans)	2	5	0	12	7	0	0	0	0	20	**0.05**
hectares in barley	21	43	0	86	4	0	0	0	0	9	**0.07**

Table 2. Farmer and farm characteristics of farms sampled in Ohio and Iowa, USA.

	LEG-N farmers					NONLEG-N farmers					t-test
	Mean	S.D.	Min	Max	n	Mean	S.D.	Min	Max	n	p-value
Age	53	15	39	76	6	56	10	30	75	20	0.29
Level of education[1]	2.7	1.0	2	4	7	3.4	1.1	1	5	20	**0.10**
Days worked off-farm in the last year[2]	1.4	0.8	1	3	7	1.4	1.0	0	5	21	0.50
Gross sales from farming in 2006[3]	4.6	1.0	3	6	7	5.0	0.9	4	6	21	0.17
Gross household income from all sources in 2006[4]	2.6	2.1	1	6	7	3.3	1.3	1	6	21	0.13

Note: [1] 1 = Less than high school diploma, 2 = High school or equivalent, 3 = Some college or vocational training, 4 = College bachelor degree, 5 = Advanced degree.

[2] 1 = None, 2 = 1 to 49 days, 3 = 50 to 99 days, 4 = 100 to 199 days, 5 = 200 days or more.

[3] 1 = Less than $10,000, 2 = $10,000 - $49,999, 3 = $50,000 - $99,999, 4 = $100,000 - $249,999, 5 = $250,000 - $499,999, 6 = $500,000 - $999,999, 7 = $1,000,000 or more.

[4] 1 = Less than $30,000, 2 = $30,000 - $59,999, 3 = $60,000 - $89,999, 4 = $90,000 - $119,999, 5 = $120,000 - $149,999, 6 = $150,000 or more.

Table 3. Importance of information sources for farmers interviewed in Ohio and Iowa, USA.

	LEG-N farmers					NONLEG-N farmers					*t*-test
	Mean	S.D.	Min	Max	*n*	Mean	S.D.	Min	Max	*n*	*p*- value
Other farmers	3.9	1.1	2.5	5	7	3.4	0.7	2	5	20	**0.07**
Fertiliser/input dealers	2.6	1.6	1	5	6	3.1	1.3	1	5	20	0.22
Independent crop consultants	1.0	0.0	1	1	6	2.4	1.5	1	5	20	**0.02**
Farmer professional organisation(s)	2.9	1.4	1	5	6	2.1	1.0	1	4	17	**0.06**
Extension agent, university specialist, or extension publications	2.3	1.3	1	4	6	3.0	1.2	1	5	20	**0.09**
Soil and water conservation agents	2.5	0.8	1	3	6	2.9	1.0	1	5	20	0.22
Buyer(s) of your crops	2.2	1.8	1	5	6	2.9	1.3	1	5	20	0.13
Landlord(s)	1.8	1.0	1	3	6	1.8	1.0	1	4.5	20	0.49
Family, friends, or neighbours	2.3	0.8	1	3	6	2.5	1.2	1	5	20	0.34

Note: Values in the table are: 1 = none, 2 = little, 3 = moderate, 4 = high, 5 = very high.

farmers" (3.9 versus 3.4) and "farmer professional organisation(s)" (2.9 versus 2.1). In the Iowa study site, a number of LEG-N farmers belonged to Practical Farmers of Iowa, a farmer-to-farmer learning organisation that prioritises on-farm experimentation, and which has had a significant impact on farmers' abilities to transition to alternative practices (Lighthall and Roberts 1995, Bell 2004). Similarly, if less well developed, farmer-to-farmer learning networks are also important to the LEG-N farmers in Ohio. One LEG-N farmer in Iowa said:

> I really enjoy talking with elderly farmers who farmed organically for a long time in their farming career before they started using chemicals. That's a lot of fun — finding out what they used to do, and how it worked. I guess I'm sort of an introvert by nature, but to be successful with organics you really have to be more of an extrovert — to gain the information you need to do it — and so that's a result of doing something really different like this. I guess that's one way I've changed.

LEG-N farmers were much less likely to value "independent crop consultants" (1.0 versus 2.4) and "extension agent, university specialist, or extension publications" (2.3 versus 3.0), compared to NONLEG-N farmers. There was consensus amongst the LEG-N farmers that independent crop consultants were of no use to them. The Cooperative Extension system of the land-grant universities were also of little use. This highlights the considerable disconnect between both the private and public knowledge production systems and the needs of agroecologically-oriented farmers in the Midwest. In addition, in an open-ended interview one LEG-N farmer noted that his approach to applying information from external sources is different:

> When I was conventional, I would make a call to the agronomist and they would say, "Oh, I think you should use this spray at this time." They'd even come out and scout the field and say, "Now it's time to spray." You know, you would rely on somebody else's expertise and then you just followed their directions and now I have to — I still use a lot of those outside resources — but I make up my own mind as to how to do it.

3.2.3. Level of various farming and management competencies

To pursue the importance of knowledge networks and knowledge production in greater depth, we asked farmers to assess themselves on a number of farming and nitrogen-related competencies, following Wolf and Primmer's conceptualisation of land management competencies (Wolf and Primmer 2006). While farmers in the two groups rated seven of the competencies similarly, there were four important differences (Table 4).

The LEG-N farmers rated themselves higher than NONLEG-N farmers for "organic production methods" (3.8 versus 2.5) and, to a lesser extent, "on-farm experimentation (formal or informal)" (4.1 versus 3.7). The LEG-N farmers were much more likely to be certified organic than NONLEG-N farmers. The difference in on-farm experimentation competence suggests the importance of learning for LEG-N farmers as they create and modify their production systems based on organic nitrogen sources in a context where such practices are rare. Compared to synthetic nitrogen fertilisers, the availability of nitrogen from organic sources is less predictable (and outcomes for crop yield can be more variable), since biotic processes regulate nitrogen mineralisation to inorganic nitrogen forms. Ecological nutrient management is therefore knowledge-intensive. For example, farmers can improve their knowledge regarding selection of legume species or cover crop mixtures to support particular ecosystem functions, or how to time legume incorporation to synchronise nitrogen mineralisation with cash crop demand. One farmer noted: "My extension agent tells me that cover crops don't work, so I am doing experiments to show otherwise," which also supports the survey results indicating that LEG-N farmers were less likely to value information from Cooperative Extension agents.

Compared to NONLEG-N farmers, LEG-N farmers rated themselves lower on "nitrogen application using precision agriculture" (2.3 versus 3.4) and "making use of government conservation programs" (2.7 versus 3.3). Precision agriculture, which enables farmers to vary their nitrogen application rates at small spatial scales to reduce nitrogen losses, is becoming an increasingly common fertiliser application method in the Midwest, given the common use of GPS-enabled tractors. Since this

Table 4. Competency assessments of farmers interviewed in Ohio and Iowa, USA.

	LEG-N farmers					NONLEG-N farmers					*t*-test
	Mean	S.D.	Min	Max	*n*	Mean	S.D.	Min	Max	*n*	*p*-value
Assessment of on-farm competencies											
Seeking out new information	3.8	0.3	3.5	4	6	4.0	0.5	3	5	21	0.29
Differentiating valid information from invalid information	4.0	0.6	3	5	6	3.7	0.6	3	5	21	0.19
Maximising commercial fertiliser use efficiency	3.6	0.8	3	5	6	3.9	0.8	2	5	21	0.20
Nitrogen application using precision agriculture	2.3	1.5	1	5	6	3.4	1.4	1	5	21	**0.07**
Nitrogen crediting[1] (including soybeans)	3.7	1.0	2	5	6	3.2	1.1	1	5	21	0.17
On-farm experimentation (formal or informal)	4.1	0.7	3	5	6	3.7	0.7	3	5	21	**0.10**
Farm economics	3.3	0.8	2	4	6	3.6	0.6	3	5	21	0.21
Making use of government conservation programs	2.7	1.0	1	4	6	3.3	0.9	1	5	21	**0.07**
Soil health or soil quality	3.8	0.4	3	4	6	3.7	0.7	3	5	21	0.28
Growing winter cover	2.9	1.2	1	4	6	2.5	0.8	1	4	21	0.20
Organic production methods	3.8	1.0	2	5	6	2.5	1.1	1	4	21	**0.01**
Average	3.5	0.4	3.2	4.2	6	3.4	0.4	3	4.5	21	0.35

(*Continued*)

Table 4. *(Continued)*

	LEG-N farmers					NONLEG-N farmers					*t*-test
	Mean	S.D.	Min	Max	*n*	Mean	S.D.	Min	Max	*n*	*p*-value
Assessment of competencies of off-farm service providers											
Seeking out new information	3.7	1.0	2	5	6	4.2	0.7	3	5	19	**0.09**
Differentiating valid information from invalid information	3.3	0.8	2	4	6	4.1	0.9	2	5	19	**0.03**
Maximising commercial fertiliser use efficiency	3.7	1.2	2	5	6	3.9	0.6	2	5	19	0.23
Nitrogen application using precision agriculture	3.3	1.6	1	5	6	4.1	0.9	2	5	18	**0.08**
Nitrogen crediting for manure and/or legumes (including soybeans)	4.0	1.2	2	5	5	4.0	1.0	1	5	16	0.48
On-farm experimentation (formal or informal)	4.0	1.1	2	5	6	4.3	0.8	3	5	19	0.26
Soil health or soil quality	4.1	1.2	2	5	5	4.1	0.8	3	5	18	0.49
Growing winter cover	3.5	1.0	2	4	4	3.2	1.0	1	5	14	0.31
Organic production methods	3.8	0.4	3	4	5	2.8	1.2	1	5	13	**0.05**
Average	3.7	1.0	2.1	4.8	6	3.7	1.0	0	4.9	21	0.48
Research on the farm											
Engaged in personal on-farm research in the last 5 years[2]	0.7	0.5	0	1	6	0.3	0.5	0	1	21	**0.08**

Note: [1]A nitrogen credit is a reduction in nitrogen fertiliser rate for a crop following a legume crop, manure, or another organic amendment.
[2]Binary variable (0 = No and 1 = Yes). Level of expertise is 1 = none, 2 = low, 3 = moderate, 4 = high, and 5 = very high.

technology applies only to synthetic nitrogen, it is predictable that NONLEG-N farmers would feel more confident in this competency. NONLEG-N farmers also have more land enrolled in the Conservation Reserve Program (CRP) and the Wetland Reserve Program (WRP), on average (14.1 hectares versus 7.3 hectares for LEG-N farmers), which translates into them feeling more confident in using government conservation programmes.

In addition, during the open-ended interviews, LEG-N farmers recounted mixed experiences with government conservation programmes. Some farmers in Iowa were enrolled in the Conservation Stewardship Program (CSP) and the Environmental Quality Incentives Program (EQIP), two working land programs within the Farm Bill that provide tiered payments for environmentally friendly practices (for CSP) and cost-sharing for various conservation practices (for EQIP). For example, two of the LEG-N farmers on the Des Moines Lobe had enrolled in a 2008 Farm Bill programme to assist farmers with transitions to certified organic production. However, farmers' access to these programmes varied by county because some county Natural Resource Conservation Service offices allocate conservation dollars towards organic or agroecological management, such as cover cropping, whereas others do not (focusing on other initiatives; for example, on no-till or riparian buffers). One farmer said: "my county decided not to pay for cover crops, or organic, or anything." When we asked him how USDA might change these programmes to increase practices like cover cropping on farms in his region he replied:

> The only way you're going to get farmers to do it is to pay them. And it has to cover the labor and the seed cost. And most of the farmers around here do not have the equipment, so they'd have to hire someone to put it on. And if you're farming 3000 acres and you've got a $2 million operating loan, and you have to get that crop harvested and in the elevator to satisfy the bank, you are not going to want to put out a cover crop. So I don't know how that's going to happen.

We asked farmers to assess their off-farm service providers on the competencies related to fertility management, too. Here, there were also four significant differences (Table 4). LEG-N farmers ranked their service providers higher for just one competency: "organic production methods" (3.8 versus 2.8). This suggests that LEG-N farmers, if they use off-farm service providers, seek out those familiar with organic agriculture. These were scarce in both sites when we did our fieldwork. For three competencies, LEG-N farmers rated their service providers lower than NONLEG-N farmers: "seeking out new information" (3.7 versus 4.2), "differentiating valid information from invalid information" (3.3 versus 4.1), and "nitrogen application using precision agriculture" (3.3 versus 4.1). While there are likely differences in the types of providers that the two groups use (especially in relation to organic production methods), these differences in ratings suggest that LEG-N farmers may be more skeptical of outside information compared to their own knowledge and that of other farmers. This interpretation is supported by the findings regarding information sources and on-farm experimentation, and perhaps makes sense given that farmers using agroecological management practices must contend with greater ecological complexity and environmental variability.

3.2.4. Marketing competencies

In the Corn Belt, LEG-N and NONLEG-N farmers have differential access to markets. LEG-N and other diversified farmers must develop new competencies and skills to successfully market their products, for which market channels that adequately value their management systems have not been developed. In the open-ended interviews with farmers who had transitioned from NONLEG-N to LEG-N management in Iowa, they described a wide range of strategies for marketing alternative products: from selling on commodity markets to, less commonly, direct marketing. Rather than hauling grain to the local grain elevator, marketing alternative products such as certified organic grains involves extensive networking — for example, with potential buyers at farming conferences — and, often, transporting the grain to longer distances than before. At conferences these farmers discuss the quantity and quality of their product with millers or other brokers. Those millers then call, offering their current price when they are buying grain. After the initial networking effort, farmers said that buyers usually stay in close contact and it is not necessary to continually seek new ones.

Several of the LEG-N farmers directly marketed a small proportion of their product. Some of the certified organic LEG-N farmers sold grains as feed to neighbouring or nearby livestock farmers. Others developed new markets for local and regional consumers. For example, one farm couple sells popcorn and flax seed, and engages in educational activities to teach potential customers how to prepare and eat their products. They have a booth at the local grocery store at which they describe the nutritional benefits of flax seed and its meal and demonstrate its use in recipes. Another farm family grinds a small proportion of their grain into flour and bakes bread to sell at their local farmers' market. Direct marketing is, of course, far more common for vegetable farmers than for grain farmers, who produce staples that typically must be processed before consumption. In addition, direct marketing activities take a great deal of time, and the benefits are largely educational and have to do with building community support for more sustainable and local production. These grain farmers are vectors for change through these efforts, although the largest proportion of their product — and income — comes from national or global-scale markets.

3.2.5. Views of civic and professional communities

The interviews asked farmers to assess their civic and professional communities according to a variety of characteristics identified by Flora and Flora (1993) as essential for supporting entrepreneurial innovation. Civic communities were defined as farmers' towns, or the nearest town if they lived in remote areas, while professional communities were the people they interacted with in their farming profession, including fellow farmers, service providers, and buyers of their goods. There were a couple of differences in the specific ratings. LEG-N farmers rated one civic community characteristic, "has relatively small differences in wealth between individuals," lower than NONLEG-N farmers (Table 5; 2.3 versus 3.0). They also rated the professional community characteristic, "promotes innovation and member learning," significantly lower than NONLEG-N farmers (3.7 versus 4.3). This is by far the largest difference between the two groups regarding their perception of their communities, which suggests that although LEG-N farmers are pioneering considerable ecological

Table 5. Ohio and Iowa farmers' ratings of civic and professional community characteristics.

	LEG-N farmers					NONLEG-N farmers					t-test
	Mean	S.D.	Min	Max	n	Mean	S.D.	Min	Max	n	p-value
Civic community characteristics											
Requires that all members think and act the same[1]	3.3	1.2	2	5	6	3.4	0.9	1	4	21	0.42
Promotes innovation and member learning	3.0	0.6	2	4	6	3.2	0.8	2	4	21	0.26
Engages in constructive criticism	3.3	1.0	2	4	6	3.5	0.8	2	4	21	0.36
Has relatively small differences in wealth between individuals	2.3	1.4	1	4	6	3.0	1.0	1	4	21	**0.10**
Invests private resources for the public good	3.6	1.1	2	5	5	3.7	0.9	1	5	21	0.44
Has closed decision-making structures dominated by a few[1]	2.8	1.0	2	4	6	3.2	1.0	1	4	20	0.21
Average	3.0	0.3	2.7	3.5	6	3.3	0.7	1.3	4	21	**0.10**
Professional community characteristics											
Requires that all members think and act the same[1]	3.7	1.0	2	5	6	3.8	0.8	2	5	21	0.40
Promotes innovation and member learning	3.7	0.5	3	4	6	4.3	0.6	3	5	21	**0.01**
Engages in constructive criticism	3.8	0.4	3	4	6	3.9	0.6	2	5	21	0.46
Has relatively small differences in wealth between individual	2.7	1.2	1	4	6	3.2	1.0	1	4	20	0.14
Invests private resources for the public good	3.8	0.4	3	4	5	3.8	0.8	2	5	21	0.49
Has closed decision-making structures dominated by a few[1]	3.8	1.0	2	5	6	3.9	0.7	2	5	21	0.42
Average	3.5	0.5	2.8	4	6	3.8	0.4	3	4.5	21	**0.05**

Note: [1] Scale reversed (i.e. 1 = strongly agree and 5 = strongly disagree) due to the question being framed in a negative way. Values in the table are: 5 = strongly agree, 4 = agree, 3 = neither agree nor disagree, 2 = disagree, 1 = strongly disagree.

management innovations in their farming systems, they do not feel particularly supported in their efforts. When looking at the differences between the averages of each of the characteristics, LEG-N farmers had more negative views of both their civic community (3.0 versus 3.3) and their professional community (3.5 versus 3.8). This suggests that LEG-N farmers are more likely to see themselves as outsiders in both their civic and professional communities.

3.2.6. Views of sustainability of their farm operation, social demands, and water quality

The interviews asked farmers to rank their farming operations on a continuum of sustainability, from 1 (the most environmentally friendly) to 10 (the most production oriented) (Table 6). There were significant differences in the averages of the rankings, with LEG-N farmers seeing themselves as more environmentally friendly than NONLEG-N farmers (3.6 versus 5.6). When asked where their farming operation would ideally be ranked on the continuum, LEG-N farmers aspired to greater environmental sustainability than NONLEG-N farmers (3.1 versus 5.5). Twenty-nine percent of LEG-N farmers wanted to be more environmentally friendly than they currently are, in contrast to only 14 percent of NONLEG-N farmers. That LEG-N farmers are more likely to strive towards even more environmentally friendly practices suggests a deep level of environmental commitment. As one noted, "I want to get as close as I can to 1 (most environmentally friendly)."

There were also large differences in how farmers perceived society's demands on them (Table 6). We asked all farmers whether society asks more of farmers now compared to 30 years ago. While 90% of NONLEG-N farmers said yes, only 40% of LEG-N farmers did so. In the open-ended discussions in the interviews, NONLEG-N farmers felt that environmental considerations had impinged considerably upon them, especially the cross-compliance requirements of CRP and WRP that farmers need to meet to continue receiving commodity payments (Swanson 1993). One NONLEG-N farmer noted: "Each generation is further from

Table 6. Views of Ohio and Iowa farmers on societal demands, sustainability, and perspectives on water quality.

	LEG-N farmers					NONLEG-N farmers					*t*-test
	Mean	S.D.	Min	Max	*n*	Mean	S.D.	Min	Max	*n*	*p*-value
Placement on a sustainability continuum[1]											
current placement on a continuum	3.6	3.1	1	10	7	5.6	2.4	1.5	10	20	**0.05**
ideal placement on a continuum	3.1	3.3	1	10	7	5.5	2.6	1	10	18	**0.04**
Societal demands											
society asks more of farmers now than 30 years ago (binary)	0.4	0.5	0	1	5	0.9	0.3	0	1	16	**0.00**
Water quality[2]											
your farm's surface water	4.5	0.5	4	5	6	4.0	0.7	3	5	19	**0.08**
your farm's ground water	4.2	0.8	3	5	6	4.2	0.6	3	5	21	0.50
local water bodies you could use (for fishing, swimming, etc.)	2.8	1.1	1	4	6	3.5	1.1	1	5	21	**0.09**

Note: [1] 10 being the most production oriented, and 1 being the most oriented to environmental sensitivity

[2] 5 = very good, 4 = good, 3 = moderate, 2 = poor, 1 = very poor

farmers, and they make the rules of how we play the game." Another noted:

> Oh yeah. Environmental regulations, livestock need permits, distance regulations, annual reports, they demand more accountability and more on the humane side. We grew up with it, went to college, and resent that outsiders tell us [what to do], like free range versus caged and gestation stalls.

In contrast, LEG-N farmers felt that societal demands were not very burdensome and did not ask much of farmers generally.

Farmers' position relative to fertility management paradigms is also related to their views on water quality. The interviews asked farmers to rate the quality of three types of water: surface water on their farms, groundwater on their farms, and local surface water bodies they might use. There was no difference in groundwater ratings, which is not surprising given that most groundwater in the Corn Belt is heavily polluted with nitrate, and the contamination of groundwater in the area occurred decades prior. However, LEG-N farmers rated their farms' surface water quality higher than NONLEG-N farmers (4.5 versus 4 of 5), and rated local water bodies significantly lower than NONLEG-N farmers (2.8 versus 3.5 of 5). This suggests that LEG-N farmers believe that their practices lead to better surface water quality, and that they are more critical of the impacts of industrial farming on nearby surface water.

4. Transition processes: Increasing the presence of ecological nitrogen management in the Corn Belt

Our results indicate that more diverse crop rotations and use of legume nitrogen sources increase the nitrogen use efficiency of grain agroecosystems, which we quantified using a nitrogen mass balance approach. For the LEG-N farmers, legume nitrogen sources comprised 50% or more of the total nitrogen inputs to their grain fields. In many cases the legumes were perennial forages, which farmers grew for livestock feed as part of mixed crop-livestock systems or sold to neighbouring livestock farmers.

Both ecological principles and a growing body of empirical evidence, including this research, provide support for the broader environmental benefits of agroecological nutrient management. However, despite their promise, these management practices are currently very rare in the USA Corn Belt.

The marginal presence of agroecological nutrient management practices is reflected in the small sample size for our study, in which we employed a purposeful sampling approach specifically seeking out these practices. It is also demonstrated by secondary data at county and state scales. For example, of all farms in Iowa and Ohio only 0.6% (or 566 farms) and 0.97% (or 578 farms) are certified organic, respectively (USDA 2008). Iowa had just 70 community supported agriculture (CSA) farms and Ohio had 107 in 2009 (Galt 2011). A mail survey of farmers in four Corn Belt states (Illinois, Indiana, Iowa, and Minnesota) reported that only 11% of farmers had planted a cover crop sometime in the last five years (Singer *et al.* 2007), although a recent national survey suggests that interest in cover cropping is increasing again (SARE and CTIC 2015).

Although interest in improving the sustainability of agriculture in the Corn Belt is growing, the socio-political and economic context creates large barriers to broader implementation of ecological nutrient management. Here, we analysed ecological and social variables, as well as the intersection of agency (actions of farmers and other individuals and communities) and structure (institutions, governance, and farm policy), for a wide management gradient, focusing on nutrient management practices. In particular, we were careful to include the practices that are rare in the region, in order to inform opportunities for increasing practices with the lowest potential for nitrogen losses.

For farmers in this study, transitioning to ecological nutrient management involved the development of several new farm-level competencies, as well as shifts in thinking. The LEG-N farmers described the importance of thinking "independently" and of applying preventative, ecological, and systems thinking to their management. For example, one LEG-N farmer who also has some livestock said of his nutrient management: "It's a totally different mind-set, I think. You do have to look at things differently — renewability and recycling nutrients, and nutrient budgeting."

The LEG-N farmers were more likely to consider their practices to be ecologically sustainable than NONLEG-N farmers, and they demonstrated a deeper ideological commitment to environmental preservation, even though fewer of them were participating in government conservation programmes. On average, the LEG-N farmers had greater agrobiodiversity (number of crops produced, and more integration of crop and livestock production), and were more likely to have organic certification, both of which likely helped to buffer against the economic risks of transitioning to ecological management in a landscape that favours farm intensification and specialisation. They had developed the competencies to manage more biodiverse, complex agroecosystems through on-farm experimentation and by developing linkages with external resources.

In terms of off-farm resources, LEG-N farmers had different knowledge networks supporting their practices. LEG-N farmers were more likely to highly value knowledge from other farmers and professional organisations that facilitate site-specific, participatory, and farmer-to-farming learning. These farmer organisations intersect with a wider sustainable agriculture movement and knowledge system (Allen 2004). The networks, then, also represent an important source of social support for an isolated group of farmers who expressed feeling like outsiders in their local civic and professional communities. The LEG-N farmers were less likely to rely on formal information networks such as state land-grant institutions or Cooperative Extension. However, even though the dominant information systems did not meet their needs by supporting agroecological practices, interviews revealed that most LEG-N farmers creatively and strategically applied information from commodity groups, land-grant universities, and Cooperative Extension. Further, they noted that slow change is also occurring within these institutions, such as a small number of Cooperative Extension positions focused on organic agriculture, and new academic programs in organic and sustainable agriculture. In this way, their relationships with institutions that have primarily supported the industrialisation of agriculture are helping to redefine and expand the limits of what counts as legitimate knowledge within these organisations (Carolan 2006).

More broadly, the Corn Belt now lacks many of the physical and technological resources — such as markets, processing facilities, and appropriately scaled machinery — for smaller-scale, diversified crop and livestock production. The LEG-N farmers were resourceful, however, demonstrating differing degrees of reliance on existing physical and market infrastructure while also building new infrastructure. For example, the broad range of marketing activities and relations described by LEG-N farmers is slowly creating new food system arrangements within the Corn Belt. Farmers who transition to ecological management, and other actors such as cooperatives and millers, are building new markets in a coevolutionary process since the dominant agricultural infrastructure does not meet their needs.

LEG-N farmers in the Corn Belt also experience lack of policies supporting their management. For instance, the farmer interviews suggested they saw themselves as less proficient at accessing government conservation programmes in part because of a disconnect between the type of conservation they practice and the type promoted by USDA programmes in their regions. At a national scale there is also a large imbalance in public and private sector support for commodity production versus conservation. Taken together, these factors reflect the path dependencies of deep investments supporting industrial agriculture in the Corn Belt. Yet the presence of LEG-N farmers in this region — in spite of the enormous challenges facing those practicing ecological management — indicates that the situation is more open than commonly portrayed. Innovation is occurring in the "interstices" or in niches within the dominant system (Geels 2011, Blesh and Wolf 2014, Galt *et al.* 2014). Importantly, this interstitial innovation is connecting bottom-up, grassroots change with top-down changes that are (slowly) occurring in existing food system institutions. By eroding the limits of the dominant regime in multiple directions — for instance, by providing real-world examples of agroecological practices that work — these pioneering efforts may eventually transform the dominant regime's seemingly locked-in components, increasing the presence of new, more sustainable and multifunctional food systems.

5. Conclusions and future research needs

Soil fertility management in temperate climates is a key dimension of farming with high stakes for global change. This is particularly the case for nitrogen, which is often readily lost to surrounding ecosystems and contributes to both climate change and water pollution. There is enormous potential for improving nitrogen management in agriculture by applying ecological principles. An agroecological nutrient management approach balances nitrogen inputs and exports, and enhances the capacity for biotic processes in soil to supply nutrients to plants, thus enabling reductions in fertiliser inputs. The USA Corn Belt is an interesting case for exploring how ecological management practices can emerge against the backdrop of a highly intensified and agro-industrialised landscape and the institutional and infrastructural support that make it possible.

Developing the motivation, skills, and competencies to transition to more sustainable management practices is not just a farm-level challenge. Analysis of agricultural innovation, therefore, needs to account for social and ecological factors from farm household to global scales. Increasing the number of farmers practicing ecological management will require both grassroots and institutional change, including within agricultural research institutions. In particular, the role of ecological knowledge should be elevated. Agroecology provides a theoretical framework with great potential to design and implement more sustainable production systems. Yet, ecological science has been undervalued and largely neglected by the dominant agricultural research and development paradigm. Even though the most proximate barriers to implementing ecological management are sociopolitical and economic, it is important to note that there is still a great deal more to learn about ecological interactions in agroecosystems and how different management systems drive different ecosystem functions.

Further, the agroecological transitions we have explored here represent complex cropping systems change ("transformation"), rather than piecemeal adoption of one or several new practices ("tweak"), or input substitution in organic agriculture (e.g. Rosset and Altieri 1997). Farmers must gain new knowledge and relationships to successfully navigate a suite of challenges and economic risks. From an ecological perspective, cross-scale interactions matter, and landscape scale diversification could enhance agroecosystem functions such as pest control and water cycling at farm and field scales (Perfecto *et al.* 2009). Such change could be supported by transformations within the many institutions and actors that support agriculture in the USA Midwest, including private service providers, technical consultants, extension agents, input suppliers, and others in the value chain. Clearly it is too simplistic to argue that Farm Bill commodity payments alone drive the current agro-industrial practices. A wide range of investments in institutions and infrastructure also support and shape Corn Belt agricultural landscapes. Yet, policy matters a great deal, too. In addition to the need to change the structure of Farm Bill subsidies, the working lands conservation programmes, especially the CSP, have some potential to support agroecological practices. However, in their current form they frequently fail to subsidise the most effective practices, as farmers in this study noted, and they have been very spatially limited due to severe underfunding by the USA Congress.

Together, farmers and scholars across many disciplines can develop new research approaches and practices that inform leverage points for food system transformation in highly industrialised contexts. Theoretical approaches linking agroecology with political ecology, socio-ecological systems analysis, or situated resilience can contribute to the development of multifunctional food systems. Critical perspectives from the social sciences, specifically, have an important role in identifying power relations that drive the intensification of agriculture, and can inform changes to public investment towards experimentation in and support of ecological agriculture.

References

Allen, P., 2004. *Together at the Table: Sustainability and Sustenance in the American Agrifood System.* The Pennsylvania State University Press, University Park, PA, USA, 272 p.

Altieri, M.A., Toledo, V.M., 2011. The agroecological revolution in Latin America: rescuing nature, ensuring food sovereignty and empowering peasants. *Journal of Peasant Studies* **38**, 587–612.

Baker, O.E. 1927. Agricultural regions of North America: Part IV — The corn belt. *Economic Geography* **3**, 447–465.

Bell, M.M., 2004. *Farming for us all: Practical Agriculture and the Cultivation of Sustainability*. Pennsylvania State University Press, University Park, PA. USA, 312 p.

Blaikie, P., 1985. *The Political Economy of Soil Erosion in Developing Countries*. Methuen, London, UK, 200 p.

Blesh, J., Drinkwater, L.E., 2013. The impact of nitrogen source and crop rotation on nitrogen mass balances in the Mississippi River Basin. *Ecological Applications* **23**, 1017–1035.

Blesh, J., Wolf, S.A., 2014. Transitions to agroecological farming systems in the Mississippi River Basin: toward an integrated socioecological analysis. *Agriculture and Human Values* **31**, 621–635.

Buttel, F.H., 2005. Ever since hightower: the politics of agricultural research activism in the molecular age. *Agriculture and Human Values* **22**, 275–283.

Buttel, F.H., Larson, O.F., Gillespie G.W. Jr., 1990. *The Sociology of Agriculture*. Greenwood Press, New York, NY, 283 p.

Caldwell, J., 2016. *Want to Raise 200-Bushel Corn*? Meredith Corporation. http://www.agriculture.com/crops/corn/wt-to-raise-200bushel-cn_136-sl16120 (retrieved January 2016).

Carolan, M.S., 2006. Social change and the adoption and adaptation of knowledge claims: Whose truth do you trust in regard to sustainable agriculture? *Agriculture and Human Values* **23**, 325–339.

Clarholm, M., 1985. Interactions of bacteria, protozoa and plants leading to mineralization of soil nitrogen. *Soil Biology and Biochemistry* **17**, 181–187.

Clark, J., 2005. The 'new associationalism' in agriculture: agro-food diversification and multifunctional production logics. *Journal of Economic Geography* **5**, 475–498.

Clark, M.S., Horwath, W.R., Shennan, C., Scow, K.M., 1998. Changes in soil chemical properties resulting from organic and low-input farming practices. *Agronomy Journal* **90**, 662–671.

Crews, T.E., Peoples, M.B., 2004. Legume versus fertilizer sources of nitrogen: ecological tradeoffs and human needs. *Agriculture, Ecosystems and Environment* **102**, 279–297.

Cronon, W., 2009. *Nature's Metropolis: Chicago and the Great West*. WW Norton & Company, New York, USA, 592 p.

Culman, S., DuPont, S., Glover, J., Buckley, D., Fick, G., Ferris, H., Crews, T., 2010. Long-term impacts of high-input annual cropping and unfertilized perennial grass production on soil properties and belowground food webs in Kansas, USA. *Agriculture, Ecosystems and Environment* **137**, 13–24.

David, M.B., Drinkwater, L.E., McIsaac, G.F., 2010. Sources of nitrate yields in the Mississippi River Basin. *Journal of Environmental Quality* **39**, 1657–1667.

Diaz, R.J., Rosenberg, R., 2008. Spreading dead zones and consequences for marine ecosystems. *Science* **321**, 926–929.

Drinkwater, L.E., Snapp, S.S., 2007. Nutrients in agroecosystems: Rethinking the management paradigm. *Advances in Agronomy* **92**, 163–186.

Drinkwater, L.E., Wagoner, P., Sarrantonio, M., 1998. Legume-based cropping systems have reduced carbon and nitrogen losses. *Nature* **396**, 262–265.

Drinkwater, L.E., Schipanski, M., Snapp, S.S., Jackson, L.E., 2008. Ecologically based nutrient management. In: Snapp, S.S., Pound, B. (Eds.), *Agricultural Systems: Agroecology and Rural Innovation for Development*. Academic Press, Burlington, MA, USA, pp. 159–208.

FAO, IFAD, WFP., 2014. The state of food insecurity in the world 2014. Strengthening the enabling environment for food security and nutrition. Rome, FAO.

Farley, J., Schmitt Filho, A., Alvez, J., Ribeiro de Freitas Jr., N., 2011. How valuing nature can transform agriculture. *Solutions* **2**, 64–73.

Fenn, M.E., Poth, M.A., Aber, J.D., Baron, J.S., Bormann, B.T., Johnson, D.W., Lemly, A.D., McNulty, S.G., Ryan, D.F., Stottlemyer, R., 1998. Nitrogen excess in North American ecosystems: predisposing factors, ecosystem responses, and management strategies. *Ecological Applications* **8**, 706–733.

Flora, C.B., Flora, J.L., 1993. Entrepreneurial social infrastructure: A necessary ingredient. *The Annals of the American Academy of Political and Social Science* **529**, 48–58.

Foley, J.A., Ramankutty, N., Brauman, K.A., Cassidy, E.S., Gerber, J.S., Johnston, M., Mueller, N.D., O'Connell, C., Ray, D.K., West, P.C., Balzer, C., Bennett, E.M., Carpenter, S.R., Hill, J., Monfreda, C., Polasky, S., Rockstrom, J., Sheehan, J., Siebert, S., Tilman, D., Zaks, D.P.M., 2011. Solutions for a cultivated planet. *Nature* **478**, 337–342.

Forsyth, T., 2003. *Critical Political Ecology: The Politics of Environmental Science*. Routledge, New York, USA.

Galloway, J.N., Aber, J.D., Erisman, J.W., Seitzinger, S.P., Howarth, R.W., Cowling, E.B., Cosby, B.J., 2003. The nitrogen cascade. *Bioscience* **53**, 341–356.

Galloway, J.N., Townsend, A.R., Erisman, J.W., Bekunda, M., Cai, Z., Freney, J.R., Martinelli, L.A., Seitzinger, S.P., Sutton, M.A., 2008. Transformation of the nitrogen cycle: recent trends, questions, and potential solutions. *Science* **320**, 889–897.

Galt, R.E., 2011. Counting and mapping community supported agriculture in the United States and California: contributions from critical cartography/GIS. ACME: *An International E-Journal for Critical Geographies* **10**, 131–162.

Galt, R.E., 2013. Placing food systems in first world political ecology: a review and research agenda. *Geography Compass* **7**, 637–658.

Galt, R.E., 2014. *Food Systems in an Unequal World: Pesticides, Vegetables, and Agrarian Capitalism in Costa Rica*. University of Arizona Press, Arizona, USA.

Galt, R.E., Gray, L.C., Hurley, P., 2014. Subversive and interstitial food spaces: transforming selves, societies, and society–environment relations through urban agriculture and foraging. *Local Environment* **19**, 133–146.

Gardner, J.B., Drinkwater, L.E., 2009. The fate of nitrogen in grain cropping systems: A meta-analysis of ^{15}N field experiments. *Ecological Applications* **19**, 2167–2184.

Geels, F.W., 2011. The multi-level perspective on sustainability transitions: responses to seven criticisms. *Environmental Innovation and Societal Transitions* **1**, 24–40.

Gliessman, S.R., 2007. *Agroecology: The Ecology of Sustainable Food Systems*. CRC Press, Boca Raton, FL, USA.

Glover, J., Reganold, J., Bell, L., Borevitz, J., Brummer, E., Buckler, E., Cox, C., Cox, T., Crews, T., Culman, S., 2010. Increasing food and ecosystem security through perennial grain breeding. *Science* **328**, 1638–1639.

Gonzalez de Molina, M., 2013. Agroecology and politics. How to get sustainability? About the necessity for a political agroecology. *Agroecology and Sustainable Food Systems* **37**, 45–59.

Gregorich, E., Drury, C., Baldock, J.A., 2001. Changes in soil carbon under long-term maize in monoculture and legume-based rotation. *Canadian Journal of Soil Science* **81**, 21–31.

Howarth, R.W., Boyer, E.W., Pabich, W.J., Galloway, J.N., 2002. Nitrogen use in the United States from 1961–2000 and potential future trends. *Ambio* **31**, 88–96.

IAASTD., 2009. *Agriculture at a Crossroads. The Global Report of the International Assessment of Agricultural Knowledge, Science, and Technology*. Island Press, Washington, D.C., USA.

IPCC., 2013. *Climate Change 2013*. The physical science basis. Summary for policymakers. United Nations, http://www.ipcc.ch/.

Jackson, L.E., Pascual, U., Hodgkin, T., 2007. Utilizing and conserving agrobiodiversity in agricultural landscapes. Agriculture, *Ecosystems and Environment* **121**, 196–210.

Kallenbach, C., Grandy, A., Frey, S., Diefendorf, A., 2015. Microbial physiology and necromass regulate agricultural soil carbon accumulation. *Soil Biology and Biochemistry* **91**, 279–290.

Kremen, C., Miles, A., 2012. Ecosystem services in biologically diversified versus conventional farming systems: benefits, externalities, and trade-offs. *Ecology and Society* **17**.

Lappé, F.M., Collins, J., Rosset, P.M., Esparza, L., 1998. *World Hunger: Twelve Myths*. Grove Press, New York, USA.

Lewontin, R., Levins, R., 2007. *Biology Under the Influence: Dialectical Essays on Ecology, Agriculture, and Health*, Monthly Review Press, New York.

Liang, C., Balser, T.C., 2010. Microbial production of recalcitrant organic matter in global soils: implications for productivity and climate policy. *Nature Reviews Microbiology* **9**, 75–75.

Lighthall, D.R., Roberts, R.S., 1995. Towards an alternative logic of technological change: insights from Corn Belt agriculture. *Journal of Rural Studies* **11**, 319–334.

Lowrance, R., Stinner, B.R., House, G., 1984. *Agricultural Ecosystems. Unifying Concepts*. John Wiley and Sons, Inc., New York, USA.

Lyson, T.A., 2004. Civic agriculture: Reconnecting rarm, food, and community. Tufts University Press, Medford, Massachusetts, USA.

Matson, P.A., Parton, W.J., Power, A.G., Swift, M.J., 1997. Agricultural intensification and ecosystem properties. *Science* **277**, 504–509.

McCarthy, J., 2005. Rural geography: Multifunctional rural geographies-reactionary or radical? *Progress in Human Geography* **29**, 773–782.

McMichael, P., 2009. A food regime genealogy. *Journal of Peasant Studies* **36**, 139–169.

MEA., 2003. *Ecosystems and Human Well-Being: A Framework for Assessment*. Island Press, Washington, D.C.

Nelson, N.O., Janke, R.R., 2007. Phosphorus sources and management in organic production systems. *HortTechnology* **17**, 442–454.

Nissen, T.M., Wander, M.M., 2003. Management and soil-quality effects on fertilizer-use efficiency and leaching. *Soil Science Society of America Journal* **67**, 1524–1532.

Odum, E.P., 1984. Properties of agroecosystems. In: Lowrance, R., Stinner, B.R., House, G.J. (Eds.), *Agricultural Ecosystems. Unifying Concepts*. John Wiley, New York, USA, pp. 5–12.

Perfecto, I., Vandermeer, J.H., Wright, A.L., 2009. *Nature's Matrix: Linking Agriculture, Conservation and Food Sovereignty*. Earthscan Publications, London, UK.

Qazi, J.A., Selfa, T.L., 2005. The politics of building alternative agro-food networks in the belly of agro-industry. *Food, Culture and Society* **8**, 45–72.

Rabalais, N.N., Turner, R.E., Scavia, D., 2002. Beyond science into policy: Gulf of Mexico hypoxia and the Mississippi River. *BioScience* **52**, 129–142.

Robertson, G.P., Vitousek, P.M., 2009. Nitrogen in agriculture: balancing the cost of an essential resource. *Annual Review of Environment and Resources* **34**, 97–125.

Rosset, P.M., Altieri, M.A., 1997. Agroecology versus input substitution: a fundamental contradiction of sustainable agriculture. *Society and Natural Resources* **10**, 283–295.

Russelle, M.P., Entz, M.H., Franzluebbers, A.J., 2007. Reconsidering integrated crop–livestock systems in North America. *Agronomy Journal* **99**, 325–334.

SARE, CTIC, 2015. 2014–2015 *Annual Report: Cover Crop Survey*. West Lafayette, IN, USA.

Schmidt, M.W., Torn, M.S., Abiven, S., Dittmar, T., Guggenberger, G., Janssens, I.A., Kleber, M., Kogel-Knabner, I., Lehmann, J., Manning, D.A., Nannipieri, P., Rasse, D.P., Weiner, S., Trumbore, S.E., 2011. Persistence of soil organic matter as an ecosystem property. *Nature* **478**, 49–56.

Shennan, C., 2008. Biotic interactions, ecological knowledge and agriculture. *Philosophical Transactions of the Royal Society of London* B **363**, 717–739.

Singer, J.W., Nusser, S.M., Alf, C.J., 2007. Are cover crops being used in the US Corn Belt? *Journal of Soil and Water Conservation* **62**, 353–358.

Smil, V., 1999. Nitrogen in crop production: An account of global flows. *Global Biogeochemical Cycles* **13**, 647–662.

Stonich, S.C., 1993. "*I am destroying the land!*": *The political ecology of poverty and environmental destruction in Honduras*. Westview Press, Boulder,

Swanson, L., 1993. Agro-environmentalism: the political economy of soil erosion in the USA. In: Harper, S. (Ed.), The Greening of Rural Policy. Bellhaven, London, UK, pp. 99–118.

Taylor, P.J., 2005. *Unruly Complexity: Ecology, Interpretation, Engagement*. University of Chicago Press, Chicago, USA.

Tomich, T.P., Brodt, S., Ferris, H., Galt, R.E., Horwath, W.R., Kebreab, E., Leveau, J.H.J., Liptzin, D., Lubell, M., Merel, P., Michelmore, R., Rosenstock, T., Scow, K., Six, J., Williams, N., Yang, L., 2011. Agroecology: a review from a global-change perspective. *Annual Review of Environment and Resources* **36**, 193–222.

Tonitto, C., David, M.B., Drinkwater, L.E., 2006. Replacing bare fallows with cover crops in fertilizer-intensive cropping systems: a meta-analysis of crop yield and N dynamics. *Agriculture, Ecosystems and Environment* **112**, 58–72.

USDA., 2008. *National Agricultural Statistics Service State and County Data.* Available at http://nass.usda.gov (retrieved August 2015). USDA, NASS, Washington, D.C., USA.

Vermeulen, S.J., Campbell, B.M., Ingram, J.S., 2012. Climate change and food systems. *Annual Review of Environment and Resources* **37**, 195–222.

Wezel, A., Casagrande, M., Celette, F., Vian, J.F., Ferrer, A., Peigné, J., 2014. Agroecological practices for sustainable agriculture. A review. *Agronomy for Sustainable Development* **34**, 1–20.

Wolf, S.A., Primmer, E., 2006. Between incentives and action: a pilot study of biodiversity conservation competencies for multifunctional forest management in Finland. *Society and Natural Resources* **19**, 845–861.

Zimmerer, K.S., 1996. Changing fortunes: Biodiversity and peasant livelihood in the Peruvian Andes. University of California Press, Berkeley, USA.

Chapter 5

A Disturbance-based Framework for Understanding Weed Community Assembly in Agroecosystems: Challenges and Opportunities for Agroecological Weed Management

Richard G. Smith* and David A. Mortensen†

*Department of Natural Resources and the Environment, University of New Hampshire, Durham, New Hampshire, USA
†Department of Plant Sciences, Pennsylvania State University, University Park, Pennsylvania, USA

1. Introduction

Weeds are a persistent feature of agriculture. Indeed, the conditions of high disturbance and high resource availability created through the practice of most forms of agriculture are the very conditions to which most plant species considered "weeds" are specifically adapted (Grime 1977, Radosevich *et al.* 2007, Smith 2015). As a consequence, among pests, weeds are the primary contributor to crop productivity and economic losses (Pimentel *et al.* 2005, Oerke 2006) and the high cost of control (Pimentel *et al.* 2005). Not surprisingly, farmers consistently rank weed control as their primary pest management challenge (Conley *et al.* 2007) and one of the greatest impediments to transitioning to organic forms of production (Barberi 2002, Martini *et al.* 2004). What is more, many of the practices employed to manage weed populations, such as tillage and application of herbicides, are often associated with non-target effects that can degrade soils, pollute water ways, and harm the health of humans and other organisms (Liebman 2001, Smith *et al.* 2011, Mortensen *et al.* 2012).

Clearly, if the science of agroecology is to contribute to the sustainability of agriculture, it must be founded in ecological principles useful for understanding agricultural weeds and their management. But what agroecological principles explicitly address weeds? Certainly, successional theory has influenced much of agroecology. Secondary succession is the change in community development over time following a disturbance (Odum 1969). This conceptual model has been used to both characterise agroecosystems relative to other types of ecosystems (Odum 1969, Hart 1980, Ewel 1999, Gliessman 2007) and as an explanation for the relative susceptibility of agroecosystems to weeds and other pest species (Gliessman 2007, Smith 2015). Fundamental to succession theory is the role that disturbance plays in creating the conditions that initiate succession. Because they are frequently disturbed, most agroecosystems can be considered to be in a perpetual state of early secondary succession (Radosevich *et al.* 2007). This critical point will be revisited in more detail later.

Community assembly theory is another body of ecological theory; however, it has not yet been as widely incorporated as a conceptual underpinning of agroecology as has succession theory. Community ecologists have used community assembly theory for some time to explain and predict the species composition of communities of organisms (see Weiher *et al.* 2011 for an overview of recent work). In this theoretical construct, a pool of species are subject to a series of filters that allow species to either pass through, thus becoming a part of a resident weed community, or be held back in a way that restricts their membership in the community (Weiher and Keddy 1999). Cornerstones of community assembly theory are the concepts of geographic, environmental, and biotic filters (Kraft *et al.* 2015). The theory posits that a regional species pool is first filtered by the organisms' ability to reach a site, thereby overcoming the geographic constraints on community membership. The species must also be adapted to the climate and soils of the site, thereby passing the environmental filter. Finally, the resulting community of organisms interact through a range of internal dynamics, the so-called "biotic filter", which affects the demography of individuals, thereby further filtering the community. These internal dynamics include species interactions such as competition, predation, herbivory, and the accumulation of plant pathogens (Belyea and Lancaster 1999, HilleRisLambers *et al.* 2012). Some aspects of community assembly theory have made their way into the agronomic weed science literature.

Booth and Swanton (2002) proposed that community assembly theory could provide a useful framework for understanding the factors that shape weed communities in agroecosystems. In particular, they highlighted how a trait-based assembly framework could be used to predict shifts in weed species composition following the implementation of new management practices or other changes to the cropping system. Additionally, they argued that because the focus is on traits, rather than species *per se*, insights from a trait-based approach would be generally applicable across cropping regions that may have very different pools of species. Subsequent research has expanded our understanding of the relative importance of particular management practices in determining patterns of weed community assembly in a variety of agroecosystems (e.g. Smith 2006, Ryan *et al.* 2010, Storkey *et al.* 2010, Garnier and Navas 2012, Perronne *et al.* 2015).

Booth and Swanton (2002) point out that while natural ecosystems and managed agroecosystems follow the same ecological principles, they also differ in a number of important ways. Despite this recognition, however, little attempt has been made to fully integrate these differences that are specific to agricultural systems into a unified community assembly framework for agroecosystems.

Here we attempt to further differentiate how managed agricultural systems are unique from other types of systems and therefore necessitate an expanded assembly framework. While assembly theory has been applied by other agricultural scientists, there are features of these so called novel ecosystems (Hobbs *et al.* 2006) that must be incorporated into this framework in order to deepen its utility within the allied disciplines of agroecology and ecologically-based weed management. Specifically, we highlight the existence of higher-order constraints on assembly filters and the assembly processes that operate in agroecosystems. We argue that there is value in recognising and better understanding the nature of these constraints, particularly if we desire to move weed management towards more environmentally-benign practices based on agroecological principles. We then contextualise cropping systems based on disturbance; specifically their position along interacting gradients of disturbance, intensity and frequency, and highlight how viewing cropping systems along these axes of disturbance illuminates the nature of the filtering processes that occur in these systems and therefore the traits of weeds that are most likely to be constrained by these filters. Finally, we describe how common agricultural management practices fit within the context of a disturbance-based assembly framework and then recommend how these theoretical constructs might be implemented in practice.

2. Differences between assembly processes in natural and agroecosystems

In natural ecosystems, plant communities assemble from a pool of species that are available to colonise a given location, i.e. the geographic species pool. Environmental filters further restrict community membership to those species that are adapted or otherwise physiologically capable of establishing under the environmental conditions specific to the location, i.e. the habitat species pool. Internal dynamics, i.e. biological filters, represent the final filter on community membership. These biological filters include competition, herbivory, and other interference interactions among species (HilleRisLambers *et al.* 2012). The resulting community thus consists of all of the species that are capable, because of their unique combination of traits, of "passing" through each of these three broad classes of filters (Figure 1(a)). Much research in ecology has attempted to

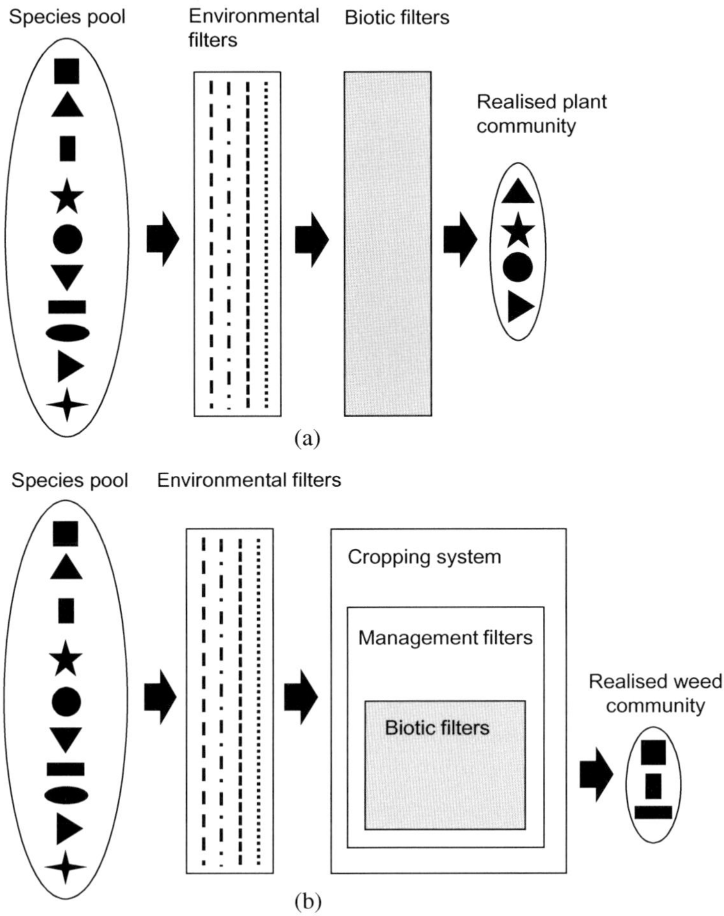

Figure 1. Conceptual model illustrating the community assembly process in (a) natural plant communities where biotic filters such as competition and other species interactions strongly shape community composition and (b) agroecosystems where biotic filters are embedded within a series of hierarchical filters that are ultimately constrained by the cropping system. Symbols represent individual weed species, each differing in specific traits. Figure (a) modified from Weiher and Keddy (1999) and Booth and Swanton (2002).

detect signals of these filters and their relative importance in ultimately determining community structure (see reviews in Weiher and Keddy 1999, Weiher *et al.* 2011, Garnier and Navas 2012, HilleRisLambers *et al.* 2012).

These same processes play out in agroecosystems, with some critical differences (Booth and Swanton 2002). In agroecosystems, the geographic and habitat species pools are still constrained by dispersal and environmental filters; however, once species pass through the environmental filters, additional filters operate that are unique and distinct from those that operate in natural ecosystems (Figure 1(b)). These filters represent constraints on the nature and magnitude of subsequent filters, including the biological filters which are often the explicit or implicit focus of many of our ecologically-based weed management strategies (Liebman and Davis 2000).

The first unique filter operating in agroecosystems is the cropping system itself (Figure 2). The cropping system incorporates the range of crops that will be grown, the range of equipment that will be used to carry out the activities and operations on the farm, and even the philosophy and mental models of the farm operator, which dictates the range of approaches that will be employed to manage the crops and associated pest populations (Eckert and Bell 2005, Radosevich *et al.* 2007, Jabbour *et al.* 2014). The second set of filters, which are constrained by the cropping system, are embodied in the management practices themselves. These management filters include the specific mechanism(s) of action by which an herbicide or mechanical implement damages or kills weeds, as well as other disturbances, such as those associated with planting or harvesting (Booth and Swanton 2002, Theisen and Bastiaans 2015). As with assembly in natural systems, biotic filters also operate in agroecosystems and include the mechanism(s) by which a cash or cover crop might exert a weed suppressive effect or by which soil-borne or foliar insects and pathogens weaken or kill weed seeds or seedlings (Liebman and Davis 2000, Davis *et al.* 2008, Mohler *et al.* 2012, Smith *et al.* 2015). The critical point here is that the nature and magnitude of these biotic filters will be strongly constrained by the management filters, which in turn, are constrained by the cropping system filter. These constraints are therefore hierarchical in that biotic filters are embedded within the context of the management filters which, in turn, are embedded within the context of the cropping system (Figure 2). Thus, as we began, the cropping system ultimately constrains the nature and magnitude of any subsequent management and biotic filtering processes.

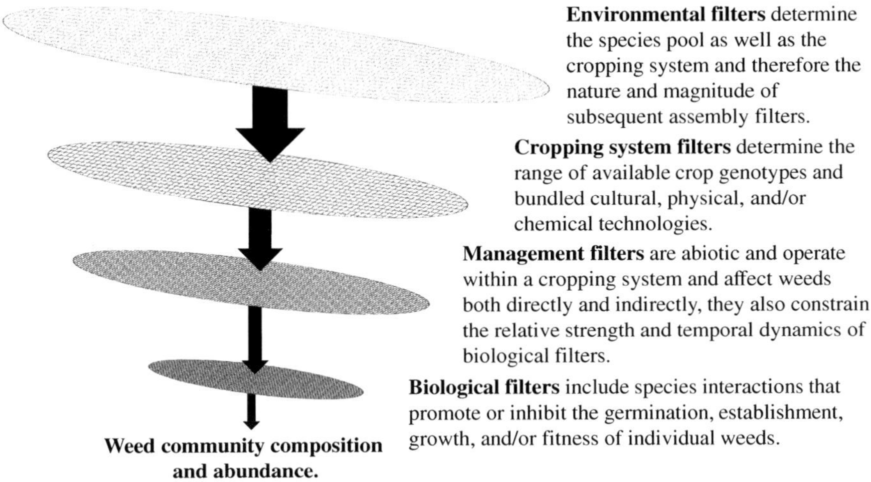

Figure 2. Hierarchy of constraints and filters that ultimately determine the species composition and abundance of weed communities in agroecosystems. These factors also determine and constrain weed management challenges and opportunities.

3. Disturbance as a unifying concept

Community assembly theory also provides a context for understanding how plant community composition and abundance might be expected to change over the course of secondary succession. Community ecologists have recognised that these two sets of theories are complementary for some time (Diamond 1975, Drake 1990, Young *et al.* 2001). Further, following colonisation, internal dynamics (i.e. biotic filters) such as competition, plant–soil feedbacks, and life-history trade-offs are the very mechanisms by which a community of species might change over time after a disturbance event. If then, disturbance is the mechanism that initiates succession, it is also then the precursor to the subsequent internal dynamics that might regulate community assembly over the course of succession. To put it another way, disturbance sets the stage but also selects the players that are available to participate in the biotic filtering that occurs later.

As we and others have stated previously, agroecosystems are unique from natural ecosystems in that they are actively managed. As such, the practice of agriculture can largely be thought of as a series of human-mediated disturbances (Freckman and Ettema 1993, Hobbs *et al.* 2008, Shennan 2008). Disturbances associated with agriculture occur in the form of tillage or herbicide application or residue burning. The act of harvesting the crop, of haying a field, or of grazing a pasture are also forms of disturbance. This simple fact suggests that rather than classifying agroecosystems based on their management philosophy or set of practices (such as organic or conventional, high versus low input) it may be more instructive in the context of the current discussion, and more accurate ecologically, to classify them in terms of their disturbance regime. This further suggests that in order for community assembly theory to have a useful application in agroecology, it must be expanded to explicitly account for the nature and magnitude of disturbances that are the defining feature of agriculture. A first step in this process is recognising (a) assembly filters operating in agroecosystems that are embedded and hierarchical starting with the cropping system filter, and (b) higher level filters such as the disturbances that make up the management filters constrain the nature and magnitude of any lower-level filters, including biotic filtering processes.

The constraining nature of disturbance-based management filters on subsequent biologically-based filters occurs because of the dynamic interplay between at least two components of the disturbance regime that makes up a particular agricultural system: specifically, intensity (i.e. how destructive a disturbance is in terms of the proportion of biomass at a site that is killed or removed; Keddy 2007) and frequency (i.e. the time interval between individual disturbance events). When conceptualised under this framework, agroecosystems can be classified based on their position along interacting gradients of disturbance intensity and disturbance frequency (Figure 3). What this implies then, is that systems characterised by frequent, high intensity disturbances, agroecosystems such as annual row crop systems or many vegetable production systems, would have weed communities structured primarily by abiotic factors (i.e. cropping system and management filters). In contrast, agroecosystems characterised by infrequent, low intensity disturbances would be structured not only by those same filters, but then also subsequent biotic filters (i.e. internal dynamics).

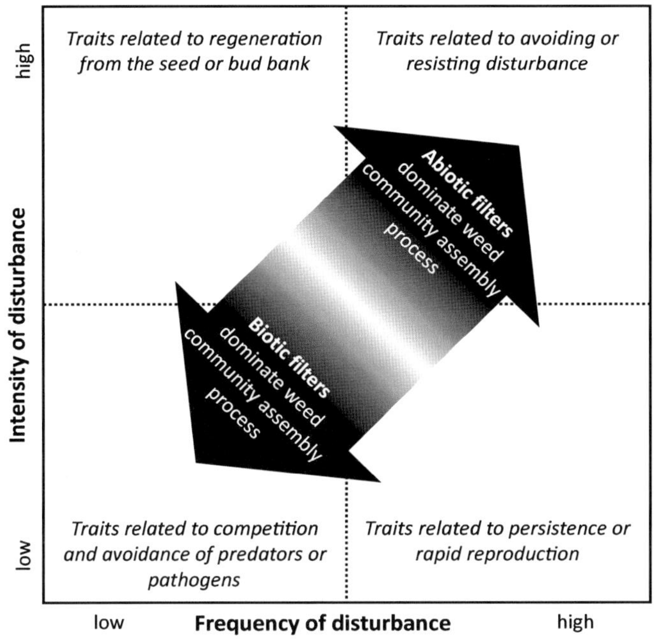

Figure 3. Conceptual model illustrating a disturbance-based framework for weed community assembly in agroecosystems. The relative importance of abiotic (i.e. management filters) versus biotic processes in structuring weed communities is dependent upon the dynamic interplay between disturbance intensity and frequency. Conditions of filtering created across these disturbance gradients will favour weed species possessing certain suites of traits over others.

The embedded nature of filters operating on weed community assembly and the role that disturbance plays in constraining these filters in agroecosystems can be illustrated with several examples. Consider two agroecosystems: a conventionally managed annual row crop system and a perennial pasture-based system. From the beginning, the fact that the cropping system is annual means weeds with germination periodicities in the warmer time of the growing season are selected for (Myers *et al.* 2004, 2005a, Smith 2006, Perronne *et al.* 2015). Further, weed communities in an annual row crop system would primarily be subjected to one or both of two strong management filters, primary tillage and/or broad-spectrum herbicide application, both of which represent intense and (because it is an annual system) frequent disturbances. Both of these filters have been demonstrated to exert strong selective (i.e. filtering) effects on the composition of weed communities (Smith 2006, Ryan *et al.* 2010, Mortensen *et al.* 2012). Other management filters operating in the annual row crop system might include those associated with planting, crop harvest or stubble mowing; however, these represent relatively less intense disturbances (Theisen and Bastiaans 2015). Because the system is annual, where the intense tillage or herbicide disturbances essentially "reset" the plant community to the initial stages of secondary succession each year, biotic filters which act over longer time scales, such as competitive exclusion, would have little time to develop and would likely have little additional effect on the resulting community dynamics of that system (Smith *et al.* 2015). Rather, disturbance would regulate the abundance of species that have managed to pass through previous filters. The literature is replete with examples of how relatively similar weed community species and functional trait composition are across annual cropping systems and how common species are repeatedly selected for through this filtering process (Radosevich *et al.* 2007).

Now, contrast the annual row crop example with the filters that might be expected to dominate in a perennial pasture-based system. In contrast to the annual system, weed communities in the perennial pasture are not reset annually by tillage or non-selective herbicide use, so biotic interactions (e.g. competition) and the lack of soil disturbance largely filters annuals from becoming a part of the final species pool. Here, management and biotic filters consisting primarily of mechanical mowing or grazing by livestock would further shape the species composition of the weed community. If the system is grazed, then palatable species may be disproportionately filtered, leaving less palatable species to dominate the community (Augustine and McNaughton 1998). Because perennial-based systems exhibit characteristics of later-successional ecosystems, other biotic filters such as plant competition, trophic interactions such as weed seed predation, and plant-soil feedback dynamics would likely be better developed and/or stronger (Odum 1969, Kardol *et al.* 2006) and therefore more important in shaping community composition.

4. Linking a disturbance-based assembly framework with species traits

The framework presented in Figure 3, offers insight into the traits that would likely be required for weed species to be successful under a disturbance regime specific to a particular cropping system. In this way, the framework results in testable predictions regarding the functional trait composition of weed communities. It also provides guidance as to how we might expect weed communities to shift following a change in the management system. For example, weed communities that assemble in cropping systems that are characterised by a low frequency of low intensity disturbances (Figure 3, bottom left) would be expected to be structured primarily by biotic filters such as competition, herbivory, and susceptibility to pathogens. As a consequence, species making up the community would tend to have traits that confer an advantage under those conditions. The community would likely be dominated by species with a perennial life history, that are relatively unpalatable, and that modify the soil environment in ways that favour their persistence (i.e. positive soil feedbacks; Kardol *et al.* 2006). Trait distributions indicative of niche-based processes, such as competitive exclusion and niche partitioning, should be particularly evident in the weed communities under these conditions (Smith *et al.* 2015). Conversely, weed communities found in systems that feature frequent and intense disturbances (Figure 3, top right) would tend to be dominated by annuals with traits conferring avoidance, resistance, or tolerance of disturbance. The agronomic literature is full of examples supporting these broad hypotheses (for a review see Buhler 1995). This framework may also help explain why in many annual cropping systems, trait distributions of the weed community appear to be structured largely by stochastic rather than niche-based processes (Petit and Fried 2012, Perronne *et al.* 2015, Smith *et al.* 2015).

Systems that feature intense disturbances that occur only infrequently (Figure 3, top left) would likely have weed communities composed of species with a different set of traits relative to the previous two situations. In this case, biotic interactions may be important during much of the assembly process; however, traits that enable species to quickly re-establish from the seed or bud bank, when disturbance does occur again, will also be important (Haas and Streibig 1982). In conditions of frequent but low intensity disturbance (Figure 3, bottom right), species with traits that make them responsive to the particular disturbance regime will likely dominate the community. It is well known that some species benefit from periodic grazing or clipping (Georgiadis *et al.* 1989). Traits related to rapid maturity and reproduction (if rcproduction occurs by seed) will likely also be selected for under these conditions.

5. Filters in the context of weed evolution

These management filters and the cropping systems that constrain them have implications for weed evolution, particularly the evolution of herbicide resistance. Particularly strong management filters may eliminate certain species from the community; however, most weed populations possess high levels of phenotypic plasticity and/or genetic variation (Neve *et al.* 2009 and references therein). Hence, some individuals within an otherwise susceptible population may be capable of passing through an otherwise "strong" management filter. Over time, if the same strong filter is applied and continues to allow only those few individuals that possess a unique set of traits (in this case a mutation that confers resistance) to pass, the weed community will become increasingly dominated by individuals with those traits. Similarly, if the diversity of filters is low, weed communities will tend to be dominated by those species with traits that allow passage through those few filters — to the point that the management filters are selecting for a recalcitrant weed community. This is why cropping systems that constrain management to only a few strong management filters (such as glyphosate) are potentially problematic.

Clearly, the nature and relative importance of the management and biotic filters are constrained by the cropping system; however, we argue that the cropping system filter is THE overarching influence on weedy plant community composition. In the most extreme cases, such as in fields that are planted to continuous maize or continuous wheat in many agroecological zones of the world, continuous planting of the same crop selects for weeds with a phenology that matches the crop (Radosevich *et al.* 2007, Fried *et al.* 2009). All else being equal, the abundance of those phenologically adapted weeds will build up over time. Additional filtering of that adapted flora via repeated use of similar herbicides and tillage regimes will further narrow the community composition. Here, as the selection pressure intensifies, the abundance of those adapted species increases with each annual iteration of the cropping systems cycle. This underscores the important point that filtering not only defines the species composition of the community, but also the abundance of the species that are particularly well adapted to that cropping system. After all, having a species-diverse weed community is not often a problem (Pollnac *et al.* 2009), and in some cases may be highly beneficial for the conservation of rare or threatened native plant species or beneficial insects (e.g. Marshall *et al.* 2003); rather, problems arise when the reproductive success of a few species that are particularly adept at passing these filters results in an uneven weed flora with a small number of particularly well adapted species occurring at very high abundance.

6. Feedback dynamics can alter the nature and magnitude of management and biotic filters

An important feature of our expanded community assembly framework is that it is dynamic and not static, iterative and not linear. In Figure 2, the filters are represented in a linear series with the strongest filter, the environmental constraints, at the top and biological constraints at the bottom. However, we know that weed communities, and the cropping systems of which they are a part, are dynamic. Weed communities may increase or decrease in response to management-induced changes in the soil environment, or feedbacks driven by biotic interactions may alter the efficacy of the management filters (see review by HilleRisLambers *et al.* 2012). We use two examples to illustrate the dynamic nature of this filtering process.

In this first example, imagine a cropping practice, such as reducing tillage, increasing the complexity of the crop rotation, or using high biomass cover crops, that results in incremental increases in the carbon content and diversity of the microbial community and nutrient pools in the soil (McDaniel *et al.* 2014, Tiemann *et al.* 2015). Here, the cropping system alters the trajectory of the soil organic matter content, which in turn alters the environmental filtering of the weed community by decreasing soil moisture stress and enhancing nutrient cycling in soils. Such effects are known to directly affect the fitness of plant species, as well as indirectly affect the interaction of those species, and therefore potentially the intensity of biotic filtering, through changes in the intensity of interspecific competition (Smith *et al.* 2010).

In the second example, changes in abundance in the realised weed community, feeds back to alter management filtering. It is known that propagule pressure drives weedy plant invasions by increasing the likelihood of recruitment success in general; however, in intensively managed agroecosystems propagule pressure can also reduce the efficacy of control practices via inverse density dependence dynamics (Hartzler and Roth 1993, Myers *et al.* 2005b). For example, it has been shown that high weed density can reduce the efficacy of some herbicides, either by making the active ingredients less bioavailable because of 'dilution' through plant uptake (Hoffman and Lavy 1978) or because some individuals within the population are physically protected from contact with the spray droplets by the canopies of other individuals (Dieleman *et al.* 1999). The efficacy of mechanical and thermal control practices can similarly be reduced with increasing density of some types of weeds (Buhler *et al.* 1992, Ascard 1995). In all cases, a build-up in abundance of weedy plant species feeds-back to reduce the severity of management filtering, thus setting up a positive feedback on the population dynamics of adapted species, where the intensity of filtering is reduced with each annual progression in the cropping cycle.

Both of these examples suggest that where a particular agroecosystem is positioned along the disturbance axes illustrated in Figure 3 can be dynamic. Hence, changes to management or feedbacks driven by biotic interactions might result in a shift in the weed community from dominance by one particular trait syndrome to another.

7. Agroecological weed management practices in the context of filters

Our discussion of filters to this point has been very theoretical. Here we provide more concrete examples of some specific agroecological weed management practices and where they fit within a disturbance-based assembly framework. Table 1 provides a list of some common agricultural practices that are aimed explicitly at managing weeds or that are likely to have suppressive or other types of effects on weed populations. The table also includes the predominant agroecological principle(s) upon which each practice is assumed to operate, as well as the type and relative intensity of filtering associated with each practice. By intensity of filtering, we mean the ability of a particular filtering process to lead to a predictable change in the trait distribution of a weed community over agriculturally-relevant spatial and temporal scales (i.e. fewer than five cropping seasons). For most of these practices, the actual intensity of the filtering effect is unknown or is likely to vary depending on the disturbance regime associated with the cropping system in which they are implemented; hence these represent our best-guess approximations.

While all weed management practices can be conceptualised as filters, not all weed management practices are necessarily predicated on agroecological principles (Table 1). It could be argued that part of the reason for this stems from differences in how weeds themselves are defined and conceptualised by those that are concerned with their presence in agricultural fields. For example, if one considers weeds to be noxious "invaders" of cropping systems, organisms that rob farms of their productivity and farmers of their livelihoods, then it is understandable that some may consider acceptable only those practices that are especially effective at killing and removing weeds. This approach to weeds and their control does not necessarily require an understanding of the biology and ecology of weeds, or of agroecology in general, but rather of which chemistry or implement does the job of killing weeds most effectively, easily, and cheaply. Herbicides and tillage are often used in this context. It is important to note, however, that herbicides and tillage can also be implemented within the context of an integrated weed management (IWM) system, often with the aims of reducing the quantity of herbicides applied to farm fields and managing for the evolution of herbicide-resistant weed populations, which does necessitate some degree of understanding of weed biology and ecology (Swanton and Weise 1991).

Both herbicides and tillage are highly effective at killing weeds. Hence, both of these practices can be considered particularly strong filters

Table 1. Common agricultural practices that directly or indirectly target weed populations and that operate as filters during the process of weed community assembly.

Practice	Agroecological principle	Type of filter	Intensity of filtering
Herbicide	None, IWM	Disturbance-based	Strong
Tillage	None, IWM	Disturbance-based	Strong
Flaming/thermal	None, IWM	Disturbance-based	Intermediate-strong
Plastic mulch	None, IWM	Physical, resource-based	Intermediate-strong
Row spacing/ seeding rate	Resource pre-emption	Biological, resource-based	Weak
Planting arrangement	Resource pre-emption	Biological, resource-based	Weak
Row orientation	Resource pre-emption	Biological, resource-based	Weak
Fertiliser placement	Resource pre-emption	Biological, resource-based	Weak
Competitive cultivars	Resource pre-emption	Biological, resource-based	Weak
Living mulch	Resource pre-emption, diversity	Biological, resource-based	Intermediate-weak
Cover cropping	Resource pre-emption, diversity	Biological, resource-based	Intermediate-weak
Intercropping	Resource pre-emption, diversity	Biological, resource-based	Intermediate-weak
Seed predation	Natural enemies	Biological, mortality-based	Intermediate-strong
Allelopathic green manures	Natural enemies	Biological, mortality-based	Intermediate-weak
Soil amendments	Natural enemies	Biological, mortality-based	Intermediate-weak
Crop rotation	Diversity, many-little-hammers	Multiple types	Variable

Note: Integrated weed management (IWM).

on the assembly of weed communities. Additionally, because they are strong filters, changes in herbicide use or tillage can result in relatively rapid shifts in the composition and abundance of weed communities, selecting for those weed species that are able to adapt to or otherwise avoid these practices. Paradoxically, because they are also disturbance-based filters, use of herbicides and/or tillage also often reinforce the very conditions that promote weed recolonisation (Smith 2015). And, as discussed previously, because these practices are disturbance-based, they also have the potential to constrain the nature and magnitude of most of the biological filtering agents that form the basis of most of our agroecologically-based weed management practices, some of which are described below.

An agroecologically-based weed management perspective views weeds with respect to their biology and ecology and within the context of the agroecosystem of which they are a part. Most weeds can be characterised as early successional species, organisms that are adapted to colonise and establish in recently disturbed sites. As such, they also tend to be highly responsive to available resources, and invest much of their energy into rapid growth and reproduction (Radosevich *et al.* 2007). Most weeds are also capable of dormancy and therefore can form relatively long-lived and persistent seed banks. This ecologically-based definition of weeds helps to shed light on why most weeds are such successful and persistent members of our agroecosystems. Our agricultural practices create the very conditions that encourage the colonisation, establishment, and growth of the plants we consider weeds; specifically, frequently disturbed soil and an abundance of available resources (Smith 2015). An ecologically-based definition of weeds as early successional plant species coupled with the realisation that our cropping practices are conducive to their success therefore provide clues for how weeds might then be managed, particularly with respect to the conditions that make agroecosystems so conducive to weeds. Hence a primary focus of agroecologically-based weed management is not so much on killing weeds, but rather changing the resource conditions of the cropping system that favour their colonisation, establishment, and/or growth.

Some agroecologically-based management practices that reduce the resources available to weeds include using cover crops before planting or after harvest of cash crops in order to suppress weeds during portions of the growing season when cash crops are not present (Hodgdon *et al.* 2016); growing living mulches between the rows of cash crops to reduce opportunities for weeds to establish there (Liebman and Dyck 1993, Warren *et al.* 2015); using narrower crop row spacing or higher seeding densities to minimise bare soil, hasten shading by the crop canopy, and improve crop competition for soil nutrients (Swanton and Weise 1991, Weiner *et al.* 2001); and planting particularly competitive crop cultivars to better compete with weeds for resources during the cropping season (Andrew *et al.* 2015). Even a practice as simple as changing the orientation of crop rows relative to the path of the sun can increase light interception by the crop canopy in some crops (thereby reducing light available to the weeds), in some cases resulting in as much as a 51% reduction in weed abundance compared to other row orientations (Borger *et al.* 2010). Crop rotation offers the opportunity to integrate a multitude of these resource-based biological filters, thereby not only potentially increasing their cumulative effect over time, but also reducing the likelihood that particular weed species will adapt or otherwise avoid filtering and become recalcitrant, and therefore problematic, over the longer term (Liebman and Gallandt 1997). These resource-based biological filters are likely much weaker in terms of their ability to drive changes in the relative trait distributions of the weed community than the strong disturbance-based filters of herbicides and tillage because they rely on species interactions that stress rather than kill weeds. The fact that the weeds and crops emerge at roughly the same time, and that competition occurs for only a portion of the total growing season, particularly in annual cropping systems, likely further weakens their filtering effect (Smith *et al.* 2015).

Agroecologically-based practices aimed at increasing weed mortality via biological agents have the potential to be stronger assembly filters than those based on resource pre-emption and include the use of green manures that have allelopathic or other phytotoxic properties (Weston 1996); use of manures, composts, or other soil quality enhancements that promote soil microbial degradation of weed seeds and growth suppression of weed seedlings (Liebman and Davis 2000, Kremer and Li 2003); and conservation or establishment of habitat in or around crop fields or other practices that promote the abundance and activities of weed seed predators

(Menalled *et al.* 2007). Weed seed predation in particular has the potential to be a strong assembly filter (because once a seed is consumed, it is dead); however, predation rates in agroecosystems are often patchy in both space and time (Westerman *et al.* 2003, Jacob *et al.* 2006), reducing the likelihood that it or other biologically-based filters would lead to dramatic shifts in weed community composition and trait distributions over agriculturally-relevant timescales, except in systems where the frequency and intensity of disturbance are both extremely low (Figure 3). Further, because most biological filters require some amount of time to develop (in the case of food-web mediated interactions such as weed seed predation) or become intense (in the case of resource-based interactions such as competitive exclusion from cover crops), periodic implementation of strong-disturbance-based filters often likely undermine the effectiveness of many biologically-based filters in most agroecosystems (Smith *et al.* 2015).

8. Putting a disturbance-based assembly framework into practice

Ideally, a disturbance-based assembly framework will not only help researchers and land managers understand how weed communities assemble but also to better manage weed problems. This framework suggests that dramatic changes in the trajectory of weed assembly are unlikely to occur in the absence of changes to the cropping system. At first this may seem obvious; however, this point is worth elaborating. A case in point is the profound glyphosate resistance problem being experienced in regions of the world where glyphosate-resistant crops have been adopted (Mortensen *et al.* 2012). The primary solutions being proposed, represent modest changes to the management filters, in this case using new transgenic crops that are resistant to additional herbicides, but not changes to the cropping system. As a consequence, these additional management filters (i.e. additional herbicide modes of action) will likely result in a two- to three-fold increase in herbicide use and select for a weed community with multiple resistance traits (Egan *et al.* 2011, Mortensen *et al.* 2012). In contrast, fundamentally altering the cropping system filter by adding diversity to the crop rotation, in the form of a winter cereal in a rotation that features spring-sown crops or a perennial forage crop in a rotation that features only annual crops, would likely reduce the need for herbicide-based and other management filters, reduce the abundance of problematic weedy species, almost certainly increase the diversity of the flora, and reduce the overall density of the resistant fraction of the seed bank (Smith *et al.* 2008, Davis *et al.* 2012, Harker *et al.* 2016). Again, none of these changes are likely to arise without a concerted change in the cropping system filter.

Secondly, ecologically-based weed management attempts to rely more heavily on biotic interactions to manage weed communities rather than the use of herbicides (Liebman and Davis 2000). If, however, the nature and strength of specific biotic filters are being undermined by disturbance-based management filters, which in turn are ultimately constrained by the cropping system, then these constraints need to be identified and addressed. In this case, it may be necessary to modify, change, or remove higher-level constraining management filters before biotic interactions can be expected to become impactful filters on the assembly process (e.g. Hulvey and Aigner 2014).

Given the large differences in functional traits being selected for across the disturbance gradient it may be that intermittent disturbance could keep weedy plant populations "off balance" in a way that local abundances remain low, and well adapted species are selected against. This is akin to the "many little hammers" concept articulated by Liebman and Gallandt (1997), which, while not initially conceived within a community assembly framework, has obvious parallels. For progress here, we need a greater understanding of these filtering effects and the degree to which seed longevity enables species to "skip" temporally dynamic filters that would otherwise select against a species' success. Ideally, hierarchical filters will be implemented in a way that reinforces filtering processes that are beneficial from the perspective of weed management rather than undermining.

Additional research aimed at understanding the relative role that biotic and abiotic filters play in agroecosystems that fall across the spectrum of disturbance gradients illustrated in Figure 3, could be a fruitful avenue for future research. Fruitful, too, will be research that better quantifies the nature, magnitude, and mechanism of action of particular disturbance-based management filters. For example, both tillage and broad-spectrum herbicides represent intense disturbance; however, each filter likely acts on very different sets of traits and therefore each likely have very different effects on the assembly process.

When advancing ecologically-based weed management, we must also recognise that in practice, farmers are constrained by features of their cropping system and abiotic environment in a way that can limit the range of possible or tractable filters. While we realise this to be the case, we also know that flexibility exists with the design of cropping system, and those that are designed based on sound ecological principles are likely to provide sufficient levels of weedy plant suppression while limiting the adverse effects that some filtering management practices can have on the environment.

9. Conclusions

As ecologists working in agroecosystems for a combined 40 years, it has become apparent to us that weedy pests are largely a product of the cropping system in which they occur. Often, weed management practices aimed simply at killing weeds fail because the underlying factors that shape weed communities and make cropping systems suitable habitats for them to persist are not well understood or addressed. Towards that end, we argue that much progress could be made in understanding the factors that shape weed communities in agroecosystems by better contextualising the disturbance associated with arable cropping within a community assembly framework. Under a community assembly framework, cropping systems can be characterised by the frequency and intensity of disturbance that occur in these systems. Disturbance can then be viewed as both a filter that directly affects the assembly of weed communities and as a constraint on the nature and magnitude of other filters that may operate in these systems. This disturbance-based assembly framework suggests that in agroecosystems, community assembly filters are hierarchical and therefore agroecological approaches to managing weeds will be most effective when they work with, rather than against higher-level filtering processes. In some agroecosystems, attempts to manipulate weed community assembly via species interactions or other biological processes (i.e. biotic filters) may be ineffective without fundamentally altering the nature of the cropping system and associated regime of disturbances. Further, this framework assumes functional traits, rather than taxonomic classification, determines whether individual species establish and persist in particular cropping systems. We hypothesise that weed communities can be characterised by four broad "trait syndromes" which allow species to persist under varying levels of agricultural disturbance. Cropping system constraints on weed and other pest community dynamics must be understood if we are to devise more robust and environmentally benign methods of management.

Acknowledgements

This work was supported by USDA NIFA AFRI Grant No. 2013-67014-21318.

References

Andrew, I.K.S., Storkey, J., Sparkes, D.L., 2015. A review of the potential for competitive cereal cultivars as a tool in integrated weed management. *Weed Research* **55**, 239–248.

Ascard, J., 1995. Effects of flame weeding on weed species at different developmental stages. *Weed Research* **35**, 397–411.

Augustine, D.J., McNaughton, S.J. 1998. Ungulate effects on the functional species composition of plant communities: herbivore selectivity and plant tolerance. *The Journal of Wildlife Management* **62**, 1165–1183.

Barberi, P., 2002. Weed management in organic agriculture: are we addressing the right issues? *Weed Research* **42**, 177–193.

Belyea, L.R., Lancaster, J., 1999. Assembly rules within a contingent ecology. *OIKOS* **86**, 402–416.

Booth, B.D., Swanton, C.J., 2002. Assembly theory applied to weed communities. *Weed Science* **50**, 2–13.

Borger, C.P.D., Hashem, A., Pathan, S., 2010. Manipulating crop row orientation to suppress weeds and increase crop yield. *Weed Science* **58**, 174–178.

Buhler, D.D., 1995. Influence of tillage systems on weed population dynamics and management in corn and soybean in the central USA. *Crop Science* **35**, 1247–1258.

Buhler, D.D., Gunsolus, J.L., Ralston, D.F., 1992. Integrated weed management techniques to reduce herbicide inputs in soybean. *Agronomy Journal* **84**, 973–978.

Conley, S.P., Krupke, C., Santini, J., Shaner, G., 2007. Pest management in Indiana soybean production systems. *Journal of Extension* **45**(4), 4RIB8, Available at: http://www.joe.org/joe/2007august/rb8.php

Davis, A.S., Schutte, B.J., Iannuzzi, J., Renner, K.A., 2008. Chemical and physical defense of weed seeds in relation to soil seedbank persistence. *Weed Science* **56**, 676–684.

Davis, A.S., Hill, J.D., Chase, C.A., Johanns, A.M., Liebman, M., 2012. Increasing cropping system diversity balances productivity, profitability and environmental health. *PLoS ONE* **7**(10), e47149. doi:10.1371/journal.pone.0047149

Diamond, J.M., 1975. Assembly of species communities. In: Cody, M. L., Diamond, J. M. (Eds.), *Ecology and Evolution of Communities*. Harvard University Press, Cambridge, MA, USA, pp. 342–444.

Dieleman, J.A., Mortensen, D.A., Martin, A.R., 1999. Influence of velvetleaf (*Abutilon theophrasti*) and common sunflower (*Helianthus annuus*) density variation on weed management outcomes. *Weed Science* **47**, 81–89.

Drake, J.A., 1990. The mechanics of community assembly and succession. *Journal of Theoretical Biology* **147**, 213–233.

Eckert, E., Bell, A., 2005. Invisible force: farmers' mental models and how they influence learning and actions. *Journal of Extension* **43**, 3FEA2, Available at: http://www.joe.org/joe/2005june/a2.php

Egan, J.F., Maxwell, B.D., Mortensen, D.A., Ryan, M.R., Smith, R.G., 2011. 2,4-Dichlorophenoxyacetic acid (2,4-D)-resistant crops and the potential for evolution of 2,4-D-resistant weeds. *Proceedings of the National Academy of Sciences* **108**(11) E38, Available at: www.pnas.org/cgi/doi/10.1073/pnas.1017414108

Ewel, J.J., 1999. Natural systems as models for the design of sustainable systems of land use. *Agroforestry Systems* **45**, 1–21.

Freckman, D.W., Ettema, C.H., 1993. Assessing nematode communities in agroecosystems of varying human intervention. *Agriculture, Ecosystems and Environment* **45**, 239–261.

Fried, G., Chauvel, B., Reboud, X., 2009. A functional analysis of large-scale temporal shifts from 1970 to 2000 in weed assemblages of sunflower crops in France. *Journal of Vegetation Science* **20**, 49–58.

Garnier, E., Navas, M., 2012. A trait-based approach to comparative functional plant ecology: concepts, methods and applications for agroecology. *A review. Agronomy for Sustainable Development* **32**, 365–399.

Georgiadis, N.J., Ruess, R.W., McNaughton, S.J., Western, D., 1989. Ecological conditions that determine when grazing stimulates grass production. *Oecologia* **81**, 316–322.

Gliessman, S.R., 2007. *Agroecology: The Ecology of Sustainable Food Systems*. CRC Press, Boca Raton, Florida, USA.

Grime, J.P., 1977. Evidence for the existence of three primary strategies in plants and its relevance to ecological and evolutionary theory. *American Naturalist* **111**, 1169–1194.

Haas, H., Streibig, J.C., 1982. Changing patterns of weed distribution as a result of herbicide use and other agronomic factors. In: LeBaron, H.M., Gressel, J. (Eds.), *Herbicide Resistance in Plants*. John Wiley and Sons, New York, pp. 57–79.

Harker, K.N., O'Donovan, J.T., Turkington, T.K., Blackshaw, R.E., Lupwayi, N.Z., Smith, E.G., Johnson, E.N., Pageau, D., Shirtliffe, S.J., Gulden, R.H., Rowsell, J., Hall, L.M., Willenborg, C.J., 2016. Diverse rotations and optimal cultural practices control wild oat (*Avena fatua*). *Weed Science* **64**, 170–180.

Hart, R.D., 1980. A natural ecosystem analog approach to the design of a successional crop system for tropical forest environments. *Biotropica* **12**, 73–82.

Hartzler, R.G., Roth, G.W., 1993. Effect of prior year's weed control on herbicide effectiveness in corn (*Zea mays*). *Weed Technology* **7**, 611–614.

HilleRisLambers, J., Adler, P.B., Harpole, W.S., Levine, J.M., Mayfield, M.M., 2012. Rethinking community assembly through the lens of coexistence theory. Annual Review of Ecology, *Evolution, and Systematics* **43**, 227–248.

Hobbs, R.J., Arico, S., Aronson, J., Baron, J.S., Bridgewater, P., Cramer, V.A., Epstein, P.R., Ewel, J.J., Klink, C.A., Lugo, A.E., Norton, D., Ojima, D., Richardson, D.M., Sanderson, E.W., Valladares, F., Vila, M., Zamora, R., Zobel, M., 2006. Novel ecosystems: theoretical and management aspects of the new ecological world order. *Global Ecology and Biogeography* **15**, 1–7.

Hobbs, P.R., Sayre, K., Gupta, R., 2008. The role of conservation agriculture in sustainable agriculture. *Philosophical Transactions of the Royal Society B* **363**, 543–555.

Hodgdon, E., Warren, N.D., Smith, R.G., Sideman, R.G., 2016. In-season and carry-over effects of cover crops on productivity and weed suppression. *Agronomy Journal* **108**, doi:10.2134/agronj2015.0419

Hoffman, D.W., Lavy, T.L., 1978. Plant competition for atrazine. *Weed Science* **26**, 94–99.

Hulvey, K.B., Aigner, P.A. 2014. Using filter-based community assembly models to improve restoration outcomes. *Journal of Applied Ecology* **51**, 997–1005.

Jabbour, R., Zwickle, S., Gallandt, E.R., McPhee, K.E., Wilson, R.S., Doohan, D., 2014. Mental models of organic weed management: comparison of

New England US farmer and expert models. *Renewable Agriculture and Food Systems* **29**, 319–333.

Jacob, H.S., Minkey, D.M., Gallagher, R.S., Borger, C.P., 2006. Variation in postdispersal weed seed predation in a crop field. *Weed Science* **54**, 148–155.

Kardol, P., Bezemer, T.M., van der Putten, W.H., 2006. Temporal variation in plant-soil feedback controls succession. *Ecology Letters* **9**, 1080–1088.

Keddy, P., 2007. *Plants and Vegetation: Origins, Processes, and Consequences.* Cambridge University Press.

Kraft, N.J.B., Adler, P.B., Godoy, O., James, E.C., Fuller, S., Levine, J.M., 2015. Community assembly, coexistence and the environmental filtering metaphor. *Functional Ecology* **29**, 592–599.

Kremer, R.J., Li, J., 2003. Developing weed-suppressive soils through improved soil quality management. *Soil and Tillage Research* **72**, 193–202.

Liebman, M., 2001.Weed management: a need for ecological approaches. In: Liebman, M., Mohler, C.L., Staver, C.P. (Eds.), *Ecological Management of Agricultural Weeds*. Cambridge University Press, Cambridge, UK, pp. 1–39.

Liebman, M., Davis, A.S., 2000. Integration of soil, crop and weed management in low-external-input farming systems. *Weed Research* **40**, 27–47.

Liebman, M., Dyck, E., 1993. Crop rotation and intercropping strategies for weed management. *Ecological Applications* **3**, 92–122.

Liebman, M., Gallandt, E.R., 1997. Ecological management of crop-weed interactions. In: Jackson, L.E. (Ed.), *Ecology in Agriculture*. Academic Press, San Diego, CA, USA, pp. 291–343.

Marshall, E.J.P., Brown, V.K., Boatman, N.D., Lutman, P.J.W., Squire, G.R., Ward, L.K., 2003. The role of weeds in supporting biological diversity within crop fields. *Weed Research* **43**, 77–89.

Martini, E.A., J.S. Buyer, D.C. Bryant, T.K. Hartz, and R.F. Denison., 2004. Yield increases during the organic transition: improving soil quality or increasing experience? *Field Crops Research* **86**, 255–266.

McDaniel, M.D., Tiemann, L.K., Grandy, A.S., 2014. Does agricultural crop diversity enhance soil microbial biomass and organic matter dynamics? A meta-analysis. *Ecological Applications* **24**, 560–570.

Menalled, F.D., Smith, R.G., Dauer, J.T., Fox, T.B., 2007. Impact of agricultural management on carabid communities and weed seed predation. *Agriculture, Ecosystems and Environment* **118**, 49–54.

Mohler, C.L., Dykeman, C., Nelson, E.B., DiTommaso, A. 2012. Reduction in weed seedling emergence by pathogens following the incorporation of green crop residue. *Weed Research* **52**, 467–477.

Mortensen, D.A., Egan, J.F., Maxwell, B.D., Ryan, M.R., Smith, R.G., 2012. Navigating a critical juncture for sustainable weed management. *Bioscience* **62**, 75–89.

Myers, M.W., Curran, W.S., VanGessel, M.J., Calvin, D.D., Mortensen, D.A., Majek, B.A., Karsten, H.D., Roth, G.W., 2004. Predicting weed emergence for eight annual species in the northeastern United States. *Weed Science* **52**, 913–919.

Myers, M.W., Curran, W.S., VanGessel, M.J., Majek, B.A., Mortensen, D.A., Calvin, D.D., Karsten, H.D., Roth, G.W., 2005a. Effect of soil disturbance on annual weed emergence in the northeastern United States. *Weed Technology* **19**, 274–282.

Myers, M.W., Curran, W.S., VanGessel, M.J., Majek, B.A., Scott, B.A., Mortensen, D.A., Calvin, D.D., Karsten, H.D., Roth, G.W., 2005b. The effect of weed density and application timing on weed control and corn grain yield. *Weed Technology* **19**, 102–107.

Neve, P., Vila-Aiub, M., Roux, F., 2009. Evolutionary-thinking in agricultural weed management. *New Phytologist* **184**, 783–793.

Odum, E.P., 1969. The strategy of ecosystem development. *Science* **164**, 262–270.

Oerke, E.C. 2006. Crop losses to pests. *Journal of Agricultural Science* **144**, 31–43.

Perronne, R., Le Corre, V., Bretagnolle, V., Gaba, S., 2015. Stochastic processes and crop types shape weed community assembly in arable fields. *Journal of Vegetation Science* **26**, 348–359.

Petit, S., Fried, G., 2012. Patterns of co-occurrence at the field and landscape level. *Journal of Vegetation Science* **23**, 1137–1147.

Pimentel, D., Zuniga, R., Morrison, D., 2005. Update on the environmental and economic costs associated with alien-invasive species in the United States. *Ecological Economics* **52**, 273–288.

Pollnac, F.W., Maxwell, B.D., Menalled, F.D., 2009. Weed community characteristics and crop performance: a neighbourhood approach. *Weed Research* **49**, 242–250.

Radosevich, S.R., Holt, J.S., Ghersa, C.M., 2007. *Ecology of Weeds and Invasive Plants: Relationship to Agriculture and Natural Resource Management.* John Wiley and Sons, Hoboken, New Jersey, USA.

Ryan, M.R., Smith, R.G., Mirsky, S.B., Mortensen, D.A., Seidel, R., 2010. Management filters and species traits: Weed community assembly in long-term organic and conventional systems. *Weed Science* **58**, 265–277.

Shennan, C., 2008. Biotic interactions, ecological knowledge and agriculture. *Philosophical Transactions of the Royal Society B* **363**, 717–739.

Smith, R.G., 2006. Timing of tillage is an important filter on the assembly of weed communities. *Weed Science* **54**, 705–712.

Smith, R.G., 2015. A succession-energy framework for reducing non-target impacts of annual crop production. *Agricultural Systems* **133**, 14–21.

Smith, R.G., Gross, K.L., Robertson, G.P., 2008. Effects of crop diversity on agroecosystem function: crop yield response. *Ecosystems* **11**, 355–366.

Smith, R.G., Mortensen, D.A., Ryan, M.R., 2010. A new hypothesis for the functional role of diversity in mediating resource pools and weed-crop competition in agroecosystems. *Weed Research* **50**, 37–48.

Smith, R.G., Ryan, M.R., Menalled, F.D., 2011. Direct and indirect impacts of weed management practices on soil quality. In: Hatfield, J.L., Sauer, T.J. (Eds.), *Soil Management: Building a Stable Base for Agriculture*. American Society of Agronomy and Soil Science Society of America, Madison, Wisconsin, USA, pp. 275–286.

Smith, R.G., Atwood, L.W., Pollnac, F.W., Warren, N.D., 2015. Cover crop species as distinct biotic filters in weed community assembly. *Weed Science* **63**, 282–295.

Storkey, J., Moss, S.R., Cussans, J.W., 2010. Using assembly theory to explain changes in a weed flora in response to agricultural intensification. *Weed Science* **58**, 39–46.

Swanton, C.J., Weise, S.F., 1991. Integrated weed management: the rationale and approach. *Weed Technology* **5**, 657–663.

Theisen, G., Bastiaans, L., 2015. Low disturbance seeding suppresses weeds in no-tillage soybean. *Weed Research* **55**, 598–608.

Tiemann, L. K., Grandy, A. S., Marin-Spiotta, E., Atkinson, E. E., McDaniel. M.D., 2015. Crop rotational diversity enhances belowground communities and functions in an agroecosystem. *Ecology Letters* **18**, 761–771.

Warren, N.D., Smith, R.G., Sideman, R.G., 2015. Effects of living mulch and fertilizer on the performance of broccoli in plasticulture. *HortScience* **50**, 218–224.

Weiher, E., Keddy, P.A., 1999. Assembly rules as general constraints on community composition. In: Weiher, E., Keddy, P.A. (Eds.), *Ecological Assembly Rules: Perspectives, Advances, Re-treats*. Cambridge University Press, Cambridge, pp. 251–271.

Weiher, E., Freund, D., Bunton, T., Stefanski, A., Lee, T., Bentivenga, S., 2011. Advances, challenges and a developing synthesis of ecological community assembly theory. *Philosophical Transactions of the Royal Society B* **366**, 2403–2413.

Weiner, J., Griepentrog, H., Kristensen, L., 2001. Suppression of weeds by spring wheat *Triticum aestivum* increases with crop density and spatial uniformity. *Journal of Applied Ecology* **38**, 784–790.

Westerman, P.R., Hofman, A., Vet, L.E.M., van der Werf, W., 2003. Relative importance of vertebrates and invertebrates in epigeaic weed seed predation in organic cereal fields. *Agriculture, Ecosystems and Environment* **95**, 417–425.

Weston, L.A., 1996. Utilization of allelopathy for weed management in agroecosystems. *Agronomy Journal* **88**, 860–866.

Young, T.P., Chase, J.M., Huddleston, R.T., 2001. Community succession and assembly: comparing, contrasting and combining paradigms in the context of ecological restoration. *Ecological Restoration* **19**, 5–18.

Chapter 6

Integration of Genetically Modified Crop Traits in Agroecological Practices in Europe: A Critical Review

Heikki M.T. Hokkanen and Ingeborg Menzler-Hokkanen

University of Helsinki, Department of Agricultural Sciences, Latokartanonkaari 5, PL 27, FIN-00014 Helsinki, Finland

1. General issues about integration of genetically modified crops in agroecological practices and IPM systems

The general approach for pest management strategies have long been dominated by quests for "silver bullet" products to control pest outbreaks after they occur, instead of improving the overall stability of the system. Managing undesired variables in agricultural ecosystems perpetuates a "quick-fix mentality" that targets symptoms and fails to address the root causes of pest problems. In the future, Integrated Pest Management (IPM) can be expected to continue to be a dominant theme in agriculture. IPM is complex, demands an ecological understanding of pest problems and can be challenging to implement. A holistic approach in IPM, where an optimal combination of pest control tactics are utilised, is the ultimate aim.

Genetically modified (GM) crops have been taken up by growers at a phenomenal rate, with about 70–80% of global plantings in some main crops such as soybean and cotton (James 2014). Theoretically, most GM crops and their new traits have the potential to significantly improve crop production especially under heavy pest, disease, and weed pressure. In this chapter, we critically review the possibility of GM crops to provide benefits for biological control and sustainable agroecological practices in Europe. We intend to identify their potential and shortcomings in their performance, which might explain the present reluctance in Europe to incorporate them as an agroecological tool.

1.1. *Reality gap in IPM*

Currently GM crops are usually seen by growers as a standalone technology for pest and disease control, without any real attempt to integrate them as a component in integrated pest management (e.g. Kos *et al.* 2009). This erodes the theoretical benefits of pest and disease resistant GM crops, which seldom seem to be realised in a sustainable way and offer only short-term, unsustainable pest control benefits. Eventually these benefits may be lost, as has been observed already in parts of the USA by the return of growers to conventional maize varieties.

This can be illustrated by considering the reality between the traditional "IPM pyramid" and the actual situation in mainstream pest management (Figure 1): ideally, pest management is based to a large extent on avoidance, and the use of chemical pesticides is just a small tip of the pyramid. In reality, the pyramid is upside down, where most of the actual pest management is done by the use of chemical pesticides. It becomes obvious that this method of pest management cannot be stable and sustainable. When GM crops are also seen by growers as a standalone pest management tool, its fate will be the same as that of chemical pesticides: unstable and unsustainable.

There are already severe signs that this is happening. For example, herbicide resistant weeds have arisen at an accelerated rate, and it is a paradox that, in herbicide tolerant cotton, growers have had to resort to hand weeding: *Amaranthus palmeri* needs to be hand weeded in

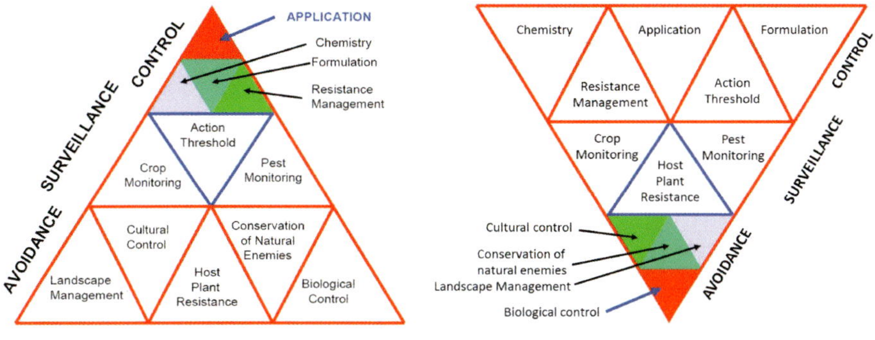

Figure 1. The "reality gap" in pest management: Ideal IPM as promoted for >50 years ((a) modified from Naranjo (2001), and current reality in mainstream pest management, (b) Hokkanen (2015)).

herbicide tolerant (HT) cotton in the USA (Sosnoskie and Culpepper 2014, Culpepper 2015). Hand weeding in HT cotton in 2000–2005 was practised by 17% of growers on a total of 5% cotton area in the USA at the cost of $2.40/acre, while in 2006–2010, 92% of growers hand-weeded 52% cotton area, at the cost of $23.70/acre. Similarly, Tabashnik *et al.* (2013) have shown how the first reports concerning field evolved resistance in *Lepidoptera* to *Bacillus thuringiensis* (*Bt*) crops appeared 10 years after commercial cultivation of these crops, and more species have been added to that list annually, so that in 2011 at least six different species had reduced sensitivity to Bt crops around the world.

The lack of integration of transgenic crops into IPM systems has been pointed out by many authors. Kos *et al.* (2009) believe that this is primarily the result of a gap, which exists between the reductionist approach in biotechnology, with focus on a single solution to pest problems, and the holistic approach in IPM, where an optimal combination of pest control tactics is the aim. Insect-resistant transgenic crops have been marketed as providing the solution to pest problems, while in reality a single solution has rarely been sufficient. Currently available transgenic crops based on Bt can at least temporarily solve some moth or beetle problems, but often secondary pests such as aphids and whiteflies can arise when competition from the main pests is relieved (Cloutier *et al.* 2008). It thus appears that transgenic plants have the greatest potential when grown as part of IPM programmes (Kos *et al.* 2009). It has already been demonstrated that currently available insect-resistant transgenic plants are compatible with other IPM approaches, such as biological, chemical, and cultural control (Smith 2005, Romeis *et al.* 2008a).

1.2. *GM crops and biological control*

The promise of GM crops has been to provide (i) higher yields, (ii) more efficient pest, disease, and weed control, (iii) lower use of pesticides, (iv) improved biological control, and (v) improved possibility for IPM. GM crops have been taken up by growers at a phenomenal rate, with about 70–80% of global plantings in some main crops such as soybean and cotton (James 2014). Theoretically, most GM crops and their new traits have the potential to significantly improve crop production especially under heavy pest, disease, and weed pressure. In particular, insecticidal and virus-resistant crops can help to keep pests and diseases in check, to lower the chemical pesticide load in the environment, and support complementary integrated pest management tactics such as increased reliance on natural control. It should be noted, however, that GM crops available for commercial use do not increase the yield potential of a variety, in fact, under low pest pressure yields, may even decrease (Gurian-Sherman 2009).

Two major traits have been commercialised so far: insect resistance and herbicide tolerance. A number of reviews (e.g. Lundgren *et al.* 2009, Tian *et al.* 2015) show that insect resistance in GM crops (i) dramatically reduces pesticide inputs, (ii) improves biological control of other pests, (iii) conserves pollination services, and (iv) improves crop productivity under high target pest pressure. On the other hand, herbicide tolerance (i) usually increases herbicide inputs, (ii) interferes with biocontrol and pollination by removing nectar and pollen resources, but (iii) improves crop productivity under high weed pressure.

Mechanisms through which natural enemies may be affected by GM crops (Lundgren *et al.* 2009) include (i) toxicity-mediated effects such as toxicity of non-prey food items (e.g. pollen) from GM crops, and toxin-accumulating prey on GM crops, (ii) changes in the crop environment, such as unintended alterations to the GM crop plant, GM crop induced changes in prey quality and abundance, and altered plant communities due to growing of HT crops, (iii) conservation tillage — facilitated by the use of HT crops — and its impacts on biological control, (iv) prey-mediated effects of GM crops (including impacts on predators, parasitoids and insect pathogens), and (v) toxic effects of herbicides used with GM crops, which can affect arthropod natural enemies.

2. The potential of GM crops as a tool for sustainable agroecological management schemes

GM crops are currently grown over 180 million ha, on all continents, but where are they integrated as a tool in agroecological or IPM management schemes? It is illustrative that at the recent conference, the 8th International IPM Symposium (IPM: Solutions for a Changing World) March 23–26, 2015 in Salt Lake City, Utah, USA, only two out of 47 separate Symposia dealt specifically with GM crops. These two symposia focussed solely on the problematics of herbicide tolerant weeds, and actually did not put them in IPM context. Therefore, in the USA and internationally, GM crops do not seem to be an issue in the discussions on IPM, nor are the crops put into IPM context even at scientific conferences.

As a first step in increasing the sustainability of the technology, and agricultural systems in general, we propose to analyse the possible contributions of GM crops and their traits to IPM relevant functions in a comprehensive manner, as outlined in Table 1. Here we provide a first estimate of the potential, or usefulness of currently available and near-market traits: to support biological control practices, to allow better use of cultural control methods (e.g. changes in soil management made possible by HT crops and possibly benefiting overwintering parasitoids and entomopathogens in the soil; Hokkanen *et al.* 1988), to improve possibilities for resistance management, to reduce pesticide use, and to allow better land use management to support vital ecosystem functions such as biocontrol and pollination (roughly following the EU–IPM principles; Barzman *et al.* 2015).

Our focus in this review is on the integration of the altered practices with other components of crop protection and crop management. We provide a first systematic study of these new, emerging IPM options for the

Table 1. Overview of current GM traits available, and their potential to contribute to IPM-relevant functions in the agroecosystem.

	Potential contribution of GM traits to IPM-relevant functions (direct and indirect)				
Traits available, or close to market	Biological control	Cultural control methods	Resistance management	Reduction in pesticide use	Land use and management
Insect resistance					
Single Lep	++/+	+	–	++	–
Single Col	+	+	–	+	–
Stacked Lep+Col	++/+	+	0/–	++	–
Bt	++/+	+	–	++	–
RNAi	+++	+	++	++	?
Herbicide tolerance					
Glyphosate	+/0/–	+	–	–	+/–
Glufosinate	+/0/–	+	–	–	+/–
Stacked	+/0/–	+	0/–	–	+/–
Disease resistance					
Fungal	+	+	+/0	++	0
Viral	+/0	+	+	+	0
Nematode	+/0	+	?	+	0
Stacked traits					
IR+HT	+/–	+	–	0	–
IR+HR+DR	+/–	+	–	+/0	–
IR+HR+DR+NR	+/–	+	–	+/0	–

Note: The authors' own judgement, based on review of related literature. +++ = high potential; – = highly problematic; 0 = no obvious connection. Combinations (e.g. +/0/–) imply that several outcomes are possible, and the contribution can be positive, neutral or negative, depending on the case. Lep = *Lepidoptera*, Col = *Coleoptera*, Bt = *Bacillus thuringiensis*, RNAi = RNA interference, IR = insect resistance, HT = herbicide tolerance, DR = disease resistance, NR = nematode resistance. Adapted from Hokkanen and Menzler-Hokkanen (2016).

GM crops and traits, which already are, or will be available in the near future. Specific provisions related to the introduction of GM crops, such as the requirement for sowing refuge areas with non-Bt cultivars in Bt crops, or coexistence-related measures, will be taken into consideration when appropriate. Additionally, we emphasise the need for flexibility in developing new management options for GM crops: they need to be tailored according to the receiving environments, and the different agroecosystems taking into account the climatic conditions (Arpaia *et al.* 2014). The pest/disease resistant GM crop may be considered as one tool in the IPM toolbox, and therefore the optimal toolbox for any region will depend on selecting the best IPM tools that work together, thus increasing the sustainability of the IPM system. A key question is: are GM crops just basic IPM tools, or do they have some features that give added value?

3. Opportunities for IPM, based on insect resistance in GM crops

There is no question that host plant genetic resistance plays an important role as a preventive measure in IPM programmes. Whether insecticidal GM crops can be placed in the same box is less clear. GM crops are sometimes considered as an IPM tool just as any other resistant cultivar (Kennedy 2008, Birch *et al.* 2011). However, as Bt-transgenic crops are "insecticidal plants", unlike conventionally bred cultivars, their use fundamentally conflicts with IPM principle number 3 (requesting pesticide applications to be based on thresholds) (Barzman *et al.* 2015). This is because, like seed dressing with insecticides, the application of the pesticide is systematic, and does not depend on assessing the pest pressure and the use of thresholds. Often the use of GM crops also conflicts with IPM principle number 7 (requiring the adoption of antiresistance strategies), although sometimes refuge areas sown with non-GM cultivar may be applied to delay resistance build-up (Meissle *et al.* 2011).

Transgenic insect resistance has many advantages over what can be achieved with conventional breeding. New varieties can be obtained much faster than by classical breeding. In addition, the level of resistance achievable in GM plants is usually higher than resistance levels occurring naturally in the gene pool of a plant species, because genes can be introduced from a wide variety of sources, and linkage drag is minimised (Smith 2005, Kennedy 2008). In GM transgenic insect resistance crops, the exposure of non-target organisms is significantly reduced, compared to that resulting from pesticide spraying (Jouanin *et al.* 1998). In addition, pests feeding on plant parts, which normally cannot be reached by pesticide sprays such as stem borers, are exposed to the transgenic toxins and can be targeted with this technology (Jouanin *et al.* 1998, Thies and Devare 2007, Romeis *et al.* 2008b). Because the toxins are continuously expressed in plant tissues, target insects are exposed throughout the infestation period, and also during the most vulnerable life stages (Jouanin *et al.* 1998, Thies and Devare 2007, Romeis *et al.* 2008b). It should also be noted that transgenic resistance in Bt-plants is much more specific for the target insect than most pesticides, because specific Bt-toxins are toxic only to a limited range of species within one or two insect orders (de Maagd *et al.* 1999). With emerging, newer techniques such as RNAi-based control, the specificity may still be increased (Ramon *et al.* 2014).

3.1. *IPM impacts of altered insecticide use in GM crops*

Changes in the insecticide spraying regime on a crop will have far-reaching consequences on several ecosystem functions, with relevance to crop protection. At least two types of possibilities may open up for improved IPM:

(i) lower and/or less frequent insecticide treatments will improve possibilities to benefit from enhanced activity of natural enemies in controlling pests, but also weeds (via seed predator activity, plus weed-specific herbivores); and
(ii) fewer spray treatments will allow a better choice of their timing so that natural enemies of key pests can be protected (more freedom in the selection of spray windows).

The benefits of growing Bt crops such as cotton, corn, potato and rice have been well described (e.g. Romeis *et al.* 2008b). Many lepidopteran pests such as the European corn borer (*Ostrinia nubilalis*) and the cotton bollworm (*Helicoverpa armigera*), have been successfully controlled in Bt crops. This has resulted in increases in crop yields, while at the same time allowing reduced use of pesticides in these crops (Shelton *et al.* 2002, Smith 2005, Fitt 2008). Brookes and Barfoot (2015) report that in all countries that have adopted the growing of Bt maize, there was a reduction in insecticide use on that crop by 52% cumulatively during 1996–2013 (Figure 2), and in Bt cotton similarly by 27%.

Natural enemies of pests have been shown to have benefited from the decrease in the application of broad-spectrum pesticides, and therefore it is expected that their contribution to natural pest control in these crops has increased (Sisterson *et al.* 2007, Romeis *et al.* 2008a). This, in turn, should reduce the number or severity of secondary pest outbreaks, and decrease the risk of resistance development. In addition, if a pest consumes sublethal levels of the insecticidal toxin, its development time may be prolonged, and its exposure to natural enemies is increased. This is hypothesised to lead to slow-growth and high-mortality (Williams 1999, Cornelissen and Stiling 2006). Active release of biological control agents in case of secondary pest outbreaks, or when resistance is emerging, could reduce the probability of pest outbreaks. While these risks are usually regarded as disadvantages of GM crops, they disappear when GM crops are used in an IPM context. When biological control of the secondary or unanticipated pests is not sufficiently fast or efficient, it is still possible to

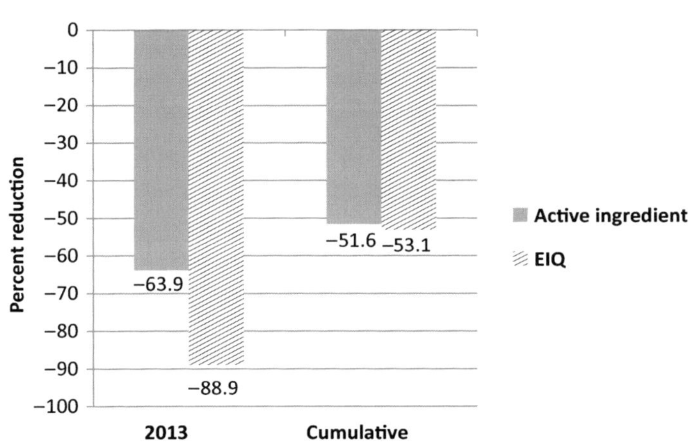

Figure 2. Reduction in insecticide use and the calculated environmental load (EIQ) from using GM insect resistant maize in adopting countries in 2013, and cumulative 1996–2013. Adapted from Brookes and Barfoot (2015).

apply specific pesticides to prevent crop loss — the rational use of pesticides is possible within IPM (Dent 1995, Koul *et al.* 2004).

However, in reality, the comparison of natural enemy activities on non-Bt and Bt crops depends greatly on the context. When the non-Bt crops are not sprayed with insecticides (which is often the case in maize growing), there is very little potential for Bt crops to reduce the environmental impact on natural enemies. In a meta-analysis of arthropod functional guilds, LaReesa Wolfenbarger *et al.* (2008) found that: (i) predators were less abundant in unsprayed Bt cotton than in unsprayed non-Bt cotton; (ii) predators and herbivores were more frequent in unsprayed Bt crops than in sprayed non-Bt crops; and (iii) no differences were found, when both Bt and non-Bt crops were sprayed. The authors concluded that although both Bt toxins and insecticides may influence arthropods, the effects of insecticides are much greater than those of Bt toxins. Similar results have been reported for Bt cotton and Bt maize (Marvier *et al.* 2007).

In some cases, the widespread adoption of GM insect resistance technology has resulted in "area-wide" suppression of key pests such as stem borers in maize. As a result, also conventional farmers (e.g. of maize in the USA) have benefited from a lower level of pest infestation, and the reduced need for insecticide treatments (Brookes and Barfoot 2015).

We are not aware of any studies, or practical cases, where the improved possibilities provided by the GM insect resistance technology, would have been exploited for actively utilising other IPM tools for crop protection, e.g. conservation biocontrol, inundative biocontrol, assessing weed seed predation levels, or making use of wider or altered treatment windows for pesticide treatments to protect key natural enemies of important remaining pests.

3.2. *Integration of GM crops with the use of trap cropping*

Genetic engineering could conceivably be helpful in utilising trap cropping for pest management by making GM "dead-end" trap plants available, such as Bt collard and Bt Indian mustard to protect cabbage from diamondback moth damage (Hokkanen 1991, Shelton *et al.* 2008). In some cases, it might be possible to employ GM trap crops for managing the pests, and to cultivate the main crop using conventional varieties. Also, the use of GM herbicide tolerant crops would make it easier to manage flower strips as beetle banks, should these strips become sources for weeds (Ratnadass and Barzman 2014).

Ribonucleic acid interference (RNAi) techniques:

RNAi-mediated silencing of gene function has been reported in many insect orders such as Lepidoptera, Hemiptera, Coleoptera, Diptera, and Hymenoptera (Price and Gatehouse 2008), and appears therefore potentially useful for the control of various insect pests. Transgenic plants expressing double-stranded RNA (dsRNA) is one possible delivery method of the dsRNA product for insect control. When the target insect feeds on the plant, messenger RNA (mRNA) levels of a selected gene in the target herbivore are reduced, and this can consequently affect the development and survival of the insect (Baum *et al.* 2007, Price and Gatehouse 2008). For example, transgenic maize engineered to express dsRNA, based on the sequence of a gene encoding an essential protein in the western corn rootworm (*Diabrotica virgifera virgifera*), resulted in a significant reduction in root damage by this species (Baum *et al.* 2007). If the introduced dsRNA is highly specific to the target insect, it will effectively eliminate negative effects on non-target organisms. It is unclear as yet whether such specificity is achievable (Lundgren 2014). This appears as a very promising approach for transgenic pest control (Zotti and Smagghe 2015). The first application utilising RNAi-transgenic crop plant is expected to reach the market in 2017 (Macron 2014).

4. Herbicide tolerance as a component tool in IPM

Herbicide tolerant (HT) crops have had a profound impact on weed management. Most of the impact has been by glyphosate resistant maize, cotton, soybean and canola. Significant economic savings, yield increases and more effective and simple weed management have contributed to the widespread adoption of the technology (Duke 2015). Initially, glyphosate resistant crops allowed to reduce tillage significantly, and reduced the overall environmental impact of weed management. Continuous use of glyphosate, however, over broad areas rapidly facilitated the evolution of glyphosate resistant weeds. This has resulted in increasing again the use of tillage, and other herbicides together with glyphosate, reducing the initial environmental benefits of glyphosate resistant crops (Duke 2015).

Transgenic crops with resistance to other active ingredients, stacked with glyphosate and/or glufosinate resistance, are expected to become available in the next few years (Duke 2015). These technologies will provide additional weed management options for farmers, but will not have all of the positive effects such as reduced cost, simplified management, lower environmental impact, and reduced tillage, that glyphosate resistant crops had initially. The industry expects that in the future, other herbicide resistant crops (including non-transgenic varieties), herbicides with new modes of action, and technologies that are currently being developed (e.g. bioherbicides, sprayable herbicidal RNAi, and robotic weeding) may affect the role of transgenic HT crops in weed management (Duke 2015).

4.1. *Adoption of HT crops, and impacts on herbicide use*

Soybeans genetically engineered with HT traits have been the most widely and rapidly adopted GM crop in the USA, followed by HT cotton. USDA survey data, reported by Benbrook (2012), show that adoption of HT soybeans rose from 17% in 1997 to 68% in 2001, and to 93% in 2010. Plantings of HT cotton rose from 10% in 1997 to 56% in 2001, and to 78% in 2010. Adoption of HT maize was slower, but reached 70% in 2010.

There are conflicting data concerning the impact of HT crops on the trends in total herbicide use in the target crops. Benbrook (2012) reports USDA data indicating that herbicide resistant crop technology led to a 239 million kg increase in herbicide use in the USA between 1996 and 2011. HT varieties increased herbicide use in soybean by 23% between 1996 and 2006, and in cotton by 42% from 1996 till 2010. Brookes and Barfoot (2015), however, report globally that in HT soybeans there has been virtually no change in the quantity of herbicide use from 1996 to 2013, while in HT maize cumulative change has been –9.2% over the same time, in cotton –7.2%, and in rapeseed –16.5%. In all crops, the calculated environmental impact quotient (EIQ) (Kovach *et al.* 1992) has cumulatively decreased by a range from –9.5% to –27.9%. The EIQ and its utility has been recently criticised, and it is currently considered as a poor measure of herbicide environmental impact (Kniss and Coburn 2015).

4.2. *Herbicide resistance in weeds*

Weed resistance to glyphosate has become a major issue affecting farmers using GM HT crops. Worldwide there are now 35 weed species resistant to glyphosate (Heap 2016). In the USA, there are currently 15 weeds exhibiting resistance to glyphosate, and in Australia, 12 species. In Argentina, Brazil and Canada, where GM HT crops are also widely grown, the number of weed species showing resistance to glyphosate are 7, 6, and 5, respectively. Some of the glyphosate resistant species, such as marestail (*Conyza canadensis*), waterhemp (*Amaranthus tuberculatus*) and palmer pigweed (*Amaranthus palmeri*) in the USA, are now widespread with the affected area being about 30–40% of the total area annually devoted to maize, cotton and soybeans (Brookes and Barfoot 2015).

Where farmers are faced with weeds resistant to glyphosate in their GM HT crops, they are increasingly advised to include other herbicides with different modes of action, in combination with glyphosate. In some cases, farmers need to revert to ploughing and/or hand-weeding in their weed management systems. These changes already are influencing the mix, total amount, cost and overall use of herbicides in GM HT crops (Brookes and Barfoot 2015).

For example, in 2013, in the USA, 64% of the HT soybean crop received an additional herbicide treatment, compared with 14% in 2006. The increase in non-glyphosate herbicide use is primarily in response to weed scientist recommendations to diversify weed management programmes (Brookes and Barfoot 2015). Since the mid 2000s, the average amount of herbicide used on HT cotton in the USA has increased, via a combination of increased use of glyphosate and additional use of other herbicides (Brookes and Barfoot 2015).

Glyphosate resistant weeds are therefore forcing farmers to respond by increasing herbicide application rates, by making multiple applications of herbicides, and by applying complementary herbicides, using deep tillage to bury weed seeds, and by resorting to manual weeding (Benbrook 2012).

In recent years, the first three of the above responses have been the most common. Each response increases the amount of herbicides applied on HR crops. All five interventions increase farmers' costs. Moreover, if the response to the current resistance problems is to develop 2,4-D and dicamba herbicide resistant maize and soybeans, there will be an increased reliance on older, higher risk herbicides. This is being currently pushed for by the industry, which foresees based on an upward scenario in the planting of 2,4-D HR maize, which should reach 55% of maize hectares by 2019, the use of 2,4-D on maize in the USA would increase over 30-fold from 2010 levels (Benbrook 2012).

4.3. *Herbicide tolerant crops and conservation agriculture*

The use of HT crops has facilitated a substantial increase in certain forms of conservation agriculture, which is an important part of the ecological intensification scenario for agriculture. On the other hand, it is difficult to consider HT crops as part of IPM, because they are based on calendar treatments with synthetic herbicides such as glyphosate (Ratnadass and Barzman 2014). The use of conservation tillage systems has increased in the USA throughout the 1980s and 1990s. Only 30% of soybean farmers applied conservation tillage in 1996, but 63% did in 2006. Similarly, 33% of maize area was under conservation tillage in 1990, but 40% in 2006 (Fernandez-Cornejo *et al.* 2013).

There is some evidence that HT crops facilitate conservation tillage, because weed control is more effective and less costly. Most studies concerning the interaction of HT crops and conservation tillage systems have found that they are correlated, but it is difficult to determine whether HT adoption causes farmers to adopt conservation tillage, or whether adoption of conservation tillage causes farmers to use HT seeds (Fernandez-Cornejo *et al.* 2013). In most cases, adoption of HT crops facilitated the use of conservation tillage, and *vice versa*. This would imply that by encouraging farmers to adopt conservation tillage, HT crops indirectly benefit the environment by reducing soil losses and erosion, runoff, fuel use, and the carbon footprint of agriculture (Fernandez-Cornejo *et al.* 2013).

Conservation tillage as such can be highly useful in facilitating the IPM tactic of conservation biological control. Hokkanen *et al.* (1988) showed how parasitoids of the pollen beetle can be effectively conserved by no-till practices after oilseed rape harvest (direct drilling of following crop). Ploughing and harrowing kills 75% of the specific parasitoids *Phradis morionellus*, which overwinter in the soil. The simple conservation measure of no-till after oilseed rape will improve biocontrol of the pollen beetle and will keep its populations at low levels (Hokkanen 2008). Conservation tillage will also help in increasing and maintaining pest-suppressive microbial biodiversity (entomopathogenic fungi and nematodes) in the field soil, as was shown in a large collaborative field study on oilseed rape (Hokkanen 2006, Zec-Vojinovic *et al.* 2006). Integrated pest management with conservation tillage significantly increased pest suppressiveness of soils already after two years of IPM (Hokkanen 2006). It should be noted, however, that some common pre-emergence (e.g. trifluralin) herbicides are strongly antagonistic to beneficial fungi in the soil (Vänninen and Hokkanen 1988).

4.4. *Effects of HT crops on biodiversity*

Negative effects of HT crops on plant and invertebrate biodiversity in ecosystems would result in the loss of associated ecosystem services such as food and refuge for natural enemies of pests, and pollination services.

The largest field study available on the effects of altered weed management with HT crops is a series of trials in the UK called "Farm Scale Evaluation" (FSE). The FSE consisted of a large number of beet, maize, and oilseed rape fields. Of each field, one half was sown with a HT GM crop, and treated with the respective herbicide (glyphosate or glufosinate). The other half was sown with a conventional cultivar of the same crop, and treated with a herbicide according to standard practice.

In the FSE study, a clear reduction in the diversity of the arable flora by HT crops was demonstrated (Heard *et al.* 2003). In HT beet, weed biomass was one-sixth of that in non-GM beets, while the production of weed seeds ("seed rain") was one third, and the seed bank in the following year was about 80% of that in the non-GM beet. In HT spring oilseed rape and in maize, similar results were obtained. The same patterns were observed in the field margins, but they were less pronounced than within the fields (Roy *et al.* 2003).

In the FSE study, the impacts on arthropods depended on the taxonomic group. Most taxa of ground surface active (ground beetles, rove beetles) and aerial arthropods had similar populations in non-GM and GM crops. The pattern of bees was similar to weed presence within the crops, and butterflies were usually more frequent in non-GM crops. Collembola, however, were always more frequent in GM crops (Haughton *et al.* 2003).The slug *Deroceras reticulatum* was in general more frequent in GM crops, but the difference was significant in only two cases (Brooks *et al.* 2003).

It was concluded from the FSE that frequent use of GM crops would reduce the biodiversity of the arable flora, with secondary effects on the fauna. Continuous use of effective herbicides is likely to reduce weed populations, weed diversity, and weed seed banks over time. Reductions in biodiversity appear unavoidable for the flora, and are likely for herbivorous and nectar-feeding arthropods. The impact on other arthropods is more difficult to predict. There is some flexibility for fine-tuning the time of herbicide application in HT crops. In some trials, early application of glyphosate resulted in very few weeds, while delayed applications resulted in higher weed abundance, without negative impacts on the yield (Strandberg and Pedersen 2002). Similarly, in sugar beet, weed management with an early band application, followed by a late overall application of glyphosate, resulted in higher weed and arthropod populations, without compromising yield (Dewar *et al.* 2003). This kind of responses could possibly be employed for IPM tactics, where some of the benefits of weeds (e.g. biodiversity, provision of pollen and nectar) could be retained without compromising on weed management.

In the context of HT crops, little attention has been paid to the critically important role that weeds play in facilitating pest control ecosystem services by providing nectar and pollen to foraging parasitoids and predators, If parasitoids lack nectar sources, their longevity and fecundity are strongly reduced. Similarly, hover flies and other predators need pollen for ovarial development and production of eggs. If the cropping system does not provide these resources, IPM principle number one (prevention) is heavily compromised, and the natural enemy complex is crippled. HT crops clearly are working against this principle.

5. Disease resistance: An essential goal in agroecological practices

Disease resistance in GM crops is not widely available. It will impact IPM on the crop at least via lower number of pesticide treatments, either fungicides against fungal diseases, antibiotics against bacterial diseases, or insecticides against vector insects. All of these can have significant non-target effects on beneficial organisms (e.g. arthropod natural enemies, or antagonistic microbes). Therefore, decreased dependence on pesticides can offer increased opportunities to other IPM tactics, such as entomovectoring (Hokkanen *et al.* 2015).

5.1. *Ecological impacts of altered chemical disease control regime*

Non-target effects of the fungicides and insecticides used in Swiss orchards were evaluated by Speiser *et al.* (2013). Of the pesticides used in the organic orchard, only sulphur at high concentrations had substantial non-target toxicity (to predatory mites). Reduction of sulphur by two-thirds had a major beneficial effect on predatory mites, but only a minor effect on other non-target organisms. Of the pesticides used in the integrated orchard, several (captane, thiacloprid, and chlorpyriphos) had severe toxicity to non-target organisms. Reduction of the fungicide captane by one-third had a major beneficial effect on hoverflies. The insecticides thiacloprid and chlorpyriphos had severe toxicity to many non-target organisms. It was concluded that the beneficial effect on hoverflies might be relatively small in comparison to the negative effects on predatory insects and bees (Speiser *et al.* 2013). These results indicate that if disease resistant GM varieties were available, allowing significant reductions in the use of fungicidal and insecticidal (against disease vectors) products on crops, their beneficial ecological impacts might be significant.

Concerning the great variety of microorganisms that live on the skin (epiphytes) and within the tissue (endophytes) of apples, the management system is likely to have a profound impact on the microbial populations (Granado *et al.* 2008). Any changes in pesticide use, particularly in the use of streptomycine, are likely to affect these microorganisms. At present, very little is known about them. The novel gene products in disease resistant GM apples are not expected to have any impact on microorganisms, as compared to traditionally bred cultivars (Speiser *et al.* 2013).

Considering the possible impacts of GM apple trees on the undergrowth vegetation, Speiser *et al.* (2013) considered that neither the transgenes nor the associated changes in the fungicides and bactericides were expected to have a direct impact on the undergrowth. They consider, however, that all reductions in the number of spray passages will favour biodiversity in the undergrowth. The improved management flexibility in the GM scenarios would allow spraying to be done when conditions were more optimal, and thus would be expected to further increase biodiversity in the undergrowth between rows. It was noted that biodiversity of the undergrowth vegetation was positively affected by scab resistance, and that with combined resistance to scab and fire blight, the effect was more pronounced (Speiser *et al.* 2013). Diverse undergrowth vegetation has been shown to promote beneficial arthropods in apple orchards (Wyss 1995).

5.2. *Trap cropping and disease resistance*

Trap cropping based on genetically engineered plants can also be effective in controlling insect-vectored pathogens, such as papaya ringspot virus (PRSV). The virus is transmitted by many aphid species in a non-persistent manner, making it exceptionally difficult to control through the use of insecticides (Gonsalves 1998). Besides using PRSV-resistant papaya as a direct control, some growers in Hawaii are using borders of PRSV-resistant papaya as a trap crop to reduce the movement of PRSV into the interior of the field, where non-genetically engineered papaya are grown. This has proven a successful tactic to allow production of both non-genetically and genetically engineered papaya simultaneously (Shelton and Badenes-Perez 2006).

6. GM impacts of simplifying rotations

Introduction of GM crops may induce changes in crop production systems over time (at field, farm, and landscape level). The Swiss study (Speiser *et al.* 2013) suggests that (i) the potential impact of GM crops depends on the production system, and (ii) that GM crops facilitate the general trend towards shorter crop rotations. A general tendency towards larger fields is an independent trend, but may be enhanced with the adoption of GM crop varieties. This will have relevance for the landscape-level processes, and may undermine some of the foreseen ecological benefits concerning, for example, biocontrol and pollination ecosystem services. Furthermore, it is worth noting that coexistence requirements favour larger field sizes as a minimum distance has to be kept between GM and non-GM crops (Hüsken *et al.* 2007).

7. Present day integration of GM crops in agroecological practice

The potential contribution of GM crops to improve agroecological practices, in relation to several other cropping techniques (Figure 3) seem to be met with scepticism. GM crops have been extremely rapidly integrated

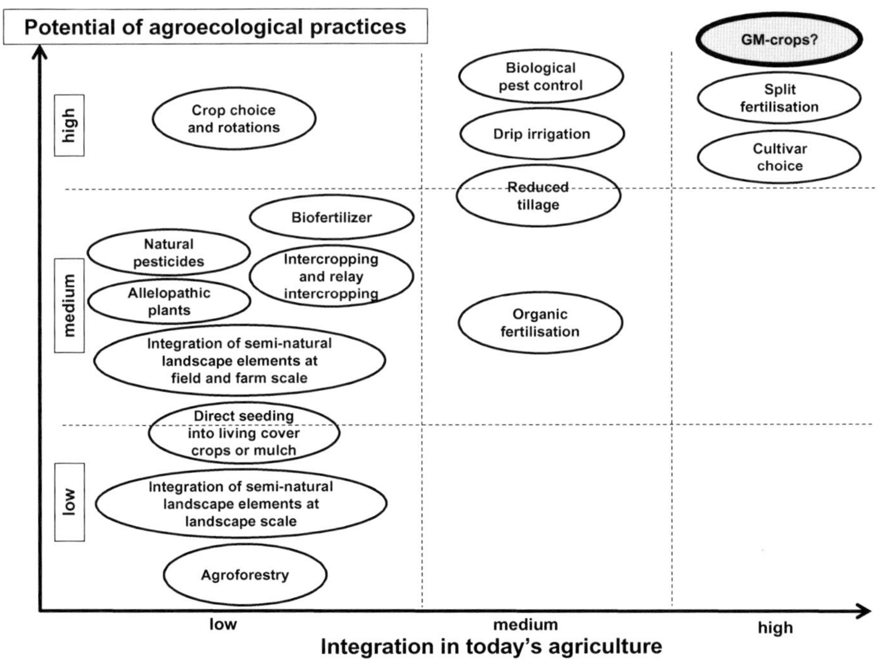

Figure 3. Potential contribution of GM crops to improve agroecological practices, in relation to several other cropping techniques (modified from Wezel *et al.* (2014)).

in today's global agriculture (where allowed), and they have a high potential to benefit the ecological processes in agricultural ecosystems. Unfortunately, due to the "Reality Gap" (Figure 1), this opportunity does not seem to be seized, and we remain pessimistic that these potentially valuable techniques will actually be used for the benefit of biological control, nor for becoming integrated in modern IPM systems and/or agroecological based agricultural practices.

8. Conclusions

GM crops have been extremely rapidly integrated in today's global agriculture (where allowed), and they carry a high potential to become an integrated part of modern agricultural systems. The possible contributions of GM crops and their traits to IPM relevant functions have been analysed in a comprehensive manner. For example, reduction in insecticide spraying should lead to improved biocontrol of other pests, and will allow for a wider selection of spray windows to maximise the beneficial effects. It may also allow applications of biocontrol products e.g. for disease control. Possible changes, which may benefit other crop protection components could include improved flexibility in tillage regimes, weed control regimes (e.g. to promote parasitoids at a particular time), planting time, or other agronomic benefits. Genetic engineering could conceivably be helpful in utilising trap cropping for pest management, by making GM "dead-end" trap plants available, such as Bt collard and Bt Indian mustard to protect cabbage from diamondback moth damage. Unfortunately, GM crops are currently mostly introduced to the growers and are seen by the growers as a standalone technology for pest and disease control, without any real attempt to integrate them as a component in integrated pest management. The theoretical benefits of pest and disease resistant GM crops are eroded by this approach. Eventually pest control benefits may be lost, as has been observed already in parts of the USA. As long as GM crops are globally not an issue in the discussions on IPM, and when GM crops are seen by growers as a standalone pest management tool, their fate will be the same as that of chemical pesticides: unstable and unsustainable technologies.

Acknowledgements

Our research has been supported by the EU-FP7 project AMIGA, "Assessing and Monitoring the Impacts of Genetically modified plants on Agro-ecosystems", Grant agreement No. 289706. We wish to thank for valuable comments and for discussions on the topic our AMIGA colleagues Salvatore Arpaia, Nick Birch, and Bert Lotz, our colleagues Alexander Wezel and Aurelie Ferrer at ISARA Lyon, as well as active student feedback and discussions at ISARA as well as at the University of Helsinki. We wish to thank Dr. William B. Walker for critical reading, commenting, and language edition of an earlier draft of the text.

References

Arpaia, S., Messéan, A., Birch, N.A., Hokkanen, H., Härtel, S., van Loon, J., Lövei, G., Park, J., Spreafico, H., Squire, G.R., Steffan-Dewenter, I., Tebbe, C.,van der Voet, H., 2014. Assessing and monitoring impacts of genetically modified plants on agro-ecosystems: the approach of AMIGA project. *Entomologia* **154**, 79–86.

Barzman, M., Bàrberi, P., Birch, N.E., Boonekamp, P., Dachbrodt-Saaydeh, S., Graf, B., Hommel, B., Jensen, J.E., Kiss, J., Kudsk, P., Lamichhane, J.R., Messéan, A., Moonen, A.-C., Ratnadass, A., Ricci, P., Sarah, J.-L., Sattin, M., 2015. Eight principles of integrated pest management. *Agronomy for Sustainable Development* **35**, 1199–1215.

Baum, J.A., Bogaert, T., Clinton, W., Heck, G.R., Feldmann, P., Ilagan, O., Johnson, S., Plaetinck, G., Munyikwa, T., Pleau, M., Vaughn, T., Roberts, J., 2007. Control of coleopteran insect pests through RNA interference. *Nature Biotechnology* **25**, 1322–1326.

Benbrook, C.M., 2012. Impacts of genetically engineered crops on pesticide use in the U.S. — the first sixteen years. *Environmental Sciences Europe* **24**, 24.

Birch, A.N.E., Begg, G.S., Squire, G.R., 2011. How agro-ecological research helps to address food security issues under new IPM and pesticide reduction policies for global crop production systems. *Journal of Experimental Botany* **62**, 3251–3261.

Brookes, G., Barfoot, P., 2015. *GM Crops: Global Socio-economic and Environmental Impacts 1996–2013*. PG Economics Ltd, Dorchester, UK.

www.pgeconomics.co.uk/pdf/2015globalimpactstudyfinalMay2015.pdf (accessed 15 November 2015).

Brooks, D.R., Bohan, D.A., Champion, G.T., Haughton, A.J., Hawes, C., Heard, M.S., Clark, S.J., Dewar, A.M., Firbank, L.G., Perry, J.N., Rothery, P., Scott, R.J., Woiwod, I.P., Birchall, C., Skellern, M.P., Walker, J.H., Baker, P., Bell, D., Browne, E.L., Dewar, A.J.G., Fairfax, C.M., Garner, B.H., Haylock, L.A., Horne, S.L., Hulmes, S.E., Mason, N.S., Norton, L.R., Nuttall, P., Randle, Z., Rossall, M.J., Sands, R.J.N., Singer, E.J., Walker, M.J., 2003. Invertebrate responses to the management of genetically modified herbicide-tolerant and conventional spring crops. I Soil-surface-active invertebrates. *Philosophical Transactions of the Royal Society B* **358**(1439), 1847–1862.

Cloutier, C., Boudreault, S., Michaud, D., 2008. Impact of Colorado potato beetle-resistant potatoes on non-target arthropods: a meta-analysis of factors potentially involved in the failure of a Bt transgenic plant. *Cahiers Agricultures* **17**, 388–394.

Cornelissen, T., Stiling, A., 2006. Does low nutritional quality act as a plant defence? An experimental test of the slow-growth, high mortality hypothesis. *Ecological Entomology* **31**, 32–40.

Culpepper, S., 2015. Presentation at the 8th International IPM Symposium [IPM: Solutions for a Changing World] March 23–26, 2015, Salt Lake City, Utah, USA.

de Maagd, R.A., Bosch, D., Stiekema, W., 1999. Bacillus thuringiensis toxin-mediated insect resistance in plants. *Trends in Plant Science* **4**, 9–1.

Dent, D.R. (Ed.), 1995. Integrated Pest Management, Chapman & Hall, London. xii + 356 p.

Dewar, A.M., May, M.J., Woiwod, I.P., Haylock, L.A., Champion, G.T., Garner, B.H., Sands, R.J.N., Qi, A., Pidgeon, J.D., 2003. A novel approach to the use of genetically modified herbicide tolerant crops for environmental benefit. *Proceedings of the Royal Society B* **270**, 335–340.

Duke, S.O., 2015. Perspectives on transgenic, herbicide-resistant crops in the United States almost 20 years after introduction. *Pest Management Science* **71**, 652–657.

Fernandez-Cornejo, J., Hallahan, C., Nehring, R., Wechsler, S., Grube, A. 2013. Conservation Tillage, Herbicide Use, and Genetically Engineered Soybeans in the U.S.A. ISB News Report April 2013, Information Systems for Biotechnology, Virginia Tech, USA, www.isb.vt.edu/news/2013/Apr/Conservation-Tillage.pdf (accessed 15 November 2015).

Fitt, G.P., 2008. Have Bt crops led to changes in insecticide use patterns and impacted IPM? In: Romeis, J., Shelton, A., Kennedy, G. (Eds.), *Integration of Insect-Resistant Genetically Modified Crops within IPM Programs*. Springer, Dordrecht, Netherlands, pp. 303–328.

Gonsalves, D. 1998. Control of papaya ringspot virus in papaya: a case study. *Annual Review of Phytopathology* **36**, 415–437.

Granado, J., Thürig, B., Kieffer, E., Petrini, L., Fliessbach, A., Tamm, L., Weibel, F.P., Wyss, G.S., 2008. Culturable fungi of stored 'golden delicious' apple fruits: a one-season comparison study of organic and integrated production systems in Switzerland. *Microbial Ecology* **56**, 720–732.

Gurian-Sherman, D., 2009. *Failure to yield. Evaluating the Performance of Genetically Engineered Crops*. Union of Concerned Scientists, Cambridge, MA, USA. http://www.ucsusa.org/sites/default/files/legacy/assets/documents/food_and_agriculture/failure-to-yield.pdf

Haughton, A.J., Champion, G.T., Hawes, C., Heard, M.S., Brooks, D.R., Bohan, D.A., Clark, S.J., Dewar, A.M., Firbank, L.G., Osborne, J.L., Perry, J.N., Rotherry, P., Roy, D.B., Scott, R.J., Woiwod, I.P., Birchall, C., Skellern, M.P., Walker, J.H., Baker, P., Browne, E.L., Dewar, A.J.G., Garner, B.H., Haylock, L.A., Horne, S.L., Mason, N.S., Sands, R.J.N., Walker, M.J., 2003. Invertebrate responses to the management of genetically modified herbicide-tolerant and conventional spring crops. II Within-field epigeal and aerial arthropods. *Philosophical Transactions of the Royal Society B* **358**, 1863–1877.

Heap, I., 2016. *The International Survey of Herbicide Resistant Weeds*. Available at www.weedscience.org, (accessed 14 May 2016).

Heard, M.S., Hawes, C., Champion, G.T., Clark, S.J., Firbank, L.G., Haughton, A.J., Parish, A.M., Perry, J.N., Rothery, P., Scott, R.J., Skellern, M.P., Squire, G.R., Hill, M.O., 2003. Weeds in fields with contrasting conventional and genetically modified herbicide-tolerant crops. I. Effects on abundance and diversity. *Philosophical Transactions of the Royal Society B* **358**, 1819–1832.

Hokkanen, H.M.T., 1991. Trap cropping in pest management. *Annual Review of Entomology* **36**, 119–138.

Hokkanen, H.M.T., 2006. Integrating insect pathogens into oilseed rape cropping systems. CD-ROM *Proceedings of the International Symposium 'Integrated Pest Management of Oilseed Rape Pests'*, Göttingen, Germany, 3–5 April 2006.

Hokkanen, H.M.T., 2008. Biological control methods of pest insects in oilseed rape. *Bulletin OEPP/EPPO Bulletin* **38**, 104–109.

Hokkanen, H.M.T., 2015. Integrated pest management at the crossroads: Science, politics, or business (as usual)? *Arthropod-Plant Interactions* **9**, 543–545.

Hokkanen, H., Husberg, G.-B., Söderblom, M., 1988. Natural enemy conservation for the integrated control of the rape blossom beetle *Meligethes aeneus F. Annales Agriculturae Fenniae* **27**, 281–293.

Hokkanen, H.M.T., Menzler-Hokkanen, I., Lahdenperä, M.-L., 2015. Managing bees for delivering biological control agents and improved pollination in berry and fruit cultivation. *Sustainable Agriculture Research* **4**(3), 2015. http://dx.doi.org/10.5539/sar.v4n3p89

Hokkanen, H.M.T., Menzler-Hokkanen, I., 2016. Potential contribution of GM crops to Integrated Pest Management. *Arthropod-Plant Interactions* (submitted).

Hüsken, A., Ammann, K., Messeguer, J., Papa, R., Robson, P., Schiemann, J., Squire, G.R., Stamp, P., Sweet, J., Wilhelm, R., 2007. A major European synthesis of data on pollen and seed mediated gene flow in maize in the SIGMEA project, *unpublished.* http://www.botanischergarten.ch/Geneflow/Huesken-Synthesis-2007.pdf

James, C., 2014. Global Status of Commercialized Biotech/GM Crops: 2014. *ISAAA Brief No. 49*. ISAAA: Ithaca, NY.

Jouanin, L., Bonadé-Bottino, M., Girard, C., Morrot, G., Giband, M., 1998. Transgenic plants for insect resistance. *Plant Science* **131**, 1–11.

Kennedy, G., 2008. Integration of Insect-Resistant Genetically Modified Crops within IPM Programs. In: Romeis, J., Shelton, A.M., Kennedy, G., (Ed.), *Integration of Insect-Resistant Genetically Modified Crops within IPM Programs* (*Progress in Biological Control*). Springer, Dordrecht, The Netherlands, pp. 1–26.

Kniss, A.R., Coburn, C.W., 2015. Quantitative evaluation of the environmental impact quotient (EIQ) for comparing herbicides. *PLoS ONE* **10**(6), e0131200. http://doi.org/10.1371/journal.pone.0131200

Kos, M., van Loon, J.J.A., Dicke, M., Vet, L.E.M., 2009. Transgenic plants as vital components of integrated pest management. *Trends in Biotechnology* **27**, 621–627.

Koul, O., Dhaliwal, G.S., Cuperus, G.W. (Eds.), 2004. *Integrated Pest Management: Potential, Constraints and Challenges*. CABI Publishing, Wallingford, UK. 336 p.

Kovach, J., Petzoldt, C., Degni, J., Tette, J., 1992. A method to measure the environmental impact of pesticides. *New York's Food and Life Sciences Bulletin* **139**, 1–8.

LaReesa Wolfenbarger, L., Naranjo, S.E., Lundgren, J.G., Bitzer, R.J., Watrud, L.S. 2008. Bt crop effects on functional guilds of non-target arthropods: a meta-analysis. *PLoS ONE* **3**, e2118.

Lundgren, J.G. 2014. Considerations for risk assessment procedures of RNAi-based crops. *EFSA International Scientific Workshop: Risk assessment considerations for RNAi-based GM plants.* http://www.efsa.europa.eu/sites/default/files/event/documentset/140604-p06.pdf

Lundgren, J.G., Gassmann, A.J., Bernal, J., Duan, J.J., Ruberson, J., 2009. Ecological compatibility of GM crops and biological control. *Crop Protection* **28**, 1017–1030.

Macron, D., 2014. *Monsanto moves RNAi-based Pest-resistant Corn into Final Stage of Development.* https://www.genomeweb.com/rnai/monsanto-moves-rnai-based-pest-resistant-corn-final-stage-development (accessed 20 November 2015).

Marvier, M., McCreedy, C., Regetz, J., Kareiva, P., 2007. A meta-analysis of effects of Bt cotton and maize on nontarget invertebrates. *Science* **316**, 1475–1476.

Meissle, M., Romeis, J., Bigler, F., 2011. Bt maize and integrated pest management — a European perspective. *Pest Management Science* **67**, 1049–1058.

Naranjo, S.E., 2001. Conservation and evaluation of natural enemies in IPM systems for Bemisia tabaci. *Crop Protection* **20**, 835–852.

Price, D.R.G., Gatehouse, J.A., 2008. RNAi-mediated crop protection against insects. *Trends in Biotechnology* **26**, 393–400.

Ramon, M., Devos, Y., Lanzoni, A., Liu, Y., Gomes, A., Gennaro, A., Waigmann, E., 2014. RNAi-based GM plants: food for thought for risk assessors. *Plant Biotechnology Journal* **12**, 1271–1273.

Ratnadass, A., Barzman, M., 2014. Ecological intensification for crop protection. *Sustainable Agriculture Reviews* **14**, 53–81.

Romeis, J. *et al.*, 2008a. Insect resistant transgenic crops and biological control. In: Romeis, J., Shelton, A., Kennedy, G. (Eds.), *Integration of Insect-Resistant Genetically Modified Crops within IPM Programs.* Springer, Dordrecht, The Netherlands, pp. 87–118.

Romeis, J., Shelton, A., Kennedy, G. (Eds.), 2008b. *Integration of Insect-Resistant Genetically Modified Crops within IPM Programs.* Springer, Dordrecht, The Netherlands.

Roy, D.B., Bohan, D.A., Haughton, A.J., Hill, M.O., Osborne, J.L., Clark, S.J., Perry, J.N., Rotherry, P., Scott, R.J., Brooks, D.R., Champion, G.T., Hawes, C., Heard, M.S., Firbank, L.G., 2003. Invertebrates and vegetation of field margins adjacent to crops subject to contrasting herbicide regimes in the Farm Scale Evaluations of genetically modified herbicide tolerant crops. *Philosophical Transactions of the Royal Society of London B* **358**, 1879–1898.

Shelton, A.M., Badenes-Perez, F.R., 2006. Concepts and applications of trap cropping in pest management. *Annual Review of Entomology* **51**, 285–308.

Shelton, A.M., Zhao, J.-Z., Roush, R.T., 2002. Economic, ecological, food safety, and social consequences of the deployment of Bt transgenic plants. *Annual Review of Entomology* **47**, 845–881.

Shelton, A.M., Hatch, S.L., Zhao, J.Z., Chena, M., Earle, E.D., Cao, J., 2008. Suppression of diamondback moth using Bt-transgenic plants as a trap crop. *Crop Protection* **27**, 403–409.

Sisterson, M.S., Carrière, Y., Dennehy, T.J., Tabashnik, B.E., 2007. Effects of transgenic Bt cotton on insecticide use and abundance of two generalist predators. *Entomologia Experimentalis et Aplícata* **124**, 305–311.

Smith, C.M. (Ed.), 2005. Plant Resistance to Arthropods: Molecular and Conventional Approaches, Springer, Dordrecht, The Netherlands.

Sosnoskie, L.M., Culpepper, A.S., 2014. Glyphosate-resistant Palmer amaranth (Amaranthus palmeri) increases herbicide use, tillage, and hand-weeding in Georgia Cotton. *Weed Science* **62**, 393–402.

Speiser, B., Stolze, M., Oehen, B., Gessler, C., Weibel, F.P., Bravin, E., Kilchenmann, A., Widmer, A., Charles, R., Lang, A., Stamm, C., Triloff, P., Tamm, L., 2013. Sustainability assessment of GM crops in a Swiss agricultural context. *Agronomy for Sustainable Development* **33**, 21–61.

Strandberg, B., Pedersen, M.B., 2002. Biodiversity in glyphosate tolerant fodder beet fields — timing of herbicide application. NERI Technical Report No. 410. National Environmental Research Institute, Denmark. http://www2.dmu.dk/1_viden/2_Publikationer/3_fagrapporter/rapporter/FR410.pdf

Tabashnik, B.E., Brévault, T., Carrière, Y., 2013. Insect resistance to Bt crops: lessons from the first billion acres. *Nature Biotechnology* **31**, 510–521.

Tian, J.-C., Yao, J., Long, L.-P., Romeis, J., Shelton, A.M., 2015. Bt crops benefit natural enemies to control non-target pests. Scientific Reports, 5, 16636. http://doi.org/10.1038/srep16636

Thies, J.E., Devare, M.H., 2007. An ecological assessment of transgenic crops. *Journal of Development Studies* **43**, 97–129.

Vänninen, I., Hokkanen, H., 1988. Effect of pesticides on four species of entomopathogenic fungi in vitro. *Annales Agriculturae Fenniae* **27**, 345–353.

Wezel, A., Casagrande, M., Celette, F., Vian, J.-F., Ferrer, A., Peigné, J., 2014. Agroecological practices for sustainable agriculture. A review. *Agronomy for Sustainable Development* **34**, 1–20.

Williams, I.S., 1999. Slow-growth, high-mortality — a general hypothesis, or is it? *Ecological Entomology* **24**, 490–495.

Wyss, E., 1995. The effects of weed strips on aphids and aphidophagous predators in an apple orchard. *Entomologia Experimentalis et Applicata* **75**, 43–49.

Zec-Vojinovic, M., Hokkanen, H.M.T., Büchs, W., Klukowski, K., Luik, A., Nilsson, C., Ulber, B., Williams, I.H., 2006. Natural occurrence of pathogens of oilseed rape pests in agricultural fields in Europe. CD-ROM *Proceedings of the International Symposium 'Integrated Pest Management of Oilseed Rape Pests'*. Göttingen, Germany, 3–5 April 2006.

Zotti, M.J., Smagghe, G., 2015. RNAi technology for insect management and protection of beneficial insects from diseases: Lessons, challenges and risk assessments. *Neotropical Entomology* **44**, 197–213.

Chapter 7

Entomovectoring: An Agroecological Practice of Using Bees for Biocontrol

Ingeborg Menzler-Hokkanen and Heikki M.T. Hokkanen

University of Helsinki, Department of Agricultural Sciences, Latokartanonkaari 5, PL 27, FIN-00014 Helsinki, Finland

1. Introduction

The use of pollinating insects in the context of biological control of insect pests and diseases has its origin in the early nineties, when honeybees were used to spread biological control agents (BCA) to the flowers instead of spraying (Peng *et al.* 1992). It was not until 2007, however, that the term 'entomovector technology' was first used (Hokkanen and Menzler-Hokkanen 2007), and a more systematic development of the pollinator-and-vector technology was established. Recent research towards functioning dispenser designs, potential vectors, and their loading was conducted to improve and further develop applications of the entomovector technology, in order to significantly improve the yield and quality in organic cropping systems. In particular, organic berry and fruit production suffers heavily from the lack of effective disease and pest management tools, and from inadequate insect pollination at times. As a consequence, the expanding demand on organic berries and fruits cannot be satisfied today (European Commission 2014). In this review, we focus on the grey mould caused by *Botrytis cinerea* on strawberry and raspberry.

The EU is the biggest producer of strawberries in the world, and of the single member countries, Spain is the number two producer after the USA (FAO 2011). Grey mould is the most important biotic threat to the crop, and conventional growing methods with fungicides usually require 3–8 treatments per season, depending on weather conditions. Currently, organic strawberry growers have no means of preventing grey mould on their crops, and consequently they occasionally loose the harvest almost entirely. Conventional growers also suffer, on an average 10–20% preharvest crop losses to grey mould are reported (Strømeng 2008) and up to 25–35% (IPM-Centers 2011), despite numerous fungicide treatments.

Early trials on biological control of *Botrytis* have relied on spraying BCA on strawberry flowers. Extensive field trials in Finland showed that three spray applications with Prestop® (Verdera Oy, Espoo, Finland) formulation at flowering time remarkably decreased the number of mouldy berries and increased the marketable yield (Lahdenperä 2006). However, despite the efficacy of *Gliocladium* (the active biological control microbe in Prestop®) sprays against *Botrytis*, the method could not be widely used because of high treatment costs. Blanket spraying could so far not be adjusted to deliver the BCA only to the inflorescence at the different developmental stages of flowers, and within the required time frame to prevent grey mould growth. In contrast, bees, as an essential component of the pollination system, can colonise the flowers with the BCA via frequent pollination visits to each inflorescence at the proper time (Figure 1), and achieve disease suppression naturally.

2. Pollinators in dissemination of biological control agents

The unique concept and the success of the entomovector approach requires different ecological components such as pollinators, biocontrol agents, and plant pathogens/insect pests to interact successfully (Kevan *et al.* 2008). The entomovector technology depends on multifaceted aspects involving primarily the interactions between the appropriate components

(a) (b)

Figure 1. (a) Strawberry flowers being visited by solitary bees and (b) by honeybees (photos: Heikki Hokkanen).

of vector, control agent, formulation, and dispenser. The entomovector technology represents a unique opportunity in exploiting ecosystem services, where pollinators achieve a dual role: control agents are directly delivered on the target location, i.e. the flowers, while the pollination needs are fulfilled (Figures 1(a) and 1(b)).

The entomovector approach forms an effective disease and pest management tool during flowering of the crop, and even during development of the fruits, because the flowers are the main point of entry for many different plant pathogens (e.g. *B. cinerea*) and insect pests (e.g. thrips). Control of infections by the entomovector technology can thereby increase the marketable fruit yield (Hokkanen *et al.* 2015, Lahdenperä 2017) and even play a role in post-harvest diseases (Nallathambi *et al.* 2009).

Optimisation of the entomovector technology involves research at all of the interaction levels. Dissemination of the BCAs depends on the foraging behaviour and range of the pollinating insects. The formulation of the microbial control agent is of crucial importance for the acquisition on the body, deposition on the flowers, and viability of the control agent. Dispensers have to be designed in order to load the vector without impairing its natural behaviour and safety to the vector. Research towards functioning dispenser designs, different carrier substances and their interactions, potential vectors and their loading etc. has been conducted during the BICOPOLL project on a pan-European level, in order to significantly improve the yield and quality of organic cropping systems (Figure 2).

3. Vector species

A crucial step towards the success of the entomovector technology is the selection of the pollinator vector. In the literature, there is an unresolved debate concerning the relative merits of honeybees versus bumble bees as crop pollinators (e.g. Willmer *et al.* 1994, Aebi *et al.* 2012, Ollerton *et al.* 2012), and this discussion has implications also to entomovectoring situations.

Since the start of employing pollinating insects as a delivery system for biological control agents, three commercially available bee genera have been reported to vector BCAs efficiently onto crops: honeybees (*Apis mellifera*), bumblebees (*Bombus terrestris* and *B. impatiens*) and the mason bee (*Osmia cornuta*). Honeybees and mason bees are suitable for field conditions (Maccagnani *et al.* 2006, Hokkanen *et al.* 2015), while bumblebees are applicable under both field and greenhouse conditions (Mommaerts and Smagghe 2011). Recently, the application of blue-banded solitary bees (*Amegilla* spp.) have been shown to be a good alternative as a pollinator in greenhouses in Australia, where no bumble bees exist on the mainland (Hogendoorn *et al.* 2007). In the BICOPOLL project, we focused on the use of honeybees (Figure 3), but also investigated the potential of bumblebees and solitary bees for entomovectoring (Hokkanen *et al.* 2015).

The choice of the right pollinator species depends on different parameters: species, crop visitation rate, and deposition capacity of the BCA by the vector (Smagghe *et al.* 2012). An important vector-species dependent factor is the natural foraging behaviour. Bumblebees can tolerate temperature fluctuations, while honeybees are shown to be more sensitive to weather conditions (temperature, cloudiness, rain), which can affect proper dissemination of the BCA onto the flowers (Maccagnani *et al.* 1999, Guerra-Sanz 2008, Goulson 2010). Also, the different foraging ranges should be taken into account as a determinant for vector choice. Mason bees and bumblebees stay in relative close proximity of their nest (respectively, 100–200 m and 800–1500 m) compared to foraging ranges up to 3 km for honeybees (Vicens and Bosch 2000, Osborne *et al.* 2008).

(a) (b)

Figure 2. (a) Organic strawberry fields in Central Finland. (b) edge of an organic strawberry field, where bee hives are kept by the farmer year-round. These fields were included in the pilot entomovectoring field trials (photos: Heikki Hokkanen).

Figure 3. Field trials on commercial strawberry fields were conducted in central Finland. The setting-up of the experiments were actively supported by the strawberry growers, who also took care of all the crop management work, including operation of the bee hives and the BCA dispensers (photo: Heikki Hokkanen).

The efficiency of disease control that can be achieved depends on the frequency of visits by the vector to the crop. However, not all flowers will be visited in similar frequency, imposing therefore a high variability in the numbers of spores, or colony-forming units (CFU) of the BCA per flower ranging between 0 and 10^4 (Kovach *et al.* 2000, Shafir *et al.* 2006, Albano *et al.* 2009). This causes a variable suppression capacity, in particular when disease pressure is high (Shafir *et al.* 2006). Generally, entomovector studies report a mean deposition of 10^3–10^4 CFU/flower, an amount shown to be sufficient to control most plant pathogens, for example *B. cinerea* (Elad and Freeman 2002). Protection can last up to 31 days after application (Escande *et al.* 2002), and even in post-harvest stocks, longer shelf-lives of strawberries has been demonstrated (Hokkanen *et al.* 2015).

The levels of success in biocontrol so far obtained by employing the entomovector technology, using different vectors and BCAs in relation to the crop considered, are given in Mommaerts and Smagghe (2011), and in Hokkanen *et al.* (2015). In general, under experimental conditions in the field and in protected crops, most entomovector systems provided good control of the target diseases and pests in a wide variety of crops, comparable to or better than what can be achieved with chemical pesticides. These include using honeybees on strawberry, blueberry, raspberry, apple, pear, sunflower, clover, and oilseed rape; while bumblebees were tested successfully on strawberry, raspberry, sweet pepper, and tomato, and mansion bees on pear (Mommaerts and Smagghe 2011). Under practical field conditions in commercial berry fields and orchards, excellent results in the control of target diseases were demonstrated over several years with honeybees as vectors in strawberry, raspberry and apple, and on strawberry with bumblebees (Hokkanen *et al.* 2015).

An important consequence of the foraging behaviour of the pollinator is the secondary dissemination of the BCA: multiple flowers will be visited during foraging flight, thereby disseminating the vectored product from the dispenser to the flowers, but also similarly depositing and transporting the BCAs from one visited flower to another (Maccagnani *et al.* 2005). Due to this secondary BCA transport, a homogeneous deposition of the vectored product will be obtained over the vectored area. Conversely, crop visitation rate by the vector and flower attractiveness can have its impact on the efficiency of deposition, especially in open field conditions.

The decision of a bee to visit the target crop is in the case of honeybees foremost a "hive-decision" (Boecking, personal communication), and other resources in the vicinity of the hive will affect the visitation rate of the target crop (Raine and Chittka 2007). Finally, targeted control of plant pathogens by use of the entomovector technology seems to be most effective when vectoring of the control agent starts before most flowers are open, and thus before infection by plant pathogens can occur (Ngugi *et al.* 2002).

4. Dispensers

To load pollinators with microbial control agents in an efficient way, the use of a suitable dispenser is crucial. Research on dispensers for honeybees started by testing simple one-way dispenser, previously used to load pollinators with pollen. In loading bees with microbial control agents, two-way dispensers, where entrance and exit are separated, were developed and improved, ultimately leading to a model that was suitable for commercial use, the BeeTreat™ dispenser (Smagghe *et al.* 2012; Figure 4, Aasatek Oy Ltd, Finland).

In order to load the vector in an appropriate manner, an efficient dispenser design is required because it determines whether a good acquisition on the vector body is achieved. The dispenser is a chamber or a

(a) (b)

Figure 4. (a) The first prototype of the BeeTreat™ dispenser, used in our first pilot studies prior to the BICOPOLL project; (b) BeeTreat™ as it was developed during the BICOPOLL project, attached to a bee hive (photos: Ingeborg Menzler-Hokkanen).

'tunnel' leading from the hive to the exit of the bee hive, containing the powdery BCA-product. When bee workers leave their hive, they have to walk through this chamber and during this time the BCA product will stick to the vector's body. As a consequence, the loaded vector will transport the BCA to the flowers of the target crop. The dispenser must be designed in such a way that the vector will be loaded with a sufficient amount of the BCA product to assure safety for the vector without interfering with its foraging behaviour, and that they can be easily refilled, preferably in intervals of one day or longer. During the last 20 years several dispenser types have been developed for honeybees, mason bees, and bumblebees, each with its specificities. In general, there are two types: the one-way dispenser and the two-way dispenser. In the first one, the entrance and exit are not completely separated. In the two-way dispenser, entering and leaving the dispenser is separated, which avoids that returning bees walking back into the hive carry the BCA product into the hive. Multiple authors report a loading of greater than 10^4 CFU/bee, and a good control of the target pest or disease. The standard two-way dispenser BeeTreat™ (Figure 5) has been successfully used by some growers for 10 years, and in recent years, by hundreds of other strawberry growers in Finland (Hokkanen *et al.* 2015).

(a) (b)

Figure 5. (a) Attaching a BeeTreat™ dispenser to a bee hive; (b) dispenser attached to a bee hive without the movable ("steering") part. The photo shows the inoculum field, the part of the dispenser, which is filled with the BCA, and over which the honeybees walk when leaving the hive ((b) photos by Ingeborg Menzler-Hokkanen).

For *B. terrestris* workers, the Mommaerts dispenser (Mommaerts *et al.* 2010) seemed to be most effective with higher vector loading and three-days refilling intervals, and without any effects on foraging activity. An optimised dispenser for bumblebees is currently commercially available at the "Flying Doctors" dispenser, built into a commercial bumblebee hive (Biobest NV, Belgium).

During flight of the bee, a significant amount of loaded product is lost, and therefore the presence of the vectored BCA is not only restricted to the flowers, but is also disseminated onto the leaves, which is of particular interest in protecting other plant structures than flowers, e.g. against foliar diseases such as powdery mildews. For instance it was shown by Kapongo *et al.* (2008) that *C. rosea*, vectored by *B. impatiens*, could be recovered on 90% and 79% of the sampled tomato and sweet pepper leaves, respectively.

5. Microbial control agent formulations

Fungi, bacteria, and viruses can be used as BCA in agriculture and horticulture against plant pathogens or pest insects. Such microbial BCAs are naturally occurring microorganisms, isolated from the environment and thus are often already present in the target ecosystem, e.g. in the soil or on the foliage. In order to be applicable via the entomovector technology, they have to be modified in such a way that they can be used as a loaded product on the pollinator (Smagghe *et al.* 2012). Therefore the vector is loaded with the BCA in the form of a powdery formulation containing the organisms such as pieces of mycelium and/or spores.

Before use as BCAs, it needs to be assured that they are genetically stable, are able to survive adverse environmental conditions, and are effective at a low concentrations. Resistance to fungicides used on strawberries and raspberries under integrated pest management is further required from a competitive BCA product; furthermore, it should be easy to store and to disseminate effectively (Droby *et al.* 2009, Sharma *et al.* 2009). Formulation of the control agent is thus a crucial element of successful entomovectoring, and this is under constant improvement to guarantee that vectors acquire a sufficient amount of viable BCAs allowing an optimal transport to the flowers. Concerning formulation and the BCA product, the key elements are the carrier material, and the microbial agent itself. To illustrate this process, we examine the most widely applied product until now in commercial entomoverctoring, the Prestop Mix biofungicide (Verdera Oy, Finland), and discuss issues concerning the carrier and the organism.

In order to ensure an optimal transport by the vectors, BCAs are not used solely but in combination with a carrier material. It is well established that in general, any dust on an insect is irritating, and for example induces grooming behaviour (e.g. Alstad *et al.* 1982, Korunic 1998, Zhukovskaya *et al.* 2013).This is particularly well known from bees (e.g. Bozic and Valentincic 1995, Macedo *et al.* 2002). It should be emphasised that the product Prestop Mix has not been originally developed for entomovectoring, but rather for mixing with the soil to control numerous plant pathogenic fungi in the soil environment. For mixing with soil, the main carrier material kaolin (part of clay) is an ideal substance, and cannot be expected to cause any problems. Clay is a natural component of soils, and insects including pollinators frequently are exposed to dusts and clay particles in the field dust, particularly in the spring at the time of farmers working their fields. It is not surprising that Karise *et al.* (2016) found impacts of pure kaolin on the bees, but puzzling that when mixed with *Gliocladium catenulatum* (current name *Clonostachys rosea* f. *catenulata*) and formulated as the product Prestop Mix, the kaolin effect disappeared. The authors have not been able to provide any explanation to this contradictory result so for.

In the BICOPOLL project, it has been established that kaolin is not an ideal formulation substance for entomovectoring because of its variable particle size, and that alternative substances should be developed. The carrier material not only needs to be optimal for entomovectoring (particle size, and size distribution), but it also has to fulfil the needs of the biocontrol agent to ensure its storability and functioning after it has been distributed to the target environment (flowers in the case of entomovectoring). There is ongoing work on new carrier substances for *G. catenulatum* specifically for entomovectoring.

Concerning the living BCA itself, the fungus *G. catenulatum* is a ubiquitous organism occurring widely in different soil types all around the world (US-EPA 2002). The indigenous wild fungal strain J1446, the active organism in Prestop Mix biofungicide, was originally isolated from Finnish field soil. The total amount of *G. catenulatum* in Finnish soil with active root growth has been estimated to be 10^6–10^7 CFU/kg. The fungus is thus part of normal soil microflora, and it is probable that the numbers of the organism after use as a BCA will decrease to natural levels due to competition by natural soil microorganisms. Thus the risk for disturbances in the environment after the recommended applications is expected to be quite small (SANCO 2004, EFSA 2007). The mode of action of *G. catenulatum* strain J1446 is hyperparasitism, production of lytic enzymes and competition for living space and nutrients. There is no indication of toxin production by *G. catenulatum* (SANCO 2004, EFSA 2007). The fungus has been shown to pose no hazard to humans (SANCO 2004): it is non-pathogenic and non-infective, non-toxic, and causes no dermal or eye irritation. *G. catenulatum* has also been shown not to present any hazard to the environment, including impacts on non-targets (US-EPA 2002, EFSA 2007). Due to these favourable characteristics, the products Prestop and Prestop Mix have been accepted for use in organic farming in the EU (EFSA 2007).

In practical use, no adverse impacts have been noticed. In entomovectoring, either managed honeybees or managed bumblebees ('Flying Doctors' concept, Biobest NV, Belgium) are used to disseminate the biocontrol agents (Hokkanen *et al.* 2015). These managed insects are of course important pollinators where they are deployed, but they are reared and managed by humans. In the case of commercial bumblebees, the hives have a very short life-time: a managed bumblebee hive has longevity of maximally 6–8 weeks, after which it perishes in any case (Keppens *et al.* 2011). Thus, considering potential harmful side effects of the carrier materials and the BCA itself on bees, it is crucial to distinguish between the bees utilised for dissemination (e.g. the commercial, short-lived bumblebee colonies), and the wild bees potentially affected via visiting flowers harbouring the disseminated BCA. In vast majority of cases, pollination of plants is carried out by wild pollinators (bumblebees, solitary bees, feral honeybees, and other pollinators) (e.g. Muniz *et al.* 2013, Vaudo *et al.* 2014, Bartos *et al.* 2015). These have very little to do with the managed bees, except during encounters at the flowers during foraging.

Professional strawberry growers in Finland have been using entomovectoring with honeybees since 2006, and currently hundreds of growers employ the method for grey mould control (Hokkanen *et al.* 2015). Many growers also have experimented with bumblebees as vectors of Prestop Mix. No reports of any kind of harmful effects to the vectoring bees have been reported. Therefore, in practice, no risks to the vectoring bees themselves have been evident.

6. Strawberry and raspberry grey mould control

The first published experiments in using bee vectored biocontrol involved honeybees and the control of grey mould on strawberries. Peng *et al.* (1992) and Yu and Sutton (1997) reported good control of grey mould in raspberries and strawberries using *Gliocladium roseum*, reducing *B. cinerea* incidence from 90% to 68%, and from 64% to 48%, respectively. To our knowledge, results collected during the BICOPOLL project represent the first successful use of entomovectoring by commercial growers over large cropping areas. A review of entomovectoring (Mommaerts and Smagghe 2011) provided a listing of numerous other studies with a wide variety of target diseases, pests, crops, and antagonistic BCA, but could not identify practical applications in crop protection — other than the case of Finland. Furthermore, our Finnish on-farm research results with raspberry confirmed that grey mould can be controlled with entomovectoring in commercial production of that crop as well.

Hokkanen *et al.* (2015) show that good efficacy in grey mould control can be achieved with *G. catenulatum* at much lower doses than what is required for equal efficacy when applying the BCA by spray treatments, thus resulting in economically competitive control (Lahdenperä 2017, Lahdenperä unpublished data). Using entomovectoring, 400 g of Prestop® Mix is disseminated/ha/season, but to achieve the same level of control by spraying, the product would require about 1–2 kg/ha applied in 3–5 spray treatments (Lahdenperä 2017, Lahdenperä *unpublished* data). We assume that at least two factors contribute to this result: (i) blanket spraying of the crop at economically feasible doses does not bring high enough numbers of antagonist spores to the strawberry flowers when needed to prevent Botrytis from developing; and (ii) as at least some 40,000 new flowers open every day in a typical strawberry field per hectare (own calculations), these remain without protection until the next spraying is carried out. Entomovectoring appears to remedy both factors: (i) bees bring high amounts of BCA spores directly to the flower (several hundred spores have typically been measured after a bee visit, e.g. Peng *et al.* (1992) and Yu and Sutton (1997) — enough to prevent the grey mould fungus from colonising the flower; and (ii) bees are active every day, and visit flowers as soon as the weather conditions allow. This provides a continuous, targeted precision biocontrol to take place, and ensures thereby good protection against *B. cinerea*.

Acknowledgements

Our research team on entomovectoring wants to thank foremost the countless bee workers whom we had to carry the biocontrol agent to the target flowers during the pilot studies and field tests. Our thanks also to the strawberry growers, who allowed us to conduct experiments on their fields. We wish to thank for support and cooperation the whole entomovectoring research team (Aino-Maija Mustalahti, Lotta Kekäläinen. Laura Pulli, Marja-Leena Lahdenperä, Heikki Hokkanen, Ingeborg Menzler-Hokkanen) and the helping crew from the strawberry farming community in the Savo region of Finland including Mika Räsänen, Mr. and Mrs. Mönkkönen, Helena Hänninen, and dozens of other pioneering berry and fruit growers. Partial support to our entomovectoring work has been obtained from NordForsk projects 70066 and 45941, and the CORE ORGANIC 2 project BICOPOLL.

References

Aebi, A., Vaissiére, B.E., vanEngelsdorp, D., Delaplane, K.S., Roubik, D.W., Neumann, P., 2012. Back to the future: Apis versus non-Apis pollination. *Trends in Ecology and Evolution* **27**, 142–143.

Albano, S., Chagon, M., de Oliveira, D., Houle, E., Thibodeau. P.O., Mexia, A., 2009. Effectiveness of Apis mellifera and Bombus impatiens as dispensers of the Rootshield biofungicide (Trichoderma harzianum, strain T-22) in a strawberry crop. *Hell Plant Prot J* **2**, 57–66.

Alstad, D.N., Edmunds, Jr G.F., Weinstein, L.H., 1982. Effects of air pollutants on insect populations. *Annual Review of Entomology* **27**, 369–384.

Bartoš, M., Tropek, R., Spitzer, L., Padyšáková, E., Janšta, P., Straka, J., Tkoč, M. Janeček, Š., 2015. Specialization of pollination systems of two co-flowering phenotypically generalized Hypericum species (Hypericaceae) in Cameroon. *Arthropod-Plant Interactions* **9**, 241–252.

Bozic, J., Valentincic, T., 1995. Quantitative analysis of social grooming behavior of the honey bee Apis mellifera carnica. *Apidologie* **26**, 141–147.

Droby, S., Wisniewski, M., Macarisin, D., Wilson, C., 2009. Twenty years of postharvest biocontrol research: Is it time for a new paradigm? *Postharvest Biology and Technology* **52**, 137–145.

EFSA, 2007. Reasoned opinion on the first establishment of Annex IV of Regulation (EC) 396/2005. European Food Safety Authority. *EFSA Scientific Report* **115**, 1–161.

Elad, Y., Freeman, S., 2002. Biological control of fungal plant pathogens. In: Kempken, F. (Ed.), *The Mycota, A Comprehensive Treatise on Fungi as Experimental Systems for Basic and Applied Research*. Springer, Heidelberg, Germany.

Escande, A.R., Laich, F.S., Pedraza, M.V., 2002. Field testing of honeybee-dispersed Trichoderma spp. to manage sunflower head rot (Sclerotinia sclerotiorum). *Plant Pathology* **51**, 346–351.

European Commission, 2014. Commission staff working document. *Impact assessment, accompanying the Document Proposal for a Regulation of the European Parliament and of the Council on Organic Production and Labelling of Organic Products*. COM (2014) 180 final. Annexes 1-8. Retrieved from http://ec.europa.eu/agriculture/organic/documents/eu-policy/policy-development/impact-assessment/impact-assessment-part2_en.pdf

FAO, 2011. FAOSTAT agricultural production statistics. http://faostat.fao.org/site/567/ and at http://faostat.fao.org/site/339/ (accessed 10 February 2011).

Goulson, D., 2010. *Bumblebees Behaviour and Ecology*. Oxford University Press, New York, USA, p 317.

Guerra-Sanz, J.M., 2008. Crop pollination in greenhouses. In: James RR, Pitts-Singer T (eds) *Bee Pollination in Agriculture Ecosystems*. Oxford University Press, New York, USA.

Hogendoorn, K., Coventry, S., Keller, M.A., 2007. Foraging behaviour of a blue banded bee, Amegilla chlorocyanea in greenhouses: implications for use as tomato pollinators. *Apidologie* **38**, 86–92. doi: 10.1051/apido:2006060

Hokkanen, H.M.T., Menzler-Hokkanen, I., 2007. Use of honeybees in the biological control of plant diseases. *Entomology Research* **37**(Suppl. 1), A62–A63.

Hokkanen, H.M.T., Menzler-Hokkanen, I., Lahdenperä, M.-L. 2015. Managing bees for delivering biological control agents and improved pollination in berry and fruit cultivation. *Sustainable Agriculture Research* **4**(3). http://dx.doi.org/10.5539/sar.v4n3p89

IPM-Centers, 2011. Crop Profile for Strawberries in Louisiana. http://www.ipmcenters.org/cropprofiles/docs/LAstrawberries.pdf, accessed on 10.2.2011

Kapongo, J.P., Shipp, L., Kevan, P., Broadbent, B., 2008. Optimal concentration of Beauveria bassiana vectored by bumble bees in relation to pest and bee mortality in greenhouse tomato and sweet pepper. *Biocontrol* **53**, 797–812.

Karise, R., Muljar, R., Smagghe, G., Kaart, T.,Kuusik, A., Dreyersdorff, G., Williams, I.H., Mänd, M., 2016. Sublethal effects of kaolin and the biopesticides Prestop-Mix and BotaniGard on metabolic rate, water loss and longevity in bumble bees (Bombus terrestris). *Journal of Pest Science.*

Keppens, W., Bollens, T., Wäckers, F.L., Yokoi, H., Yoneda, M., Aoki, T., 2011. *Bumblebee Nest Box.* Biobest Belgium NV, Patent WO2011089248 A1, http://www.google.com/patents/WO2011089248A1?cl=en#backward-citations

Kevan, P.G., Kapongo, J.-P., Al-Mazra'awi, M., Shipp, L., 2008. Honey bees, bumble bees and biocontrol. In James, R.R., Pitts-Singer, T. (Eds.), *Bee Pollination in Agriculture Ecosystems.* Oxford University Press, New York, USA.

Korunic, Z., 1998. Diatomaceous earths, a group of natural insecticides. *Journal of Stored Products Research* **34**, 87–97.

Kovach, J., Petzoldt, R., Harman, G.E., 2000. Use of honey bees and bumble bees to disseminate Trichoderma harzianum 1295-22 to strawberries for Botrytis control. *Biological Control* **18**, 235–242.

Lahdenperä, M.-L., 2006. Gliocladium catenulatum as an antagonist against grey mold on strawberry. NJF seminar No. 389: *Pest Disease and Weed Management in Strawberry.* 8–9. Nov. 2006, Finland. Abstract in NJF Report No. 10, 2006.

Lahdenperä, M.-L., 2017. Use of Prestop® Mix biofungicide in entomovectoring on apple against storage rot diseases. In: Hokkanen, H.M.T., Boecking, O., Menzler-Hokkanen, I., Smagghe, G. (Eds.), *Entomovectoring for Precision Biocontrol and Enhanced Pollination of Crops: Exploiting Synergy of Ecosystem Services.* Springer, Dordrecht, The Netherlands (in the press).

Maccagnani, B., Mocioni, M., Gullino, M.L., Ladurner, E., 1999. Application of Trichoderma harzianum by using Apis mellifera as a vector for the control of grey mold of strawberry: first results. *IOBC Bulletin* **22**, 161–164.

Maccagnani, B., Mocioni, M., Ladurner, E., Gullino, M.L., Maini, S., 2005. Investigation of hive-mounted devices for the dissemination of microbiological preparations by Bombus terrestris. *Bull Insectol* **58**, 3–8.

Maccagnani, B.B.C., Biondi, E., Tesoriero, D., Maini, S., 2006. Potential of Osmia cornuta as a carrier of antagonist bacteria in biological control of fire blight: a comparison with Apis mellifera. *Acta Hortic* **704**, 379–386.

Macedo, P.A., Wu, J., Ellis, M.D., 2002. Using inert dusts to detect and assess Varroa infestations in honey bee colonies. *Journal of Apicultural Research* **41**, 3–7.

Mommaerts, V., Put, K., Vandeven, J., Jans, K., Sterk, G., Hoffmann, L., Smagghe, G., 2010. Development of a new dispenser for bumblebees and evaluation to disseminate microbiological control agents in strawberry in the greenhouse. *Pest Management Science* **66**, 1199–1207.

Mommaerts, V., Smagghe, G., 2011. Entomovectoring in plant protection. *Arthropod-Plant Interactions* **5**, 81–95.

Muniz, J.M., Pereira, A.L.C., Valim, J.O.S., Campos, W.G., 2013. Patterns and mechanisms of temporal resource partitioning among bee species visiting basil (Ocimum basilicum) flowers. *Arthropod-Plant Interactions* **7**, 491–502.

Nallathambi, P., Umamaheswari, C., Thakore, B.B.L., More, T.A., 2009. Post-harvest management of ber (Ziziphus mauritiana Lamk) fruit rot (Alternaria alternata Fr. Keissler) using Trichoderma species, fungicides and their combinations. *Crop Protection* **28**, 525–532.

Ngugi, H.K., Scherm, H., Lehman, J.S., 2002. Relationships between blueberry flower age, pollination, and conidial infection by Monilinia vaccinii-corymbosi. *Phytopathology* **92**, 1104–1109.

Ollerton, J., Price, V., Armbruster, W.S., Memmott, J., Watts, S., Waser, N.M., Totland, Ø., Goulson, D., Alarcón, R., Stout, J.C., Tarrant, S., 2012. Overplaying the role of honey bees as pollinators: a comment on Aebi and Neumann (2011). *Trends in Ecology and Evolution* **27**, 141–142.

Osborne, J.L., Martin, A.P., Carreck, N.L., Swain, J.L., Knight, M.E., Goulson, D., Hale, R.J., Sanderson, R.A., 2008. Bumblebee flight distances in relation to the forage landscape. *Journal of Animal Ecology* **77**, 406–415.

Peng, G., Sutton, J.C., Kevan, P.G., 1992. Effectiveness of honey-bees for applying the biocontrol agent Gliocladium roseum to strawberry flowers to suppress Botrytis cinerea. Canadian *Journal of Plant Pathology — Revue Canadienne De Phytopathologie* **14**, 117–129.

Raine, N.E., Chittka, L., 2007. The adaptive significance of sensory bias in a foraging context: floral colour preferences in the bumblebee Bombus terrestris. *PLoS ONE* **2**(6), e556. doi:10.1371/journal. pone.0000556

SANCO, 2004. Review report for the active substance Gliocladium catenulatum. Finalised in the Standing Committee on the Food Chain and Animal Health at its meeting on 8 October 2004 in view of the inclusion of Gliocladium catenulatum in Annex I of Directive 91/414/EEC. European Commission, Health and Consumer Protection Directorate General.

Shafir, S., Dag, A., Bilu, A., Abu-Toamy, M., Elad, Y., 2006. Honeybee dispersal of the biocontrol agent and Trichoderma harzianum T39: effectiveness in suppressing Botrytis cinerea on strawberry under field conditions. *European Journal of Plant Pathology* **116**, 119–128.

Sharma, R.R., Singh, D., Singh, R., 2009. Biological control of postharvest diseases of fruits and vegetables by microbial antagonists: A review. *Biological Control* **50**, 205–221.

Smagghe, G., Mommaerts, V., Hokkanen, H., Menzler-Hokkanen, I., 2012. Multitrophic Interactions: The Entomovector Technology. In: Smagghe, G., Diaz, I. (Eds.), Arthropod-plant interactions: Novel insights and approaches for IPM, Springer, The Netherlands, Dordrecht, pp. 127–157.

Strømeng, G.M., 2008. *Aspects of the biology of Botrytis cinerea in strawberry* (Fragaria x ananassa) *and alternative methods for disease control.* Philosophiae Doctor (PhD) Thesis 2008, 56.

US-EPA, 2002. Biopesticides registration action document. Gliocladium catenulatum strain J1446. U.S. Environmental Protection Agency, Office of Pesticide Programs, Biopesticides and Pollution Prevention Division. PC Code 021009. 12, November 2002. 26 p.

Vaudo, A.D., Patch, H.M., Mortensen, D.A., Grozinger, C.M., Tooker, J.F., 2014. Bumble bees exhibit daily behavioral patterns in pollen foraging. *Arthropod-Plant Interactions* **8**, 273–283.

Vicens, N., Bosch, J., 2000. Pollinating efficacy of Osmia cornuta and Apis mellifera (Hymenoptera: Megachilidae, Apidae) on 'Red Delicious' apple. *Environ Entomology* **29**, 235–240.

Willmer, P.G., Bataw, A.A.M., Hughes, J.P., 1994. The superiority of bumblebees to honeybees as pollinators: insect visits to raspberry flowers. *Ecological Entomology* **19**, 271–284.

Yu, H., Sutton, J.C., 1997. Effectiveness of bumblebees and honeybees for delivering inoculum of Gliocladium roseum to raspberry flowers to control Botrytis cinerea. *Biological Control* **10**, 113–122.

Zhukovskaya, M., Yanagawa, A., Forschler, B.T., 2013. Grooming behavior as a mechanism of insect disease defense. *Insects* **4**, 609–630.

Chapter 8

Agroecology and Participatory Knowledge Production and Exchange as a Basis for Food System Change: The Case of the Community Agroecology Network

Stephen R. Gliessman*, Heather Putnam† and Roseann Cohen†

**Environmental Studies, University of California, Santa Cruz, California, USA*
†*Community Agroecology Network, Santa Cruz, California, USA*

1. Introduction

Food insecurity especially in the form of seasonal hunger is persistent in smallholder coffee growing communities in Mexico and Central America (Bacon *et al.* 2014). At the same time, industrialised agriculture has supplanted centuries-old traditional food systems and agricultural practices, resulting in ecological degradation, a weakening of local economies as farmers become dependent on few cash crops, the loss of traditional environmental management knowledge, and families increasingly vulnerable to economic and ecological shocks (Gliessman 2015).

This chapter seeks to apply a model of agroecological transformation and change to evaluate the case study of a long-term partnership model between an international non-profit organisation, local organisations in Nicaragua and Mexico, and smallholder coffee farming families. The partnership model was based on agroecological principles, and utilised participatory action research (PAR) processes to jointly identify problems leading to food insecurity and vulnerability, formulate solutions rooted in local, existing knowledge and practices, and reflect on, evaluate, and adjust strategies together in annual iterations. The case study demonstrates how agroecology is a mutual, horizontal learning process among partners, and that it can reduce vulnerability among farmers who depend so much on a market commodity like coffee while ensuring food security. It also shows how the change process is an evolutionary process, not one of simple adoption.

2. Background: Smallholder coffee farmer vulnerability and agroecology as a solution

Food insecurity among smallholder coffee farmers is a common reality that has only recently been recognised, and it is still not well understood (Putnam and Brown 2013, Bacon *et al.* 2014). The most common manifestation of food insecurity among coffee smallholders is seasonal hunger, although transitory hunger also occurs as the result of ecological or economic shocks that affect a family's ability to access food (Bacon *et al.* 2014).

While access to markets and income generation tend to be high on the agendas of mainstream development organisations for alleviating food insecurity, smallholder coffee farmers solely dependent on one cash crop and the market face significant vulnerability for a number of reasons. First, volatile coffee markets combined with cyclical maize and beans markets often leave farmers facing high food prices when they are receiving low coffee prices. On top of this, climate variability has seen severe droughts, inundations, and increased pest infestation affecting both food and coffee production, in turn impacting household access to food and income. Farmer cooperative organisations have played a role in mitigating the impacts of these factors by increasing access to assistance, securing higher or more stable prices for their members' coffee through specialty markets like fair trade or organic, but these have not proven to actually reduce smallholder coffee farmer vulnerability nor significantly enhance their livelihoods (Bacon 2004, Méndez *et al.* 2010). Second and related to the first challenge, is lack of broader access to specialty coffee markets, which can theoretically lead to reduced incentive for a farmer to invest in improving their soils.

A third factor impacting farmer vulnerability is a high rate of emigration at the household level, with no discernible difference seen among families benefitting from coffee certifications (Méndez *et al.* 2010). Emigration is especially common among young household members, who increasingly see coffee farming as an unviable livelihood. The exodus of young future farmers from the farming livelihood breaks down the social fabric of a community and in fact threatens the future of smallholder coffee production. A fourth factor exacerbating smallholder vulnerability is the ongoing shift away from traditional crops and varieties for household consumption and local trade, to cash crops for export (Ghosh 2010), combined with the influx of imported — and often highly processed foods into rural communities (Friedmann 2005). This trend has in turn led to the loss of traditional agricultural knowledge and practices and the subsequent weakening of the ecological integrity of the farming system as higher-impact practices, including chemical fertilisers, become more predominant. A secondary effect of this trend is a decrease in soil fertility and increased vulnerability of coffee and food crops to disease outbreaks.

The dominant response at the state level to the vulnerability of smallholder coffee farmers is production input or seed subsidies, even as some governments recognise that cookie-cutter recipes that are productivist in spirit are bound to be ineffective, and are exploring various forms of innovative farmer engagement, including the formation of local food security councils (Putnam *et al.* 2016). At the same time, international development organisations have focused on production diversification for market as their main strategy to address smallholder vulnerability. But given the lack of understanding of the complexity of the causality of food insecurity and seasonal hunger in coffee growing communities, and the tendency to implement similar strategies in different contexts, questions remain as to how or if these strategies actually mitigate the diverse set of factors impacting household food security.

A long-term partnership model has sought to answer the question of how smallholder coffee farmers can utilise an agroecological approach to transform their food system and alleviate food insecurity and seasonal hunger. The Community Agroecology Network (CAN) is an international non-profit organisation committed to sustaining rural livelihoods and environments through collaborative research and action (Jaffe and Bacon 2008). It focuses on agroecological capacity-building and locally informed sustainable development strategies. It is a network of affiliated researchers, partner organisations, men and women farmers, and community youth leaders doing agroecology (see www.canunite.org).

CAN is committed to combining research and action for social change. CAN's action research initiatives study the interrelated problems of food security, rural outmigration, agricultural sustainability, and the lack of opportunities for young people in rural areas. CAN researchers partner with communities and local organisations to determine research agendas and methods so that research findings will be relevant and useful. We work together to jointly identify problems and potential solutions, monitor their implementation, reflect on their efficacy and challenges, and make adjustments in many iterations over time to "create agroecological landscapes" (Méndez *et al.* 2013). We recognise and value different ways of knowing, and promote intercultural exchange of knowledge and practices as a key tool to achieving food system transformation.

CAN works towards achieving food system sustainability, food security, and food sovereignty using three ideas or principles:

(i) Markets without household resiliency increase vulnerability
 Resiliency means a household is able to confront ecological shocks (like floods, droughts, and infestations associated with climate

change) and economic shocks (like cyclical commodity prices, as in the case of coffee). CAN has learned that household resiliency is achieved when families have diverse production for consumption and sale, not relying on one sole cash crop for income; this way, if one fails, the farmer and her family can rely on the rest. It has also become apparent that resiliency is strengthened with increased soil health where crops with adequate nutrition are robust and can resist pests, drought, and other extreme conditions that threaten a farmer's livelihood. Finally, resiliency is strengthened when the entire family is empowered as producers, consumers, and marketers of household production in the household economy — they have seen that when women have control over part or all of the household income, families have improved nutrition and food security.

(ii) Agroecology is a mutual learning process
It must include all stakeholder voices in a process where all ways of knowing — local, practical, academic, and professional — are valued equally as part of a larger toolbox wherein the farmer or community can choose what is most appropriate for their context, adapt practices to their own local conditions, and create their own culturally-specific toolbox from what is available. Similarly, to achieve maximum learning, it must be assured that women, men, and youth perspectives must be included throughout the process as distinct voices.

(iii) Agroecology enhances food security and sovereignty
While ecologically-sound production practices are at the core of agroecology, they are not viable without vibrant local food economies that foster fair market channels and farmer control of local food system resources, from seeds to the land. Agroecology improves soil health, water usage, pest and disease management, and the social relations of trade, ensuring robust food and cash crops, minimising pollution, enhancing biodiversity and multifunctional agrobiodiverse environments. The result is, diverse foods are reliably available at a lower cost to farmers without sacrificing their or their environment's health.

Putting these principles into place is not done by merely prescribing sets of agroecological practices or strategies that can be easily replicated. Instead, smallholder farming families and communities are engaged with CAN and local organisations in collaborative research processes that build capacity and identify solutions connected to their particular context and aspirations. Food security takes on an expanded meaning with agro-ecology, simultaneously emphasising availability, access, affordability, utilisation, stability, nutrition, social justice, and appropriateness (Putnam *et al.* 2014).

3. Engaging communities in a participatory learning process

The vulnerability of smallholder coffee farmers in Mexico and Central America became exceedingly clear during the coffee crisis of the early 2000s when coffee commodity market prices fell to their lowest levels in history, making the cost of growing and harvesting coffee higher than the price received (Bacon *et al.* 2008). CAN affiliated researchers working in these communities showed that income-based solutions (e.g. certified fair trade or organic premiums) do not reduce vulnerability. Even as prices picked up again, CAN found that farmers continued to face seasonal hunger and food insecurity. This was most serious during the "thin months" (*los meses de las vacas flacas*), a period of at least four months per year when people struggle to feed their families. People are paid for their coffee at the end of the harvest early in the year, but the income from this coffee usually runs out by May, when the rainy season begins and maize and beans are planted. Since they don't begin harvesting grain until August or September, the staple food grain market peaks at a time when this scarcity is most intense. Families often cannot afford to buy sufficient food given the high prices. The thin months from May to August are the inevitable result and when seasonal hunger occurs (Bacon *et al.* 2014).

A primary focus of CAN's work with smallholder coffee farmers and their communities is on improving food security and building food sovereignty (Figure 1). What this has led to over time has been a broad approach focused on creating healthy communities, vibrant local economies, and

Figure 1. A diverse multistoried coffee agroecosystem in northern Nicaragua. Apart from producing coffee for sale, this multifunctional system provides benefits ranging from food security to environmental services (photo: by Steve Gliessman).

landscapes where children experience improved nutrition and gain pride in their rural livelihoods, where youth and women play a stronger role in the creation of diverse household livelihoods, where diversification of coffee systems leads to improved soil fertility, and local people regain control over the availability of, and access to food in their communities.

Community engagement occurs through PAR (Bacon *et al.* 2005, Kingdon *et al.* 2007). PAR is an approach where researchers and non-researcher partners participate in an integrated, iterative, process of research (e.g. data collection), reflection (e.g. discussing findings), and action (e.g. using findings to resolve a problem), which leads to capacity building and ability to share experiences and learning with other communities. This mutual learning process means referring back to the roots of agroecology that were developed by early agroecologists observing and learning from farmers, trying things out together and communicating across knowledge systems that linked academic science and people's science (Gliessman 2016). PAR also means democratising knowledge production by making science work for the people. It focuses on capacity-building with community partners, and eventually community mobilisers learn to conduct research in their own communities, and make data driven decisions for change for themselves. There is also a strong focus on sharing knowledge, where the learning together happens, both in terms of analysing the results of initial baseline studies and analysing the implementation of actions directed at resolving community-identified problems. Finally, a key part of the PAR cycle is action, when knowledge and experimentation is converted or adapted into changed agricultural practices, social structures, and economic activities that better address the challenges the community is facing. The most effective PAR projects are a continual cyclical process of knowledge exchange, experimentation, and innovation.

The basic steps for completing a successful PAR cycle by CAN for agroecology are the following:

Step 1. Work with partner organisations and their farmer members to identify needs, develop questions, and use a participatory approach to developing the initial baseline study;

Step 2. Train community promoters and youth leaders in data collection and analysis, then do the collection;

Step 3. Analysis of data and interpretation of results involving all stakeholders;

Step 4. Share results with communities in workshops to refine results and determine action plans based on participatory data gathering and review;

Step 5. Implement action plans.

Steps beyond: Monitor and evaluate results, compare to baseline study, share feedback on future actions, and make changes in action plans based on results of implemented strategies. PAR is an iterative process of sharing and reflecting how a particular strategy is going, how can it be better suited to community needs, and where all stakeholders are involved in a mutual learning process. After completing the first five steps in the PAR cycle, and implementation of action plans takes place, the cycle can repeat itself in the iterative process of linking research, reflection, and sharing of knowledge and experience.

4. CAN case study project: Youth leadership and food sovereignty in San Ramón, Nicaragua and Veracruz, Mexico

The Youth Leadership and Education for Sustainable Agriculture and Food Sovereignty Project was launched in 2011 as a collaborative initiative of CAN, the Union of Cooperatives, San Ramon (UCA, San Ramon) in Nicaragua, and the NGO Vinculación y Desarrollo Agroecológico en el Café (VIDA) in Mexico. The project aimed to build community-based climate change resiliency, alleviate food insecurity and seasonal hunger during the thin months, strengthen livelihoods among 234 coffee farming families, and build sustainable local food systems in 12 coffee-growing communities in Nicaragua and Mexico through education and the empowerment of local youth leaders and women in these cooperatives. The first phase of the project focused on the design of a participatory monitoring and evaluation system, implementation of a comprehensive baseline study of food insecurity and household livelihoods in the communities, the consolidation of the project's youth- and women-centred methodology and training of youth leaders, capacity building in agroecological food production, the establishment of homegardens, and reforestation with fruit, wood, and fuel trees. The second phase built upon these foundations and incorporated lessons learned from the first phase, by strengthening the focus on women's income generation, youth capacity building and leadership training, nutrition education, water access for consumption and irrigation during the dry season, soil health, access to quality seed, and access to staple foods during the thin months.

The beginning of Phase 2 in 2013, coincided with the onset of the crisis of the severe coffee rust disease known as the *la roya*, which was followed in 2014 by a two-year drought that severely impacted food production in Northern Nicaragua, and reduced rainfall levels from normal in Veracruz, Mexico. These compounding shocks clarified understanding of the positive mitigating effects of the strategies being promoted, and crystalised the vision of this project towards climate change resiliency and sustainable livelihoods. Food security and sovereignty align with this vision by sharing a focus on the empowerment of families and communities to ensure for themselves availability and access to food at all times of year in any year. Similarly, the lens of climate change resiliency allowed everyone to look at the entire panorama of factors affecting food security — from climate to political structures to the quality of the soil. The project took a holistic approach to improving household climate change resiliency, food security, and livelihoods, seeking to promote personal empowerment of women and youth as change agents in their families and communities alongside increased access to and availability of basic and more diverse foods, income diversification for women and youth, nutrition education, improved water access for consumption and food production, and agroecological farming practices.

After going through the steps of the PAR cycle listed above, community/cooperative action plans were developed. Working with coffee farming families in eight communities and their cooperatives in San Ramón, Nicaragua, and four community groups in Veracruz, Mexico, the following action plans emerged. It is important to note that the action plans were considered to be living plans that left room for change response to emergent situations like drought, and to lessons learned along the way.

Action plan: Production diversification for household dietary diversity improvement and income diversification for women.

Diversification strategies included homegardens, diversification of coffee shade structure with fruit, wood, and fuel trees, diversification of patios with root and tuber crops, and diversification into protein (chicken and eggs) production (Figure 2). Women were the primary agents implementing these diversification strategies. The homegardens initially faced many challenges. First, many women stated that they did not know how to grow vegetables, as they "had not done this since their grandmother's time", as one woman stated. Basic capacity building led by trained local youth leaders commenced, focused on basic gardening and food production techniques, including seed selection, germination, transplanting, soil building and improvement technologies, and associative and rotational planting. After two years, most of the women felt comfortable and confident in their gardens and had even begun selling at monthly farmers markets or out of their homes.

A second challenge arose during the first growing season of the gardens. Through the annual survey, it was revealed that many families were

Figure 2. An example of the diversity of products being grown in homegarden agroecosystems in the community of Guzmantla, municipality of Ixhuatlan del Café, Veracruz, Mexico (photo: Heather Putnam).

not eating the food they were growing, instead they were feeding most of the vegetables to their pigs and other small livestock. In a follow-up workshop with the women gardeners, they informed us that they did not know many different ways to prepare the suddenly wide variety of vegetables available to them in their gardens. The action that followed was a series of nutrition workshops led by local women and youth in which traditional recipes were shared and tested with locally available plants, and the production of a cookbook of traditional and innovative recipes to promote dietary diversification. Another problem arose in the second year when it was learned that the women gardeners were either not saving seed or rootstock from their garden plants, or they were not able to save seed from certain plants like carrots or onions. A collective response to this problem was to refocus garden production on the plants whose seeds the women were able to save, especially plants that were locally traditional. An additional response was to train the women and youth in household seed saving techniques for vegetable and fruit seed and reproductive material.

In the years that followed, families also developed a vibrant informal seed exchange system within and among communities. When it was found that there was no existing formal market for the vegetables produced in the homegardens, the cooperatives developed regular monthly farmers markets in the nearby municipal centre of San Ramon, and a group of women established and ran a café to sell directly to consumers coffee drinks prepared from their own coffee, and excess vegetables and fruits from the homegardens associated with the project.

Action plan: Relocalise access to staple foods (grain and seed)

The initial diagnostic study in 2011 revealed two troubling trends. First, families often had to purchase maize and bean (and sometimes sorghum for animal feed) during the thin months when the price of food was highest and when their income from the previous coffee harvest had already run out. This meant that some families could not purchase maize and beans, and resorted to eating cheap but less nutritious foods like boiled bananas. Even if they did have the money to purchase grain, they often had to buy it from middlemen or go into the city to buy it as there was no local distribution facility, adding even more cost to the food. Second, it was found that families often resorted to eating their seed put aside for the next season, which meant that they were unable to replant unless they bought more seed, resulting in a vicious cycle where access to basic foods became worse. A cooperative response to this problem of a lack of local access to staple food and seed for staple food, was the establishment of grain banks and seedbanks. This resulted in a total of seven grain and seedbanks in San Ramon, Nicaragua, and five seedbanks in Veracruz, Mexico, focusing primarily on maize and beans. The banks were led by elected commissions who were trained in their management and administration, and who subsequently led efforts to produce grain and seed for the banks, to make more available to the wider community. It is a prime example of collective action and social structure finding solutions to a systemic problem by creating a new system parallel to the existing system.

Action plan: Alternative market development for export coffee

Through the formation of a collaborative marketing process that involved the key stakeholders of a much shortened coffee commodity chain, CAN has developed a branded agroecological coffee called AgroEco® Coffee. This has involved building an alternative market that is focused on agroecological production. It is not a certification, but rather a commitment to transitioning from the industrial monoculture model of production to an agroecological model. It also involves consumer education so coffee drinkers make conscious choices when buying their coffee, and including everyone in the commodity chain of coffee, from the farmers, their cooperative, the exporter, the roaster, and the researchers make at the negotiating table when the price for the coffee is being set each year. For the 2013/2014 harvest, this moved more than 12,000 kg of coffee beans from the farmers in Nicaragua to consumers, and 4,800 kg of coffee from Mexico.

Farmers received a base price for their coffee that exceeded even the premium given for fair trade and organically certified coffee. In addition to this higher return to the farmer, an additional 5% of the price paid by the coffee roaster contributes to a "Sustainable Agriculture Fund" to directly support the transition process by funding coffee innovation projects decided upon by the communities themselves. As an example, in one case growers in one cooperative knew that building organic soil fertility was a major issue, and considered supporting a composting project to produce compost from local resources. CAN and the community faced a very challenging moment with this project. Farmers, who are mostly men, with very low participation by women, are very risk averse. Despite doing farmer to farmer exchanges with farmers from communities that were already engaged in similar soil building projects, they were still hesitant. They were also facing another crisis — the outbreak of the coffee rust disease that was killing a large percentage of their coffee plants. To choose a different course of action to the conventional high-input model was seen as very difficult.

At the same time CAN, the UCA San Ramon, and VIDA wanted to strengthen women's access to capital. Progress had been made with the homegardens and farmers markets directed primarily at women, but it was evident that women were contributing more labour to coffee production (and other agricultural and household labour) than they were being compensated for. At the bargaining table, with buyer, importer, cooperative leaders, CAN, and men and women farmers all present, a second fund was created, the Women's Unpaid Labour Fund, which added another 4% to the price of the coffee to be invested in an economic incentive initiative decided and controlled by the women themselves. Something interesting happened in Nicaragua. The women's group stepped forward and decided to invest their fund in the agroecological renovation of 0.5 hectares each in their family coffee parcels. Since then, the women have expanded their agroecological coffee plots to a little over a total of 10 hectares, and men and women gather every other Saturday to make organic fertilisers collectively to support their ongoing agroecological coffee renovation. And in Mexico, the women's group decided to invest their 4% in the development of a branded coffee they call FEMCAFE, and will be selling it locally and in other specialty markets in Mexico.

Action plan: Improve coffee agroecosystems with agroecological practices

After the hesitation to the compost plan, CAN researchers, youth leaders, and cooperative extensionists in Nicaragua convinced the whole group to invest the sustainable agriculture fund in buying natural materials and inputs (raw flour, molasses, rock minerals, etc.) for making artisanal fertilisers, as well as materials to make foliar sprays from culturing fungi in leaf litter from the mountains above their communities for disease suppression. At a calculated cost of about one tenth of conventional fungicides and fertilisers, the coffee plants are showing recovery from the disease outbreak and resistance to it from new plants. The efficacy of these agroecological soil health and plant nutrition techniques became evident after one year of application to vegetable gardens and to coffee being replanted with seedlings: the plants were robust and even starting to fruit after only 17 months in the ground. CAN and the UCA San Ramon then facilitated a process of horizontal exchange so that the other seven cooperatives engaged in the project could learn the same techniques. The learning exchanges included capacity building in the elaboration of nine different soil and foliar applications, including compost, worm compost, effective microorganisms, biofertilisers, and mineral foliar applications for both food and coffee production areas. Investments were made in barrels and other equipment to allow groups to produce the fertilisers and preparations collectively where appropriate.

A similar process unfolded in the Central Highlands of Veracruz, Mexico where CAN researchers and VIDA performed a full diagnosis of the impact of *la roya* on the coffee parcels of 151 farmers in 2014, as well as an inventory of agroecological practices already being used. They subsequently identified 12 soil fertility improvement applications and other agroecological techniques including mineral foliar sprays, and began implementing farmer-to-farmer learning exchange workshops. Since *la roya* fully hit the Veracruz region approximately 18 months after it hit Nicaragua, farmers are still in the process of recovering and are currently (as of January 2016) implementing renovation efforts accompanied by the implementation of agroecological soil building and plant nutrition practices to ensure the seedlings are robust and resilient against the rust and other infestations like anthracnose.

The process of "doing" agroecology in these communities has come full circle, yet continues to move round the iterative PAR spiral of development and action. This case study is an example of how and why to incorporate the multiple aspects and stakeholders of the food system in a change process. It links aspects such as household nutrition, local food production, building alternative markets (locally and globally), diversification, improving natural soil fertility, and empowering community members as agents of agroecological change. It also links change across all the five Levels of transformative food system change presented by Gliessman (2015, 2016) and presented in Table 1.

4. The five levels of agroecological change

Agroecology is a way of redesigning food systems, from the farm to the table, with a goal of achieving ecological, economic, and social sustainability. Through transdisciplinary, participatory, and change-oriented research and action (Méndez *et al.* 2013), agroecology links together science, practice and movements (Wezel *et al.* 2009, Gliessman 2015) focused on social change. But what are the steps that must be taken to transform food systems toward the ultimate goal of sustainability, and away from the mounting evidence of the negative impacts on the environment and society caused by modern industrial agriculture and purely commercial food systems?

Table 1. The levels of transition and the integration of the three components of Agroecology (i.e. science, practice, and movements for social change) needed for the transformation to a sustainable global food system. Adapted from Gliessman (2015, 2016).

		Role of agroecology's three aspects		
Level	Scale	Agroecological research	Farmer practice and collaboration	Social change
1 Increase efficiency of industrial practices, while lessening their environmental impacts.	Farm	***Primary***	***Important*** Lowers costs and lessens environmental impacts	***Minor***
2 Substitute alternative agroecological practices and inputs.	Farm	***Primary***	***Important*** Supports shift to alternative practices	***Minor***
3 Redesign whole agroecosystems.	Farm, region	***Primary*** Develops indicators of sustainability	***Important*** Builds true sustainability at the farm scale	***Important*** Builds enterprise viability and societal support
4 Re-establish connections between growers and eaters, develop alternative food networks.	Local, regional, national	***Supportive*** Interdisciplinary research provides evidence for need for change and viability of alternatives	***Important*** Forms direct and supportive relationships	***Primary*** Economies restructured; values and behaviours changed
5 Rebuild food systems globally so that they are sustainable and equitable for all.	Local to global	***Supportive*** Transdisciplinary research and action promotes the change process and monitors sustainability	***Important*** Offers the practical basis for the needed paradigm shift	***Primary*** Food systems fundamentally transformed

The case study presented in this chapter also requires specific theories of transition to be drawn upon, both in regard to broader political transitions, and to the particular set of factors surrounding agroecological transitions. In particular, the present case study can be analysed with reference to the *five levels of change* identified by Gliessman (2015, 2016). The change process is not necessarily linear, since it depends on the starting point for each agroecosystem and the confluence of factors, issues, and histories faced. It also depends on the scale of food system operation, from a local farm to the global food system, or if the farm is located in a developing country where smallholder farms are the norm versus developed countries where simplified large-scale monoculture farms have often become the norm (Gliessman 2015). At each level of change, agroecological research contributes to changing farming practice and the development of important indicators of change. Farmer experimentation, practice, and collaboration are key to promoting innovation, understanding, and value shifts. Consumer understanding and wider social support for alternative food systems are also essential elements in supporting change. Finally, a supportive policy environment is vital for incentivising, and even regulating, a true paradigm shift in food and agriculture systems.

These factors correspond to the different levels of change. The first three levels describe the steps farmers can actually take on their farms for converting from industrial or conventional agroecosystems. Two additional levels go beyond the farm to the broader food system and the societies in which they are embedded. The five levels taken together are used in this chapter to assess and evaluate the progress the case study presented here is making towards the transformation to sustainable food systems, as well as pointing out key areas for promoting further reform. The five levels of change are described as follows and summarised in Table 1:

Level 1. *Increase the efficiency of industrial/conventional practices in order to reduce the use and consumption of costly, scarce, or environmentally damaging inputs.*

The primary goal of change at this level is to use industrial inputs more efficiently so that fewer inputs will be needed and the negative impacts of their use will also be reduced. Most conventional agricultural research has taken place at this level, through which considerable modern agricultural technologies, inputs, and practices have been developed. This research has helped farmers maintain or increase production through practices such as improved seeds, optimum planting density, more efficient pesticide and fertiliser application, and more precise use of water. So-called "precision agriculture" is a recent focus of research at Level 1. Although this kind of research has reduced some of the negative impacts of industrial agriculture, they do not help break its dependence on external human inputs and monoculture practices.

Level 2. *Substitute alternative practices for industrial/conventional inputs and practices.*

The goal of this level of transition is to replace external input-intensive and environmentally-degrading products and practices with those that are more renewable, based on natural-products, and more environmentally sound. Organic and bio-dynamic agriculture are examples of this approach. They employ alternative practices that include the use of nitrogen-fixing cover crops and rotations to replace synthetic nitrogen fertilisers, the use of natural controls of pests and diseases, and the use of organic composts for fertility and soil organic matter management. However, at this level, the basic agroecosystem is not usually altered from its more simplified form, hence many of the same problems that occur in industrial systems also occur in those with input substitution.

Level 3. *Redesign the agroecosystem so that it functions on the basis of a new set of ecological processes.*

At this level, fundamental changes in overall system design eliminate the root causes of many of the problems that continue to persist at Levels 1 and 2. The focus is on prevention of problems before they occur, rather than trying to control them after they happen. Research on whole-system conversions has provided an understanding of key yield-limiting factors. Agroecosystem structure and function is better understood, and appropriate changes in design can be implemented. Problems are recognised, adjustments made in internal site- and time-specific design and management approaches, instead of solely by the applications of external inputs. A good example is the reintroduction of diversity in farm structure and management through such actions as ecologically-based rotations, multiple cropping, agroforestry, and the integration of animals with crops.

Level 4. *Re-establish a more direct connection between those who grow food and those who consume it.*

Food system transformation occurs within a cultural and economic context, and this transformation must promote the transition to more sustainable practices. At a local level, this means those who eat must value food that is locally grown and processed, and support with their food dollars, the farmers who are attempting to move through Levels 1–3. This support becomes a kind of "food citizenship" and can be seen as a force for food system change. Communities of growers and eaters can form alternative food networks around the world where a new culture and economy of food system sustainability is being built. Food once again must be grounded in direct relationships. An important example is the current food "re-localisation" movement, with its growing networks of farmers' markets, community supported agriculture schemes, consumer cooperatives, and other more direct marketing arrangements that shorten the food chain.

Level 5. *On the foundation created by the sustainable farm-scale agroecosystems achieved at Level 3, and the new relationships of sustainability of Level 4, build a new global food system, based on equity, participation, democracy, and justice, that is not only sustainable but helps restore and protects earth's life support systems that we all depend upon.*

By thinking beyond Levels 1–4, Level 5 involves change that extends from local to global and reaches beyond the food system to the nature of human culture, civilisation, progress, and development. The depth of change is more than mere conversion or transition, and enters into the realm of full reform or transformation. With Level 5 thinking and action, agroecology provides ways to build upon farm-scale and farmer-driven change processes to a full rethinking of how we all relate to each other and to the earth that supports us. Basic beliefs, values, and ethical systems change. The expanding awareness that is part of this process, then extends to other facets of environmental and social relationships beyond food, bringing about a paradigm shift focused on how the agriculture and food systems of the future can help reduce our ecological footprint, recognise that there are limits to growth, and what it really means to live sustainably. The important role that food systems can and must play in mitigating and adapting to climate change as a global issue is one example of the value of Level 5 thinking. The growing food justice movement, where everyone in the food system enjoys the benefits of equity, justice, security, sovereignty, and sustainability, is another.

What will food systems look like when Level 5 thinking and action guides the changes that need to take place? What are the incentives needed to stimulate these changes? Can this thinking bring about essential changes in policy, support systems, funding, and choice? Is the change process actually a stepwise, longitudinal one, starting at Level 1 and ending with Level 5, or should Level 5 thinking determine where the process begins depending on where the farmer or the food system is at the moment. The five levels of practice and action for change provide a very useful framework for guiding and evaluating this transformational process.

5. Evaluation of the community-based CAN case study

Has CAN's partnership model with these communities made a difference? Has the needle on food security and sustainability been moved? Some answers can be found in Tables 2 and 3.

6. Discussion and evaluation of the change process

After five years of participatory interaction, learning, and monitoring, agroecology in the partner communities of San Ramon, Nicaragua and Veracruz, Mexico has moved the needle of change in several important ways. Most importantly, it has reduced the "thin months" considerably. Ideally, the goal is to eliminate them completely. Using the *five levels of change* framework, it would appear that the change process is still in its early stages. What began as a project focused on improving the coffee production systems and finding alternative markets that provided a better price for the coffee, mainly as a response to the coffee crisis of the early 2000s, was forced to expand its focus with the outbreak of the coffee rust

Table 2. Summary of selected components of food system change ranked using the "five levels of change" in coffee growing communities in the region of San Ramón, Nicaragua, and Veracruz, Mexico. Data based on 95 households in San Ramón and 135 households in Veracruz, and are the cumulative outputs from 2011–2015 as reported in Putnam *et al.* 2016 and the final project report to Keurig Green Mountain (CAN 2015b).[1]

Selected outputs of food system change	San Ramon, Nicaragua	Veracruz. Mexico
Level 1		
Number of households with garden irrigation systems installed	35	3
Number of water catchment systems installed	9	46
Level 2		
Number of soil fertility and conservation practices implemented[2]	9	13
Number of households adopting Best Agricultural Practices[3] promoted by the project	95	139
Level 3		
Number of wood, fuel, or fruit tree seedlings planted in the community	28,403	2,995
Number of fruit trees planted	5,560	1,875
Number of households diversifying with fruit trees and vine crops, and vegetable crops	95	135
Number of species of heirloom vegetable seed being saved by households	20	18
Area for diversified vegetable production in homegardens or patios in manzanas[4] per household	0.73	—
Level 4		
Percent of households consuming more than six food groups daily over period 2013–2015 (%)	82–100	84–100
Number of food storage and distribution centers (CADA[5]) at a cooperative serving 200 families	7	0
Pounds of maize and beans distributed as food by CADAs	beans: 2415 maize: 2400	0
Number households with functioning homegardens	95	139
Number of farmers markets 2011–2015	21	25

(*Continued*)

Table 2. (*Continued*)

Selected outputs of food system change	San Ramon, Nicaragua	Veracruz. Mexico
Level 5		
Community Seedbanks[6] at a cooperative serving 100 families	7	5
Pounds of corn and bean seed distributed for planting	6600	300
Number of home water filters installed and in use	40	44
Number of home chicken coops constructed	62	94
Number of value added rural enterprises for women or youth established	2	5
Number of women selling in rural enterprises and markets	30	59
Number of youth selling in rural enterprises and markets	10	31
Number of youth leaders participating in project activities	12	8
Number of school gardens functioning and supported by youth leaders	3	3
Number of community or cooperative-based commissions functioning and running food security and sovereignty infrastructure	7	4
Number of community workshops, trainings and exchanges in 2015	16	44

[1] Data for 95 households in San Ramon and 139 households in Veracruz. In most outputs the initial value was at or close to zero.

[2] For example, cover cropping, composting, reduced tillage, etc.

[3] Based on community-based experiences developed by CAN (2015a).

[4] In Nicaragua, approximately 0.7 ha or ~7000 m^2.

[5] CADAs (Spanish acronym) are Food Storage (especially maize and bean grain for consumption) and distribution centres of basic household goods purchased in bulk and housed at the cooperative for cooperative members at a reduced price.

[6] Community Seed banks are part of CADA's, where local seeds for planting are properly stored and made available to cooperative members through a 2:1 exchange and return system.

intensive production of coffee to alternative, agroecologically-based best practices.

Simultaneously, work at Level 4 focused on linking coffee farmers with coffee drinkers, both locally and on the international market. A more

Table 3. Quantitative impacts of change over the five years of collaboration in eight cooperatives and their communities in San Ramón, Nicaragua and four communities in Veracruz, Mexico. Based on 95 households in Nicaragua and 139 households in Mexico.

Indicators of Change	2011	2012	2013	2014	2015
San Ramón					
Months of adequate provisioning	7.4	7.3	8.1	9.8	9.7
Length of "thin months"	4.6	4.7	3.9	2.2	2.3
Dietary diversity score[1]	—	6.6	7.8	7.1	7.4
Percent of households consuming more than 6 food groups daily	12	83	100	83[2]	82[2]
Coping strategies index (CSI)[3]	16.8	11.0	17.5	15.1	8.1
Veracruz					
Months of adequate provisioning	—	10	10	9	10
Length of "thin months"	—	2	2	3	2
Dietary diversity score	—	6.9	6.9	7.3	8.5
Percent of households consuming more than six food groups daily	—	100	84	94	100
Coping strategies index	—	20.0	24.4	9.3	4.4

[1] Dietary diversity scores based on the number of food groups consumed during a given time period, with 12 food groups used as a foundation (based on methods reported in Swindale and Bilinsky (2006)).

[2] These slightly lower numbers can be explained by the drought, which affected some families in the sample who had less access to water to irrigate their gardens.

[3] CSI = Coping Strategies Index = this score measures the variety of behaviours that people implement to cope with scarcity. Lower score means the usage of less severe strategies or less frequency use of strategies, hence less scarcity. Strategies based on activities described in Levels 1–4.

disease that devastated coffee production beginning in 2012. What had begun primarily as a project focused on moving farmers from Level 1 to Levels 2 and 3, quickly pushed beyond to Levels 4 and 5. Early on, multiple important steps were initially taken to transition conventional, input just and democratic process for setting the price of coffee to the growers was established, grounded as much in shortening the commodity chain as it was on bringing all of the voices to the table during price negotiations. But is also was clear that the market for coffee was not the only answer. The coffee crisis of the early 2000s had made it very clear that coffee farmers and coffee growing communities were the most vulnerable when prices fell, and that informed and conscious consumers and roasters operating at Level 5 in their awareness of the injustices and inequalities that the market imposed were needed to make change happen.

By the time the project began in 2010, CAN had already established a close relationship with the farmers, their communities, and the cooperatives to which the farmers belonged. Levels 3, 4, and 5 began to interact together as the need to reduce the thin months became a primary goal. Diversification of the farms was the Level 3 component, which provided additional food for local consumption. But income to support vibrant local livelihoods could only come about at Levels 4 and 5. Home gardens, for example, diversified the farm landscape at Level 3 (Tables 2 and 3), but farmers, most of whom were women, needed the help and incentive of Level 4 mark*et al*ternatives in order to expand their use and diversification. Local farmers markets, exchange within their communities, and the reduction in the need to purchase food have all been coming together.

It is at Level 5, however, where longer-term, more permanent change comes about. Creating opportunity for women to take the lead, and be rewarded, in livelihood diversification, be it installing a homegarden, planting women-managed coffee plots, or building a café and farm store in a nearby urban centre, required a change in culture and custom. There is a story to tell behind this process, with beliefs, values, and actions taking place at Level 5 making it happen. There is also a story to tell about how until some of the Level 5 changes occurred, many of the changes that happened at Levels 3 and 4 would have never been possible. School gardens, youth leadership training, creating awareness of local food customs, expanding understanding of opportunities and needs in communities, and local youth becoming leaders in local projects and other activities that provide options for staying in their communities and becoming part of the change process, are all Level 5 actions that promote and encourage change at all levels in the system.

CAN has been a partner in this change process, operating from a Level 5 perspective that change could occur, and it could result in a major paradigm shift from the very beginning. But rather than push this approach on the communities it was working with, an open, participatory approach was used from the beginning. As agroecologists, CAN researchers had an understanding of sustainability and alternative farming practices and designs. As farmers, the community members had generations of farming experience and local understanding of their environment. These "ways of knowing" came together in the transdisciplinary respect for each other's knowledge and experience. Through broad participation, transparency and trust developed. Experiments were initiated. Observations and monitoring were done jointly, and local capacity among youth to both take data and analyse it was carried out. Findings were shared, discussed, and used to take new steps, try alternative management practices (Level 2), and begin diversifying and redesigning their farms (Level 3). As results (both positive and negative) were generated, more ownership for the change process began to appear as well. This is probably most commonly called "empowerment", and has begun taking place in all members of the communities — men, women, and youth (Level 5).

This "ownership" of the change process is too often very slow to come about. With the 5 Levels of change process, it can happen. An organisation like CAN will not be able to accompany the change process in the communities it works with forever. Eventually, the communities must guide it themselves. But getting fully to Level 5 is a challenging process, with resistance to change all too common. When we do, the paradigm shifts. The local social organisations, capacity building, and sharing of knowledge across social networks then come together to form the resilience rural smallholder communities need. This case study is intended to demonstrate how CAN's approach to agroecology is a mutual learning process with community partners, and that it can reduce vulnerability among farmers who depend so much on a market commodity like coffee while ensuring food security. It also shows how the change process is an evolutionary process, not one of simple adoption.

As a farmer said during an exchange about halfway through this 5-year project:

> "When you first came to our community, you said that we had in our own hands what we needed to change. We did not understand what you meant at that time. But now we do. And we are making the change happen." (Don Pedro, La Pita)

Acknowledgement

The research reported in this chapter has been guided by coauthor Associate Director Heather Putnam, who has been the primary force behind developing the partnerships and coordinating the research with the communities in Nicaragua and Mexico reported here. We are also extremely grateful, for the trust and participation from the dedicated leaders and staff at VIDA AC and the UCA San Ramon, youth leaders in the cooperatives and communities, and all of the farming families with whom we are engaged. We would also like to recognise the contributions of Suraya Arslan, Wendy Godeck and Igor Fridman in supporting data collection over the years.

A preliminary version of this chapter was included in the report "The Agroecology Transition to Food System Sustainability", a working paper of the International Panel of Experts on Sustainable Food Systems (www.ipes-food.org). Permission to adapt parts of this report to this chapter is gratefully acknowledged.

References

Bacon, C.M., 2004. Confronting the coffee crisis: Can fair trade, organic, and specialty coffees reduce small-scale farmer vulnerability in Northern Nicaragua? *World Development* **33**(3), 497–511.

Bacon, C.M., Méndez, V.E., Brown, M., 2005. *Participatory Action Research and Support for Community Development and Conservation: Examples from Shade Coffee Landscapes of El Salvador and Nicaragua.* Center for Agroecology and Sustainable Food Systems, University of California, Santa Cruz, CA, USA, 22 p.

Bacon, C.M., Méndez, V.E., Gliessman, S.R., Goodman, D., Fox, J.A. (Eds.), 2008. *Confronting the Coffee Crisis: Fair Trade, Sustainable Livelihoods, and Ecosystems in Mexico and Central America.* MIT Press, Cambridge, MA, USA.

Bacon, C.M., Sundstrom, W.A., Flores-Gomez, M.A., Méndez, V.E., Santos, R., Goldoftas, B., Dougherty, I., 2014. Explaining the 'hungry farmer paradox': Smallholders and fair trade cooperatives navigate seasonality and change in Nicaragua's corn and coffee markets. *Global Environmental Change* **25**, 133–149.

Community Agroecology Network. 2015a. *Best Practices Guide* (In Spanish) http://www.canunite.org/newsite/wp-content/uploads/2015/03/buenas_practicasWholeGuide.pdf

Community Agroecology Network. 2015b. Phase 2 Year 3-Annual Report. Project: *Youth Leadership and Education for Sustainable Agriculture and Food Sovereignty.* http://www.canunite.org/our-work/projects-2/youth-leadership-food-sovereignty-2/

Friedmann, H., 2005. From colonialism to green capitalism: social movements and the emergence of food regimes. In: Buttel, F.H., McMichael, P. (Eds), *New Directions in the Sociology of Global Development: Research in Rural Sociology and Development*, Volume 1. Elsevier, Amsterdam, The Netherlands.

Ghosh, J., 2010. The unnatural coupling: food and global finance. *Journal of Agrarian Change* **10**(1), 72–86.

Gliessman, S.R., 2015. Converting to ecologically based management. *Agroecology: The Ecology of Sustainable Food Systems.* 3rd Edn. CRC Press/Taylor and Francis Group, Boca Raton, Florida, USA, pp. 277–285.

Gliessman, S.R., 2016. Transforming food systems with agroecology. *Agroecology and Sustainable Food Systems* **40**, 188–190.

Jaffe, R.M., Bacon C. M., 2008. From differentiated coffee markets toward alternative trade and knowledge networks. In: Méndez, V.E., Bacon, C., Cohen, R., Gliessman, S.R. (Eds.), *Agroecology: A Transdisciplinary, Participatory, and Action-oriented Approach.* CRC Press/Taylor & Francis Group, Boca Raton, FL, USA, pp. 311–336.

Kingdon, S., Pain, R., Kesby, M. (Eds.), 2007. *Participatory Action Research Approaches and Methods.* Routledge, Oxon, England, UK.

Méndez, V.E., Bacon, C.M., Olson, M., Petchers, S., Herrador, D., Carranza, C., Trujillo, L., Guadarrama-Zugasti, C, Cordon, A., Mendoza A., 2010. Effects of Fair Trade and organic certifications on small-scale coffee farmer households in Central America and Mexico. *Renewable Agriculture and Food Systems* **25**(3), 236–251.

Méndez, V.E., Bacon, C.M., Cohen, R., 2013. Agroecology as a transdisciplinary, participatory, and action-oriented approach. *Agroecology and Sustainable Food Systems* **37**(1), 3–18.

Putnam, H., Brown, J.C., 2013. The importance of process in achieving food sovereignty: Participatory Action Research (PAR) in coffeelands of Nicaragua. In: Schanbacher, W.D. (Ed.), *The Global Food System: Issues and Solutions*. Praeger, Santa Barbara, California, USA, pp. 243–265.

Putnam, H., Godek, W., Kissmann, S., Luckson Pierre, J., Alvarado Dzul, S.H., Calix de Dios, H., Gliessman, S.R. 2014. Coupling Agroecology and PAR to Identify Appropriate Food Security and Sovereignty Strategies in Indigenous Communities. *Agroecology and Sustainable Food Systems* **38**, 165–198.

Putnam, H., Cohen, R., Jaffe, R.M., 2016. Agroecology as a food security strategy in coffee-growing communities: Opportunities and challenges in San Ramon, Nicaragua. In: Méndez, V.E., Bacon, C., Cohen, R., Gliessman, S.R. (Eds.), *Agroecology: A Transdisciplinary, Participatory, and Action-oriented Approach*. CRC Press/Taylor & Francis Group, Boca Raton, FL, USA, pp. 193–216.

Swindale, A., Bilinsky, P., 2006. Household dietary diversity score for measurement of household food access: Indicator guide (v. 2). Food and Nutrition Technical Assistance Project, Academy for Educational Development, Washington D.C., USA. www.fantaproject.org/publications/hdds_mahfp.shtml.

Wezel, A., Bellon, S., Dore, T., Francis, C., Vallod, D., David, C., 2009. Agroecology as a science, a movement and a practice. A review. *Agronomy for Sustainable Development* **29**(4), 503–515.

Chapter 9

Assessing the Impact of Management Interventions in Agroecological and Conventional Cropping Systems Using Indicators of Sustainability

Cathy Hawes

James Hutton Institute, Dundee, Scotland, UK

1. Introduction

There is an increasing pressure on the agricultural industry to provide large quantities of high quality food while maintaining or improving environmental and ecological sustainability. The intensification of agricultural production since the 1950s has resulted in increases in yield, but only at heavy environmental cost: increased greenhouse gas emissions, increased water pollution through loss of agro-chemicals from fields, loss of arable plant and animal biodiversity with negative effects on ecosystem function and services, and degraded physical structure of the soil causing soil erosion and compaction.

Agricultural land in the EU now covers almost half of the total land area with over 100 million hectares under arable cultivation, 65 million hectares permanent grassland, and 12 million hectares of perennial crops (Henle *et al.* 2008). The negative impact of intensification over such a large proportion of the EU has raised serious concerns about the long-term impacts on biodiversity, ecosystem service provision, and the sustainability of food production systems. This is of particular concern since the need to increase food production in the next 50 years to match the demands of rising human population (Defra 2011), is in direct conflict with the need to conserve biodiversity and protect long term sustainability.

In an effort to achieve a balance between these potentially conflicting goals, many 'sustainable' management practices have been developed that aim to reduce the reliance on non-renewable inputs, improve the efficiency of crop production, and enhance the functioning of agroecological systems. Improvements to the agricultural environment in Europe are implemented primarily through agri-environment schemes (AESs) which were developed following the CAP reform in the late 1980s and early 1990s (Carey *et al.* 2003). These measures are designed for example to halt or reverse the decline in farmland biodiversity and encourage farmers to protect and enhance the farmland habitat by providing subsidies to adopt environmentally-friendly farming techniques. The overall objectives are to counteract the negative effects of modern agriculture by reducing pollution from nutrient and pesticide losses, protecting biodiversity, restoring landscapes and preventing rural depopulation (Kleijn and Sutherland 2003, Herzog *et al.* 2005).

Examples of agri-environment schemes include:

- Environmentally favourable extensification of farming, based on the principles of integrated farm management and organic agriculture. The United Nations Environment Programme proposed targets to increase organic agriculture to 5–10% of the EU agricultural area by 2020 as a means to reduce emissions, create carbon sinks, preserve soil organic matter, and increase biodiversity by not using mineral fertiliser or chemical pesticides. However, organic arable agriculture tends to be relatively low yielding and may therefore conflict with food security targets to meet the growing demand. Agroecological practices such as leaving stubble over winter, sowing cover crops for wild bird food, leaving uncropped strips and conservation headlands are also options for improving biodiversity and system function, regardless of organic status.

- Loss of habitat diversity is a key factor driving declines in farmland biodiversity (Vickery *et al.* 2004). Management options to rectify this include conservation of high-value habitats and their associated biodiversity, particularly semi-natural grasslands and low intensity pasture systems, wetlands, and woodlands. However, in many cases, this is still insufficient to support viable populations of many farmland species, so other options include the creation of new wildlife habitats for enhanced biodiversity through field margin management, beetle banks, uncropped headlands, hedgerows, riparian buffers and set aside (Hof and Bright 2010).
- Preservation of landscape and historical features such as hedgerows, ditches, woods, and meadows (Herzog *et al.* 2005). These historical landscapes are highly valued by society as is the farmland wildlife they support. The aim is to generate a diverse mosaic of habitats in the landscape and increase the connectivity between them.

Agri-environment schemes, of one form or another, now cover approximately 25% of the farmland in Europe (Taylor and Morecroft 2009) and have cost around €20 billion, or 22% of the EU expenditure for rural development between 2007 and 2013 (EU 2015). Given the high level of public expenditure on these schemes, they need to be accountable and demonstrate good value for money. There is, therefore, an urgent need to assess their efficacy to either justify their continuation in their current form, or to improve on them in the development of new schemes that are better able to deliver the desired goals.

Systematic evaluations of agri-environment schemes are relatively rare in the peer reviewed literature and most impact assessment studies have demonstrated mixed effects on species populations (positive, negative, and neutral, relative to the background rate of decline). The direction of effect depends largely on the specific options that were selected for study, the target habitat and the landscape context (Kleijn *et al.* 2004, Bengtsson *et al.* 2005, Knop *et al.* 2006). For example, conversion of farms to organic production, reduction in arable production, and increase in permanent pasture and grass leys has been shown to have a positive effect on grassland plant species diversity, moth populations and carabid abundance (Taylor and Morecroft 2009) and positive outcomes have also been reported for grassy field margins and hedgerows in supporting hedgehog populations (Hof and Bright 2010). At a farm level, agroecological practices in general can be associated with increases in numbers of birds (Vickery *et al.* 2002), bumblebees (Pywell *et al.* 2006) and small mammals (Tattersall *et al.* 2002). However, in an evaluation of Dutch agri-environment schemes in 78 paired fields, Kleijn *et al.* (2004) showed no effect on in-field and field margin plant species richness or bird numbers, but a positive outcome for the species richness of pollinators (bees and hoverflies). Positive effects were only detected for half of the species groups that were assessed (Kleijn *et al.* 2006). Carvell *et al.* (2007) also tested a range of Environmental Stewardship Schemes options in England designed to restore foraging and nesting habitats for bumblebee species. Not all options were equally valuable, and the main intervention to have a significant impact on bumblebee numbers was uncropped margins sown with a legume wildflower mix, which performed better than natural regeneration, conservation headlands, or sown grass mixes. Marshall *et al.* (2006) compared 21 farms with 6 m grass margins at the edges of arable fields with 21 control sites. Positive effects were detected for field margin herbaceous flora, bees and orthoptera, but there were no effects on birds, liniphiid spiders, or carabids. Similarly, Herzog *et al.* (2005) demonstrated a beneficial effect of wetland Ecological Compensation Areas in Switzerland on birds and plants but no effect for hedgerow compensation areas. Fuentes-Montemayor *et al.* (2011) also found no effect of hedgerows on moth numbers, but increasing the percentage of semi-natural habitat (species rich grasslands in particular) at a local scale (within 250 m) had a significant positive effect on moths. As a final example, the grey partridge has received a lot of attention in assessment studies as an important game bird and one of the 19 indicator species used by the UK government to assess progress with the target to reverse the decline in farmland bird populations by 2020. For this particular indicator, conservation headlands and beetle banks have been shown to be valuable management options in increasing local populations (Ewald *et al.* 2010), but other studies have shown that these specific options tend not to have any positive effect on other indicator groups, such as hedgehogs, moths, or beetles. One 'recipe' cannot therefore benefit all taxa and it is evident from these mixed results that what is optimal for one purpose may be suboptimal or even counter-productive for another (Donald and Evans 2006, Carey *et al.* 2005).

Although examples can be found that demonstrate positive effects of specific conservation strategies for individual target taxa, there is very little empirical evidence in the literature that agri-environment schemes in general have any national scale benefit to farmland wildlife or ecosystem services. For example, Perkins *et al.* (2011) showed a positive effect of targeted agri-environment schemes (those specifically designed to benefit a particular species or habitat) on the corn bunting, but no change under entry level schemes (those designed to enhance farmland habitats in general, SEERAD 2003). This suggests that positive outcomes are only likely where the management option has been specifically designed for the target indicator, as in the grey partridge example above. Most conservation management options taken up by farmers are those that cause least disruption to normal cropping practice rather than those specifically designed to conserve a particular taxonomic group. This results in a wide range of different options being selected for adoption by farmers volunteering to take part in the scheme, and a very mixed and inconsistent impact. This highlights the potential for conflicting goals in farmland management where the requirements necessary for the conservation of different components of farmland biodiversity are at odds with each other and may be very different from management recommendations designed to protect environmental health or enhance food security (Henle *et al.* 2008). Carey *et al.* (2005) tested a multidisciplinary approach for assessing agroecological practices across 500 farms, involving landscape architects, social scientists, and ecologists and found that positive outcomes according to experts in one discipline may be negatively rated by experts in another.

There is therefore an urgent need to design evaluation studies for the introduction of any new agricultural practice or management strategy to provide an assessment of the impact on the whole system, rather than just targeting specific elements in isolation. Monitoring in this way can provide a feedback mechanism for improving the outcome of future management for environmental and conservation benefit. Indicators for monitoring impact should therefore be selected that represent all key components of agricultural systems. In this chapter, we present a systems approach for monitoring impact that takes account of ecosystem functions and the trade-offs between them. This whole-systems approach not only deals with biodiversity related indicators that have historically been the focus for agri-environment schemes and agroecological practices in general, but also takes into account the wider impacts of a change in management or intervention on all components of the system that relate to sustainability, including economic and environmental functions as well as ecological processes.

2. A systems approach for impact assessment

A framework for system design, implementation and impact assessment has been established as part of the James Hutton Institute's Centre for Sustainable Cropping (CSC) at Balruddery Farm, Dundee, Scotland (www.hutton.ac.uk/csc); a long-term experimental platform for interdisciplinary research on arable sustainability. The structure is outlined in Table 1 and will be used here as a case study to illustrate the selection of indicators on the basis of the ecosystem processes or services that they represent. The framework is based on a whole-systems approach so that trade-offs and complementarities between system components can be identified and accounted for. The first step was to define the overall objectives and requirements of the production system: (i) enhancing biodiversity for conservation and ecosystem service provision such as nutrient cycling, pollination, and pest control, (ii) reducing environmental footprint by minimising losses through runoff, erosion, and greenhouse gas emissions, and (iii) maintaining or enhancing yield and product quality for economic sustainability and food security.

Once the goals were defined, the next step was to identify a suite of management interventions that can be put together an integrated cropping system to achieve the targets as set. These were selected on the basis of evidence from published data and field trials for significant positive effects on one or more of the desired outcomes. Finally, the cropping

Table 1. The framework for designing a new cropping system established at the CSC, Balruddery Farm, Scotland. Ecological, environmental, and economic objectives were set and management changes to meet each were proposed, along with a set of indicators for monitoring efficacy of the sustainable system in achieving the desired outcomes.

Objective	Management change	System function	Indicator sets
(1) *Ecological Sustainability:* Enhancing biodiversity for provision of ecosystem services (nutrient cycling, pollination, pest control)	Reduced herbicide and specific targeting of competitive weeds to achieve a 5–10% cover of beneficial broadleaved weeds in-field. Reduced/threshold pesticide application to minimise non-target effects. Conservation headlands with no fertiliser or herbicide application. Multifunctional sown margins for increasing resource availability for natural enemies and pollinators.	(1) Weeds supporting arable food webs	(A) Arable weed seedbank diversity and species composition. (B) Emerged weed and margin flora composition and abundance.
		(2) Regulation of pest and disease populations	(C) Natural enemy activity and abundance in relation to relative abundance of functional groups in arable invertebrate food webs. (D) Crop/soil borne pest and disease incidence and spread.
		(3) Pollination	(E) Pollinator abundance and activity.
		(4) Decomposition and nutrient cycling	(F) Earthworm abundance and microbial diversity, in relation to litter decomposition rates.

(*Continued*)

Table 1. (*Continued*)

Objective	Management change	System function	Indicator sets
(2) *Environmental Sustainability*: Minimising losses through erosion, runoff, leaching, and GHG emissions	Non-inversion tillage to improve soil physical structure Stubble and organic matter incorporation Controlled traffic to reduce compaction Cover cropping to retain nutrients overwinter for subsequent crop Variable rate agrochemical inputs tailored to soil nutrient supply and disease incidence Engineered riparian buffers and multifunctional margins to reduces losses from field	(5) Erosion control (6) Enhancing soil water quality and reducing losses (7) Air pollution mitigation	(G) Soil physical structure, water holding capacity, carbon content, sediment loss. (H) Leaching of nutrients and agrochemicals into soil water, runoff, plant nutrient concentrations in soil. (I) Carbon footprint calculations from inputs and machinery use. (J) Direct measurement of greenhouse gas emissions from soil and plants.
(3) *Economic Sustainability*: Maintaining or improving yield, food security, and human health	Alternative sources of nutrients to compensate for reduced mineral fertiliser Enhancing uptake of existing plant nutrients through improved soil structure Reducing loss of available nutrients through improved soil structure IPM strategies to compensate for reduced crop protection chemical applications	(8) Primary productivity and production efficiency	(K) Crop health, productivity, yield, and product quality. (L) Production efficiency and financial margins.

system was implemented in a split field design across six fields in a crop rotation of potato, winter wheat, winter barley, winter oilseed, beans, and spring barley. Direct comparisons are then made between the new integrated cropping system in one half of the field and current conventional practice in the other. At the CSC, the sustainable system differs from conventional practice by including measures to: (i) improve soil biophysical structure (non-inversion tillage, controlled traffic, addition of municipal green waste compost and straw incorporation); (ii) reduce losses of soil, water and nutrients (cover cropping, tied-ridging, tramline management, riparian buffers); (iii) provide alternative sources of plant nutrients (biological nitrogen fixation through clover undersowing) and; (iv) enhance farmland biodiversity (wildflower margins, reduced herbicide inputs and threshold crop protection applications). The conventional system follows standard commercial practice (conventional plough, straw baled and removed, standard fertiliser and crop protection inputs). The effect of the two cropping systems on the overall goals is assessed through measurement of a suite of indicators, selected to represent all key components of the agricultural system. Evaluation of the success in terms of whether the overall goals have been achieved can then be made, with a view to improving on the outcome through iterative cycles of implementation, monitoring and testing.

It is clear that the first two objectives on biodiversity and environmental protection are in conflict with the third on maintaining high yields (at least when viewed in the short-term), and a balance must therefore be struck between managing for the environment (e.g. through agroecological practices) and managing to maximise yield output. Environmentally sensitive management may have negative consequences for some elements of the system and positive effects on others. Most of the previous work on impact assessment of a new management intervention (outlined above) focusses on single indicators, generating mixed results depending on which part of the system is represented by the chosen indicator. There is therefore a need to apply a whole systems approach for assessing the impact of any change in management. By adopting a systems approach, trade-offs between different components of the system can be quantified and optimal solutions identified.

3. Indicators of arable sustainability and system function

The selection and use of arable sustainability indicators (indicator sets A – L), as developed for the CSC and related studies, is described here with reference to the system functions (1–8) that they represent (see Table 1). These system processes and components contribute directly and indirectly to a range of ecosystem services, and are grouped into indicators of ecological, environmental and economic aspects of sustainable production systems.

3.1. *Ecological sustainability*

Ecological sustainability is defined here as the resilience and long-term persistence of viable populations of organisms that interact through competition, facilitation, mutualism, and predation within a given habitat. Experimental and theoretical studies of the relation between the diversity of organisms, the strength of the interactions between them and properties of the system of which they are part, have indicated that diversity enhances a wide range of ecosystem services including nutrient cycling, population stability, system productivity, and resilience to disturbance or changing environmental conditions. The focus is on three main functions provided by the interaction between organisms in their environment: primary production supporting resilient arable food webs, regulation of crop pest and disease populations, pollination of insect pollinated crops to enhance yield, and nutrient cycling by invertebrate and microbial decomposers.

3.1.1. System function 1: Weeds supporting arable food webs

Reduction in the availability and quality of plant food resources (Biesmeijer *et al.* 2006) through the combined effects of habitat loss, habitat fragmentation, and intensive agricultural practices (Kremen *et al.* 2002) has a negative impact on the diverse array of associated herbivores, predators, and parasitoids that depend on them for food and shelter (Siemann *et al.* 1998, Altieri *et al.* 1999, Koricheva *et al.* 2000, Haddad *et al.* 2001, Hawes *et al.* 2003). Weeds, and dicotyledon weeds in particular, represent a highly valuable resource to primary consumers within arable fields, supporting up to 10 times the biomass of herbivores per unit plant mass compared to crop plants (Karely *et al.* 2011). Diversity in this weed layer results in increased diversity of herbivore consumers (Marshall *et al.* 2003), and some studies have shown that this effect is propagated through the food web to the third trophic level (e.g. Hawes *et al.* 2003, 2010, Taylor *et al.* 2006). Declining weed diversity is therefore likely to have a detrimental effect on the composition of arable food webs with serious consequences for ecosystem function (Loreau 1998, Hector *et al.* 2001, Scherer-Lorenzen *et al.* 2003). A balance of different functional types of organisms in the arable food web is necessary to maintain the long-term functioning and sustainability of that system.

Indicator set A: Arable weed seedbank

Within arable systems, the weed seedbank is the primary source of biological diversity and crucial to the functioning of arable systems. The seedbank is therefore a valuable reference in studies of ecological impact (Firbank *et al.* 2003, Heard *et al.* 2003, Perry *et al.* 2003, Squire *et al.* 2009) and in the conservation of arable plants. In particular, the persistence of seeds in the seedbank confers a degree of resilience to the arable weed community. However, the weed flora has declined over the past 50 years, along with other indicators of diversity, the main cause being the increased use of herbicide and the competitive suppression of weeds by autumn-sown crops (Robinson and Sutherland 2002, Marshall *et al.* 2003). Comparisons of records back to the start of the century (e.g. Brenchley 1918) indicate that the most abundant have remained widespread, but the rarer arable weeds have declined, even to extinction in some parts of the UK (e.g. Preston *et al.* 2002). Although more resistant to change than the emerged weed flora, the seedbank can increase rapidly if control is relaxed over several years, or decline by up to 50% a year if seed return is severely reduced or prevented (Brenchley and Warrington 1933, Roberts 1958, 1962, Wilson and Lawson, 1992). Such responses imply that the intensification of management has reduced the general population density of seedbanks over a wide area, but populations, except of the rarest arable weeds, can recover under suitable management. At the CSC, the arable weed seedbank is sampled every year following standardised protocols described in Heard *et al.* (2003). Principle Component Analysis on the number of seeds per species sampled from each half field (sustainable cropping versus conventional) over a five-year period shows distinct differences in seedbank species composition between fields (Figure 1). Principle Component scores show six clusters representing the samples taken from each of the six fields each year. This indicates stronger similarity between samples taken from the same field in different years than samples taken from the same crop in different fields, or the same cropping system in different field halves. This trend may alter over longer time scales, but indicates the high degree of resilience of the arable weed seedbank to change in management over a five year period.

Indicator set B: Emerged weed flora

Weeds can be significant components of the vegetation in arable systems and are usually managed intensively to reduce crop–weed competition

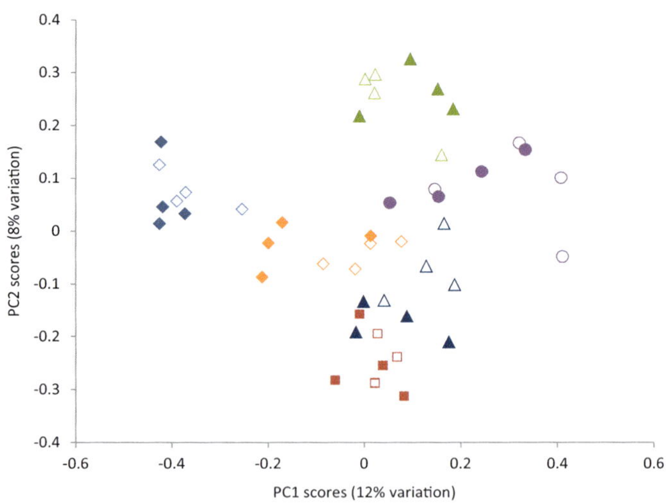

Figure 1. Principle component analysis of the arable weed seedbank sampled in each field (coded by colour) every year for four years at the Centre for Sustainable Cropping long-term experiment. Closed symbols = field half managed under a sustainable cropping system; open symbols = field half managed conventionally.

and increase crop yield. High weed population densities can lead to a serious reduction of crop yield (indicator set K), but the buried seeds and emerged plants are a direct source of energy and nutrition for arable food webs populated by species of decomposers, saprophytes, herbivores and predators that are not sustained by "cleaned" crop monocultures. These plants can also stabilise soil and reduce leaching by taking up excess fertiliser or nutrients released by decaying crop matter. Arable weeds provide a greater variety of form, composition, and function than the few crop species that dominate arable land (Norris and Kogan 2000, Hawes *et al.* 2003) and are therefore important in maintaining diverse farmland habitats. At the CSC, herbicide inputs are reduced to half of those under conventional management to allow a greater abundance of weeds, aiming at 5–10% ground cover. Where threshold levels are reached, competitive weed species are specifically targeted. Preliminary data indicate significantly higher weed densities under sustainable management and significant difference between crops: spring barley and broadleaved crops had a greater diversity and abundance of understorey weeds than potato or winter cereals (wheat and barley). There was no detectable effect of weed biomass on crop yield, suggesting that weed management is maintaining levels below economic threshold.

3.1.2. System function 2: Regulation of pest and disease populations

The presence of specific weed species can stabilise pest and beneficial natural enemy populations through provision of a greater range of microhabitats and sustained resource availability throughout the growing season. This can occur through a reduction in host apparency (herbivores are less likely to find and remain on their host plants in mixed stands compared to monocultures) and an increase in abundance of natural enemies due to greater availability of alternative prey providing continuity of resource through the season (Root 1973, Andow 1991).

Indicator set C: Predation and parasitism rates

Functional groups of invertebrates are essential for maintaining ecosystem services, particularly natural enemies that can help control populations of crop pests, and detritivores that aid the breakdown of organic matter and release of nutrients in a form suitable for uptake by plants. Natural enemy abundance is assessed as part of the monitoring programme for these arable foodwebs and their abundance can be related to the densities of alternative prey and potential competitors with which they interact.

Invertebrates living in and around arable fields all contribute to the resilience and functioning of the arable system and can prove sensitive indicators of habitat quality. A pilot study for their sensitivity to detect relatively rapid ecological changes, suggested that the most useful indicator groups for detecting direct effects of management change in arable fields are plant bugs (Heteroptera), springtails (Collembola), carabid beetles (Carabidae), and the larvae of butterflies and moths (Lepidoptera), and sawflies (Haughton personal communication). At the CSC platform, invertebrate indicators are surveyed according to standardised protocols using pitfall trapping (Brooks *et al.* 2003) and vortis suction sampling (Haughton *et al.* 2003).These data show a strong association between resource and consumers and indicate two distinct though interacting groups within arable invertebrate foodwebs: one with strongest regression relations between detritivores, omnivores and generalist predators found in winter sown cereals where grass weeds tend to dominate; and another between sap feeding invertebrates and their more specialist predator and parasitoid natural enemies, based on dicotyledonous weeds that are more abundant in spring sown crops (Figure 2, adapted from Hawes *et al.* 2010).

Whether or not, natural enemy abundance patterns translates to a reduction in pest pressure on the crop, is measured using bait cards for predation rates by generalist predators (Tolonen 1995) or caged and uncaged aphid infested plants to estimate parasitism rates (Safarzoda *et al.* 2014).

Indicator set D: Crop pathogens — Disease incidence and spread

Integrated pest and disease management (IPDM) strategies, including disease forecasting for potato blight and pest suppressive soils through strategic use of cover crops, are being introduced to the sustainable cropping system to enhance disease control without the need for increased use

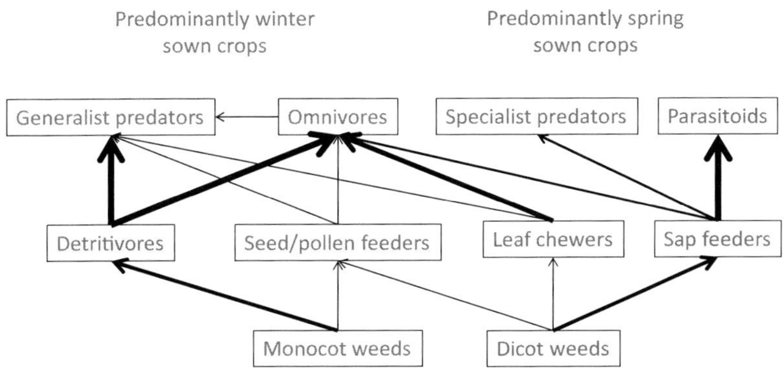

Figure 2. Invertebrate foodwebs in UK arable fields (from Hawes *et al.* 2010) showing two main interacting groups based on the strength of the regression relation between each resource-consumer pair: arrows indicate the movement of resource from food to consumer; line width represents the strength of the interaction as indicated by the variance ratio, $p < 0001$). These plant and invertebrate groups are sensitive to changes in management, disturbance and habitat quality and are therefore valuable indicators of biodiversity.

of crop protection chemicals. Effective deployment of IPDM strategies require close monitoring of pathogen populations. Disease scoring for *Rhynchosporium* in spring and winter barley crops during the 2013–2015 growing seasons indicated significantly higher scores in the sustainable management compared to conventional treatments, however scores well below critical thresholds and therefore not yield limiting. Populations of soil borne potato pathogens (*Spongospora subterranea* (powdery scab), *Rhizoctonia solani* (black scurf), *Helminthosporium solani* (silver scurf), and *Colletotricum coccodes* (black dot)) are also monitored throughout the rotation to investigate the interaction between pathogens, agronomic practice, and potato cultivar. Once inoculum is detected in the soil, its persistence and spread is being tracked through the rotation. There are some early signs of a build-up of inoculum in the sustainable cropping system, possibly due to increased organic matter content of the soil. As part of the iterative development of the sustainable cropping system, IPDM strategies to mitigate these increases in pathogens will be investigated and the most effective will be incorporated into the management package. Trends in pathogen populations will continue to be monitored throughout this process.

3.1.3. System function 3: Pollination

Declining insect pollinator numbers in the UK and worldwide (Goulson *et al.* 2008) have the potential to seriously affect the productivity of insect-pollinated crops (Kevan and Phillips 2001). For example, the annual global economic value of insect pollination is estimated to be $153 billion (Gallai *et al.* 2009) and loss of pollinator services could reduce worldwide crop production up to 8%, necessitating greater agricultural intensification (Aizen *et al.* 2009). Decreasing pollinator numbers could have consequences not only for the productivity of agricultural crops but also for population sizes of wild plant species. In Britain, parallel declines in pollinators and insect-pollinated plants have been demonstrated (Biesmeijer *et al.* 2006), and a reduction in pollination and seed set of fragmented native plant populations has been reported (Rathcke and Jules 1993). Pollinators are often resource-limited and major declines are driven by both land-use and climate change (Kremen *et al.* 2002, Williams *et al.* 2009). A major cause of pollinator decline is likely to be the reduction in the availability and quality of plant food resources (Biesmeijer *et al.* 2006). Alternative low intensity management systems that increase resource availability and quality for pollinators are therefore likely to promote pollinator diversity through niche differentiation. However, there is little quantitative data on the impact of alternative management strategies in the arable environment to inform policy and to advise landowners and conservation groups on improving insect pollinator diversity, abundance and pollination services.

Indicator set E: Pollination activity

Pollinators are monitored using standard transect walks along field margins and into flowering crops to compare activity levels under different field and management conditions in relation to resource availability (Haughton *et al.* 2003). Caged and uncaged flowering bait plants can also be used to relate activity levels to pollination rates (Ne'eman *et al.* 2010).

3.1.4. System function 4: Decomposition and nutrient cycling

The diversity and abundance of decomposers (microorganisms and macroinvertebrates) in arable soil has an impact on a range of system functions, particularly carbon turnover and nitrogen dynamics (Marschner *et al.* 2003), and is therefore important in low input arable systems where a sustainable nutrient supply for crop growth is essential for maintaining yields. Positive correlations can often be seen between soil microbial biomass and the organic matter content of field soils. Similarly, mineral nitrogen pools are found to increase with soil microbial biomass-carbon content of soils, maintaining a constant C:N ratio. Agricultural systems which receive high inputs of organic matter have augmented labile-C pools. Considering this, and the observed correlations, greater microbial activity is likely in soils with organic matter amendments and it may be predicted that this leads to increased N-supplying power of soil.

Indicator set F: Litter decomposition rate in relation to microbial diversity and detritivore abundance

Significantly increased soil microbial biomass carbon and microbial activity have been detected in the sustainable cropping system at the CSC where there are elevated levels of organic matter. Litter decomposition is measured as reduction in dry weight of barley straw, buried in mesh bags in the top 20 cm of soil after sowing in April and removed just before harvest in August (Knacker *et al.* 2003). Litter decomposition within a single year (2014) was significantly greater in the sustainable cropping system ($p < 0.001$) and significantly correlated with soil organic matter content ($p < 0.001$) (Figure 3).

3.2. *Environmental sustainability*

Environmental sustainability in arable systems may be defined as the ability of the production system to reliably maintain yields over long periods without requiring increased rates of agrochemical or non-renewable inputs. This can only be achieved by minimising the loss of resources (soil, nutrients, water) from the system and managing the agroecosystem to allow maximum efficiency in conversion of resources to product. Losses of soil, water, and nutrients from the system occur through erosion, leaching and gaseous emissions.

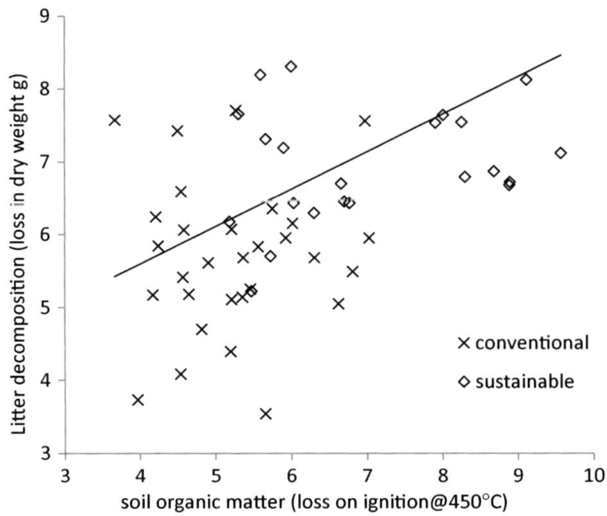

Figure 3. Relation between soil organic matter and litter decomposition across sustainable and conventional cropping systems at the CSC within the 2014 growing season (April–August).

3.2.1. System function 5: Erosion control and nutrient retention

The average annual rate of soil loss by erosion on arable land across Europe is estimated within the range of 3–40 t ha^{-1} yr^{-1}, but variability is high and extreme events exceeding 100 t ha^{-1} yr^{-1} have been recorded (Verheijen *et al.* 2009). Average nitrogen losses through soil erosion across Europe have been estimated at 60 kg N ha^{-1} yr^{-1} (calculated from Pimentel *et al.* 1995). This raises concerns about on-site nutrient depletion, decreased soil aggregation and loss of productivity, and off-site impacts through sedimentation and eutrophication of downstream water bodies (Rowan *et al.* 2012).

Indicator set G: Soil physical structure and carbon content

Management interventions to improve soil structure at the CSC include: organic matter incorporation from straw residue, municipal waste compost and cover crops, reduced or non-inversion tillage, tied ridging between potato drills to improve water infiltration (Ngetich *et al.* 2014), and limiting compaction through tramline management and controlled traffic systems. Soil strength and carbon content are used as indicators of soil physical structure to monitor the effect of management on erosion risk.

Soil strength is assessed at the CSC using the *in situ* shear vane test, where higher readings indicate greater resistance to root growth. Preliminary data suggest no significant difference in shear strength between ploughed and non-inversion tillage soils. Although data from more than a single year is required before conclusions can be drawn, these initial measurements suggest that organic matter incorporation in the sustainable cropping system may help ameliorate any increase in compaction that may result from reduced tillage. The carbon content of the soil is significantly higher ($p < 0.001$) under sustainable management (3.3% C/W) after four years of organic matter incorporation (green waste municipal compost applied at 35 t ha^{-1} yr^{-1}) compared to conventional management with no organic matter addition other than crop root residue (2.2% C/W). The effect of improvements in soil structure are illustrated in Figure 4, showing differences in water logging between conventional and sustainable cropping systems after heavy rain.

(a) (b)

Figure 4. Two halves of the same field: increased carbon input to the soil, together with non-inversion tillage as part of a sustainable management package in one field half (a) at the CSC reduces water logging relative to conventional management in the other half (b) of a winter wheat field (crop just emerging). Photo was taken on 13 November 2015 after 80 mm rain in the previous 14 days.

After four years of management under the sustainable cropping system, winter sown crops show faster rates of development relative to the conventional system. This is likely to be due to improved soil physical conditions resulting in warmer soil temperatures at establishment and in the early part of the growing season. This is an example of a trade-off between different system components, where a negative impact on the crop health through increased populations of soil borne pathogens (indicator set D) may be offset by the positive effect of increased carbon on soil physical structure and plant growth.

3.2.2. System function 6: Enhancing soil water quality

Average nitrogen losses through leaching from arable land across Europe are estimated to be around 36 kg N ha^{-1} yr^{-1} (Ngetich *et al.* 2014) resulting in a decrease in crop production efficiency and increases in water pollution through eutrophication. Movement of nitrogen and other agrochemicals to watercourses may be reduced by margin vegetation management, such as sown buffers to capture runoff and subsurface flow (Stutter *et al.* 2012). Plant material from the buffers can then be cut and used as a green manure to recycle trapped nutrients. Studies are underway at the CSC and elsewhere to test this approach and to quantify the effective recovery of plant nutrients. Initial trends suggest that different components of the buffer system have varying abilities to take up and store nutrients. Engineered buffers have also been established comprising ditches to trap sediment loss and slow the movement of water out of the system (Figure 5). Willow and alder trees are planted between the ditch and natural water course to take up excess nutrients. These provide added biodiversity benefit to the farm and can be harvested for biofuel thereby recycling nutrients otherwise lost from the system.

Indicator set H: Soil water quality

Ceramic suction cup lysimeters to depths of 10 cm, 30 cm, and 50 cm are used to sample soil water within the cropped area and either side of the buffer zone to assess the extent to which applied mineral fertilisers and other agrochemicals are lost from the system (Webster *et al.* 1993). This information can be related to timings of farm operations and rainfall events.

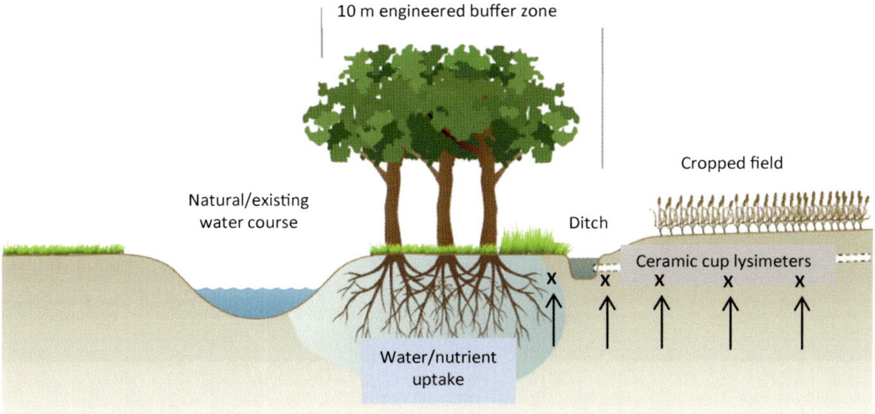

Figure 5. General design principles of the engineered buffer strips (courtesy of Marc Stutter, BUFFERTECH).

3.2.3. System function 7: Air pollution mitigation

The latest UK greenhouse gas inventory estimates the proportion of the nation's overall carbon footprint due to agriculture to be around 8%, 75% of which is directly related to fertiliser use from production, application and direct nitrous oxide emissions (Choudrie *et al.* 2008). Losses of nitrogen through gaseous emissions can amount to 52.5 kg N ha^{-1} yr^{-1} on average across European arable crops (Leip *et al.* 2008). In a survey of organic, integrated and conventional commercial farms throughout the east of Scotland, no differences were detected in the carbon footprint of crop production after the differences in nitrogen use were taken into account (Hillier *et al.* 2009). This study highlighted key opportunities for carbon mitigation, particularly through reductions in the use of mineral nitrogen fertiliser.

Indicator set I: Carbon Footprint

There are a number of online tools for calculating carbon footprint. We use The Cool Farm Tool version 2.0 (coolfarmtool.org: Unilver UK Ltd, University of Aberdeen and the Sustainable Food Lab) which calculates the carbon footprint in terms of kg CO_2 equivalents on the basis of fertiliser inputs, direct emissions of nitrous oxide, application of crop protection chemicals, crop residue management, tillage, cover cropping and fuel use, taking into account soil properties and energy required for the production of agrochemicals (particularly N from the Haber Bosch process). Output is analysed as Carbon Equivalents per hectare and per tonne of harvested product to account for any differences between cropping systems in crop yield. Off farm transport is not included because the platform is an experimental set-up comparing two management systems on adjacent field halves, and does not therefore reflect normal commercial practice beyond the field gate. Although there is not sufficient replication to analyse these data statistically until the end of the first full crop rotation, initial calculations suggest that the carbon footprint of crop production under the sustainable cropping system is on average 46% lower per kg product than under the conventional cropping system for winter sown crops and 72% lower for spring crops.

Indicator set J: Direct measures of GHG emissions

Gas samples of methane (CH_4), carbon dioxide (CO_2), and nitrous oxide (N_2O) were collected using gas exchange boxes within the cropped area to compare emissions from the soil and crop plants in the two contrasting cropping systems. Soil temperature, conductivity and moisture content were also recorded at each sample location and data were collected throughout winter, spring, and early summer. This data can be combined with data from lysimetre measurements (Indicator set H) and the N content of plants and soil to produce an estimate of the total N budget for comparison across management systems (indicator set K).

3.3. *Economic sustainability*

Economic sustainability may be defined as the ability of a farmer or the farming community to maintain a viable business indefinitely. This requires sufficient high yields and product quality, good sale prices, and low cost of production. Large quantities of nitrogen fertiliser are used to maintain high yield in most commodity crops across the world. However, fertiliser input, mainly nitrogen, is the single most important contributor to the environmental impact of arable crop production (Hillier *et al.* 2009) and can lead to widespread pollution and environmental degradation (Goulding 2000). The requirements for maintaining ecological and environmental sustainability must therefore be off set against economic considerations. Options for improving ecological and environmental sustainability without jeopardising yields or economic sustainability, are: (i) to identify alternative, renewable sources of plant nutrients to replace or reduce the use of mineral fertilisers; (ii) to minimise the loss of nutrients and agrochemicals from the system (see Section 3.2); and (iii) to maximise the efficiency with which crop plants convert resources to harvestable product.

3.3.1. System function 8: Primary productivity and production efficiency

Improvements in the efficiency of crop production have been made since the Haber–Bosch manufacture of nitrogen provided a new and relatively cheap source of nitrogen. This, together with an increase in the proportion of winter crops and crop breeding to improve resource use efficiency, has led to a steady increase in yield, reaching a peak in the 1990s (Bingham *et al.* 2012, Knight *et al.* 2013). Since then, the amount of nitrogen fertiliser applied to arable crops has remained constant in the UK, largely due to the Nitrates Directive (91/676/EEC) that restricts the application of nitrogen within Nitrate Vulnerable Zones (> 50 mg L^{-1} nitrate in the ground water). Yields have also reached a plateau due to soil conditions that limit root growth and restrict the availability of water and nutrients to the crop (McKenzie *et al.* 2009, White *et al.* 2013), or to inefficient resource use which can vary significantly with crop variety (Bingham *et al.* 2012). Maintaining yields at reduced rates of mineral fertiliser application is a major challenge for improving the sustainability of arable cropping.

Indicator set K: Nutrient availability, plant biomass production, yield and product quality

Biological nitrogen fixation (BNF) by legume crops and undersown clover can provide a renewable source of nitrogen, allowing yields to be maintained at reduced rates of mineral fertiliser. Data on faba beans from the first three years at the CSC indicated nitrogen fixation rates of over 200 kg ha^{-1} yr^{-1} (Iannetta and James, personal communication). Although a lot (ca. 80%) of this fixed nitrogen was incorporated into the grains and removed at harvest, a substantial amount (ca. 50 kg nitrogen ha^{-1}) was still left in the senescent shoots and nodulated roots. This can be considered as a net gain of nitrogen into the system, and would potentially be available for use by the succeeding non-legume crop provided that measures are put in place (e.g. cover crops) to prevent losses over winter.

BNF by undersown clover can also contribute to the nitrogen supply in arable crop rotations (Nykanen *et al.* 2008), though there is little data to quantify fixation rates and difficulties can be experienced in the establishment and maintenance of clover-cereal intercrops (Jones and Clements 2008). Spring barley crops at the CSC platform are undersown with clover and the transfer of the fixed nitrogen to concurrent and subsequent crops will be quantified in future cropping seasons. The overall target is to achieve a nitrogen supply to subsequent crops of >50 kg ha^{-1} yr^{-1} from BNF of both clover and beans and to seek other renewable sources of nitrogen to supplement mineral nitrogen fertiliser requirements.

Nitrogen inputs can also be reduced with minimal impact on crop yields by increasing efficiency using precision agriculture. Soil nutrient availability and pH is measured across the CSC platform every year and will be used both to tailor the nutrient application rates to an appropriate level, and to contribute to the calculation of nitrogen budgets along with data on soil water nitrate levels (indicator set H) and nitrate and ammonia emissions (indicator set J).

Nutrient availability correlates with plant biomass production and ultimately crop yield (Figure 5). Plant biomass production is quantified at the CSC in the month before harvest as total dry weight, % carbon and % nitrogen in weeds (separately for monocots and dicots) and crop plants (separately for crop stems and grain). Weed biomass production was greater under sustainable cropping (showing a benefit for arable biodiversity (indicator sets A and B)), but did not correlate with crop production, suggesting that weed densities were not yield limiting.

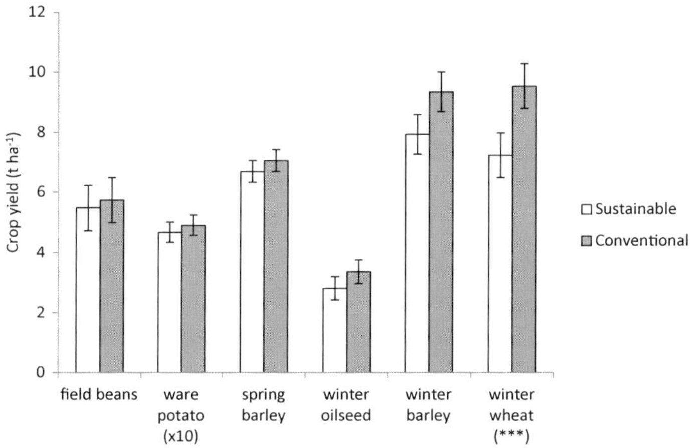

Figure 6. Crop yield averages over a six course rotation under a conventional and sustainable cropping system at the CSC. Bars = standard errors, $^{***}p < 0.001$ (general ANOVA).

Indicator set L: Production efficiency and gross financial margins

The economic impact of sustainable management systems can be assessed through a basic comparison of input costs, fuel use, tractor time and sale price, taking into account differences in quality of the harvested product, e.g. through calculations based on marketable yield (Figure 6).

Under the sustainable cropping system, yield is lower in the winter cereals, but there is no effect of cropping system on the yields of winter oilseed rape or the spring sown crops. Input costs are also reduced and this may represent sufficient compensation to maintain financial margins over the course of the rotation. Even if the sustainable cropping system incurs an economic cost in the short-term, there are likely to be benefits to longer term yield stability and system health that are difficult to translate into financial terms, although some authors have offered economic assessments of natural capital by estimating financial losses in the absence of different components of natural systems (e.g. Costanza *et al.* 1997). The CSC platform will provide long-term datasets on economic and environmental indicators to allow these trade-offs to be quantified.

4. Overall assessment of system indicators

Here, 12 key indicators of arable sustainability have been identified, relating to eight system functions. The response of each indicator to a management intervention can be assessed separately, but this does not allow trade-offs between system components to be accounted for and may not therefore provide a true representation of the response of the whole system to a change in management. The overall whole-system response may be assessed using modelling or decision software such as DEXi — a multicriteria, hierarchical decision tree that can combine qualitative and quantitative data into a single modelling framework (Bohanec *et al.* 2008, Pelzer *et al.* 2012). Overall sustainability is divided successively into smaller, more quantifiable components, each of which is characterised by one or more indicators as listed above. For example, overall sustainability will be broken down to economic and environmental factors. Economic sustainability then breaks down to profit, which comprises gross margins and salary costs. These are made up of a set of indicators including sale price, tractor time, fuel-use and input costs each of which can be measured directly as described here. The indicators are organised hierarchically allowing evaluation of the cropping system according to many potentially conflicting goals. This assessment has not yet been carried out for the case study described here since the long-term rotation is still in progress and full datasets are not yet available. However, once complete, modifications to the cropping system based on the whole system assessment can then be implemented iteratively to improve on the desired goals during subsequent rotations. These costs and trade-offs must be demonstrable and supported by hard evidence if more sustainable practices are to be adopted by the farming industry.

Acknowledgements

This work is funded by the Scottish Government and supported by the James Hutton Institute Agroecology Group.

References

Aizen, M.A., Garibaldi, L.A., Cunningham, S.A, Klein, A.M., 2009. How much does agriculture depend on pollinators? Lessons from long-term trends in crop production. *Annals of Botany* **103**, 1579–1588.

Altieri, M.A., 1999. The ecological role of biodiversity in agroecosystems. *Agriculture, Ecosystems and Environment* **74**, 19–31.

Andow, D.A., 1991. Vegetational diversity and arthropod population response. *Annual Review of Entomology* **36**, 561–586.

Bengtsson, J., Ahnstrom, J., Weibull, A.-C., 2005. The effects of organic agriculture on biodiversity and abundance: a meta-analysis. *Journal of Applied Ecology* **42**, 261–269.

Biesmeijer, J.C., Roberts, S.P.M., Reemer, M., Ohlemüller, R., Edwards, M., Peeters, T., Schaffers, A.P., Potts, S.G., Kleukers, R., Thomas, C.D., Settele, J., Kunin, W.E., 2006. Parallel declines in pollinators and insect-pollinated plants in Britain and the Netherlands. *Science* **313**, 351–354.

Bingham, I.J., Karley, A.J., White, P.J., Thomas, W.T.B., Russell, J.R., 2012. Analysis of improvements in nitrogen use efficiency associated with 75 years of barley breeding. *European Journal of Agronomy* **42**, 49–58.

Bohanec, M., Messean, A., Scatasta, S., Angevin, F., Griffiths, B., Krogh, P.H., Znidarsic, M., Dzeroski, S., 2008. A qualitative multi-attribute model for economic and ecological assessment of genetically modified crops. *Ecological Modelling* **215**, 247–261.

Brenchley, W.A., Warington, K., 1933. The weed seed population of arable soil. II. Influence of crop, soil and methods of cultivation upon the relative abundance of viable seeds. *Journal of Ecology* **21**, 103–127.

Brenchley, W.E., 1918. Buried weeds seeds. *Journal of Agricultural Science* **9**, 1–31.

Brooks, D.R., Bohan, D.A., Champion, G.T., Haughton, A.J., Hawes, C., Heard, M.S., Clark, S., Dewar, A.M., Firbank, L.G., Perry, J.N., Rothery, P., Scott, R.J., Woiwod, I.P., Birchall, C., Skellern, M.P., Walker, J.H., Baker, P., Bell, D., Browne, E.L., Fairfax, C.M., Garner, B.H., Haylock, L.A., Horne, S.L., Hulmes, S.E. Mason, N.S., Norton, L.R., Nuttal, P., Randle, Z., Rossall, M.J., Sands, R.J.N., Singer, E.J., Walker, M.J., 2003. Invertebrate responses to the management of genetically modified herbicide-tolerant and conventional spring crops. I. Soil-surface-active invertebrates. *Philosophical Transactions of the Royal Society of London B* **358** (1439), 1847–1862.

Carey, P.D., Short, C., Morris, C., Hunt, J., Proscott, A., Davis, M., Finch, C., Curry, N., Little, W., Winter, M., Parkin, A., Firbank, L.G., 2003. The multidisciplinary evaluation of a national agri-environment scheme. *Journal of Environmental Management* **69**, 71–91.

Carey, P.D., Manchester, S.J., Firbank, L.G., 2005. Performance of two agri-environment schemes in England: a comparison of ecological and multi-disciplinary evaluations. *Agriculture, Ecosystems and Environment* **108**, 178–188.

Carvell, C., Meek, W.R., Pywell, R.F., Goulson, D., Nowakowski, M., 2007. Comparing the efficacy of agri-environment schemes to enhance bumble bee abundance and diversity on arable field margins. *Journal of Applied Ecology* **44**, 29–40.

Choudrie, S.L., Jackson, J., Watterson, J.D., Murrells, T., Passant, N., Thompson, A., *et al.* 2008. UK Greenhouse Gas Inventory, 1990 to 2006, Annual Report for submission under the Framework Convention on Climate Change, DEFRA. http://www.airquality.co.uk/archive/reports/cat07/0804161424_ukghgi-90-06_main_chapters_UNFCCCsubmission_150408.pdf

Costanza, R., d'Arge, R., de Groot, R., Farber, S., Grasso, M., Hannon, B., 1997. The value of the worlds ecosystem services and natural capital. *Nature* **387**, 253–260.

Defra, 2011. Foresight. The Future of Food and Farming. Final Project Report. The Government Office for Science, London, UK.

Donald, P.F., Evans, A.D., 2006. Habitat connectivity and matrix restoration: the wider implications of agri-environment schemes. *Journal of Applied Ecology* **43**, 209–218.

EU, 2015. European Commission Environment and Agriculture. http://ec.europa.eu/environment/agriculture/index.htm

Ewald, J.A., Aebischer, N.J., Richardson, S.M., Grice, P.V., Cooke, A.I., 2010. The effect of agri-environment schemes on grey partridges at the farm level in England. *Agriculture, Ecosystem and Environment* **138**, 55–63.

Firbank, L.G., Barr, C.L., Bunce, R.G.H., Furse, M.T., Haines-Young, R., Hornung, M., Howard, D.C., Sheail, J., Sier, A., Smart, S.M., 2003. Assessing stock and change in land cover and biodiversity in GB: an introduction to Countryside Survey 2000. *Journal of Environmental Management* **67**, 207–218.

Fuentes-Montemayor, E., Goulson, D., Park, K.J., 2011. The effectiveness of agri-environment schemes for the conservation of farmland moths: assessing the importance of a landscape-scale management approach. *Journal of Applied Ecology* **48**, 532–542.

Gallai, N., Salles, J-M., Settele, J., Vaissiere, B.E., 2009. Economic valuation of the vulnerability of world agriculture confronted with pollinator decline. *Ecological Economics* **68**, 810–821.

Goulding, K., 2000. Nitrate leaching from arable and horticultural land. *Soil Use and Management* **16**, 145–151.

Goulson, D., Lye, G.C., Darvill, B., 2008. Decline and conservation of bumble bees. *Annual Review of Entomology* **53**, 191–208.

Haddad, N.M., Tilman, D., Haarstad, J., Ritchie, M., Knops, J.M.H., 2001. Contrasting effects of plant richness and composition on insect communities: a field experiment. *American Naturalist* **158**, 17–35.

Haughton, A.J., Champion, G.T., Hawes, C., Heard, M.S., Brooks, D.R., Bohan, D.A., Clark, S.J., Dewar, A.M., Firbank, L.G., Osborne, J.L., Perry, J.N., Rothery, P., Roy, D.B., Scott, R.J., Woiwod, I.P., Birchall, C., Skellern, M.P., Walker, J., Baker, P., Browne, E.L., Dewar, A.J.G., Garner, B., Haylock, L., Horne, S.L., Mason, N.S., Sands, R.J.N., Walker, M.J., 2003. Invertebrate responses to the management of genetically modified herbicide-tolerant and conventional spring crops. II. Within-field epigeal and aerial arthropods. *Philosophical Transactions of the Royal Society of London B* **358**(1439), 1863–1878.

Hawes, C., Haughton, A.J., Osborne, J.L., Roy, D.B., Clark, S.J., Perry, J.N., Rothery, P., Bohan, D.A., Brooks, D.R., Champion, G.T., Dewar, A.M., Heard, M.S., Woiwod, I.P., Daniels, R.E., Young, M.W., Parish, A.M., Scott, R.J., Firbank, L.G., Squire, G.R., 2003. Responses of plant and invertebrate trophic groups to contrasting herbicide regimes in the Farm Scale Evaluations of genetically-modified herbicide-tolerant crops. *Philosophical Transactions of the Royal Society of London B* **358**(1439), 1899–1913.

Hawes, C., Squire, G.R., Hallett, P.D., Watson, C.A., Young, M., 2010. Arable plant communities as indicators of farming practice. *Agriculture, Ecosystems and Environment* **138**, 17–26.

Heard, M.S., Hawes, C., Champion, G.T., Clark, S.J., Firbank, L.G., Haughton, A.J., Parish, A.M., Perry, J.N., Rothery, P., Scott, R.J., Skellern, M.P., Squire, G.R., Hill, M.O., 2003. Non-crop plants in fields with contrasting conventional and genetically modified herbicide-tolerant crops.1. Main effects of treatments. *Philosophical Transactions of the Royal Society of London B* **358**(1439), 1819–1832.

Hector, A., Dobson, K., Minns, A., Bazeley-White, E., Lawton, J.H. 2001. Community diversity and invasion resistance: an experimental test in a grassland ecosystem and a review of comparable studies. *Ecological Research* **16**, 819–831.

Henle, K., Alard, D., Clitherow, J., Cobb, P., Firbank, L., Kull, T., McCracken, D., Moritz, R.F.A., Niemelä, J., Rebane, M., Wascher, D., Watt, A., Young, J., 2008. Identifying and managing the conflicts between agriculture and biodiversity conservation in Europe. A review. *Agriculture, Ecosystems and Environment* **124**(1), 60–71.

Herzog, F., Dreier, S., Hofer, G., Marfurt, C., Schupbach, B., Spiess, M., Walter, T., 2005. Effect of ecological compensation areas on floristic and breeding bird diversity in Swiss agricultural landscapes. *Agriculture, Ecosystem and Environment* **108**, 189–204.

Hillier, J., Hawes, C., Squire, G.R, Hilton, A., Wale, S., Smith, P., 2009. The Carbon Footprints of Food Crop Production. *International Journal of Agricultural Sustainability* **7**, 107–118.

Hof, A.R., Bright, P.W., 2010. The value of agri-environment schemes for macro-invertebrate feeders: hedgehogs on arable farms in Britain. *Animal Conservation* **13**, 467–473.

Jones, L., Clements, R.O., 2008. Development of a low input system for growing wheat (*Triticum vulgare*) in a permanent understorey of white clover (*Trifolium repens*). *Annals Applied Biology* **123**, 109–119.

Karley, A.J., Hawes, C., Valentine, T., Johnson, S., Toorop, P., Squire, G.R., Young, M., and Iannetta, P. 2011. Can arable weeds contribute to ecosystem service provision? Functional diversity in Shepherd's purse (Capsella bursa-pastoris L. Medik.) *Aspects of Applied Biology* **109**, 31–38.

Kevan, P.G., Phillips, T.P., 2001. The economic impacts of pollinator declines: an approach to assessing the consequences. *Conservation Ecology* **5**(8), www.consecol.org/vol5/iss1/art8/

Kleijn, D., Sutherland, W.J., 2003. How effective are European agri-environment schemes in conserving and promoting biodiversity? *Journal of Applied Ecology* **40**, 947–969.

Kleijn, D., Berendse, F., Smit, R., Gilissen, N., Smit, J., Brak, B., Groeneveld, R., 2004. Ecological effectiveness of agri-environment schemes in different agricultural landscapes in the Netherlands. *Conservation Biology* **18**, 775–786.

Kleijn, D., Baquero, R.A., Clough, Y., Diaz, M., De Esteban, J., Fernandez, F., Gabriel, D., Herzog, F., Holzschuh, A., Johl, R., Knop, E., Kruess, A., Marshall, E.J.P., Steffan-Derwenter, I., Tscharntke, T., Verhulst, J., West, T.M., Yela, J.L., 2006. Mixed biodiversity benefits of agri-environment schemes in five European countries. *Ecology Letters* **9**, 243–254.

Knacker, T., Forster, B., Rombke, J., Frampton, G.K., 2003. Assessing the effects of plant protection products on organic matter breakdown in arable fields – litter decomposition test systems. *Soil Biology and Biochemistry* **35**, 1269–1287.

Knight, S., Kightley, S., Bingham, I., Hoad, S., Lang, B., Philpott, H., Stobart, R., Barnes, T.J.A., Ball, B., 2013. Desk Study to Evaluate Contributory Causes of the Current 'Yield Plateau' in Wheat and Oilseed Rape. HGCA Project Report 502. HGCA, Stoneleigh, Warwickshire: London, UK.

Knop, E., Kleijn, D., Herzog, F., Schmid, B., 2006. Effectiveness of the Swiss agri-environment scheme in promoting biodiversity. *Journal of Applied Ecology* **43**, 120–127.

Koricheva, J., Mulder, C.P.H., Schmid, B., Joshi, J., Huss-Danell, K., 2000. Numerical responses of different trophic groups of invertebrates to manipulations of plant diversity in grasslands. *Oecologia* **125**, 271–282.

Kremen, C., Williams, N.M., Thorp, R.W., 2002. Crop pollination from native bees at risk from agricultural intensification. *Proceedings of the National Academy of Science USA* **99**, 16812–16816.

Leip, A., Marchi, G., Koeble, R., Kempen, M., Britz, W., Li, C., 2008. Linking an economic model for European agriculture with a mechanistic model to estimate nitrogen and carbon losses from arable soils in Europe. *Biogeosciences* **5**, 73–94.

Loreau, M. 1998. Biodiversity and ecosystem functioning: a mechanistic model. *Proceedings of the National Academy of Science* **95**, 5632–5626.

Marshall, E.J.P., Brown, V.K., Boatman, N.D., Lutman, P.J.W., Squire, G.R., Ward, L.K., 2003. The role of weeds in supporting biological diversity within crop fields. *Weed Research* **43**, 77–89.

Marshall, E.J.P., West, T.M., Kleijn, D., 2006. Impacts of an agri-environment field margin prescription on the flora and fauna of arable farmland in different landscapes. *Agriculture, Ecosystem and Environment* **113**, 36–44.

Marschner, P., Kandeler, E., Marschner, B., 2003. Structure and function of the soil microbial community in a long term fertiliser experiment. *Soil Biology and Biochemistry* **35**, 453–461.

McKenzie, B.M., Bengough, A.G., Hallett, P.D., Thomas, W.T.B., Forster, B., McNicol, J.W., 2009. Deep rooting and drought screening of cereal crops: A novel field-based method and its application. *Field Crops Research* **112**, 165–171.

Ne'eman, G., Jürgens, A., Newstrom-Lloyd, L., Potts, S.G., Dafni, A., 2010. A framework for comparing pollinator performance: effectiveness and efficiency. *Biological Reviews* **85**, 435–451.

Ngetich, K.F., Diels, J., Shisanya, C.A., Mugwe, J.N., Mucheru-Muna, M., Mugendi, D.N., 2014. Effects of selected soil and water conservation techniques on runoff, sediment yield and maize productivity under sub-humid and semi-arid conditions in Kenya. *Catena* **121**, 288–296.

Norris, R.F., Kogan, M., 2000. Interactions between weeds, arthropod pests, and their natural enemies in managed ecosystems. *Weed Science* **48**, 94–158.

Nykanen, A., Granstedt, A., Jauhiainen, L., 2008. Residual effect of clover-rich leys on soil nitrogen and successive grain crops. *Agriculture and Food Science* **17**, 73–87.

Pelzer, E., Fortinoa , G., Bockstallerd, C., Angevina , F., Laminea, C., Moonenf, C., Vasileiadis, V., Guérina , D., Guichard, L., Reaub , R., Messéan, A., 2012. Assessing innovative cropping systems with DEXiPM, a qualitative multi-criteria assessment tool derived from DEXi. *Ecological Indicators* **18**, 171–182.

Perkins, A.J., Maggs, H.E., Watson, A., Wilson, J.D., 2011. Adaptive management and targeting of agri-environment schemes does benefit biodiversity: a case study of the corn bunting *Emberiza calandra. Journal of Applied Ecology* **48**, 514–522.

Perry, J.N., Rothery, P., Clark, S.J., Heard, M.S., Hawes, C., 2003. Design, analysis and power of the Farm-Scale Evaluations of genetically modified herbicide-tolerant crops. *Journal of Applied Ecology* **40**, 17–31.

Pimentel, D., Harvey, C., Resosudarmo, P., Sinclair, K., Kurz, D., McNair, M, Crist, S., Shpritz, L., Fritton, L., Saffouri, R., Blair, R., 1995. Environmental and economic costs of soil erosion and conservation benefits. *Science* **267**, 1117–1123.

Preston, C., Telfer, M., Arnold, H., Carey, P., Cooper, J., Dines, T., Hill, M., Pearman, D., Roy, D., Smart, S., 2002. The Changing Flora of the UK. Defra, London, UK.

Pywell, R.F., Warman, E.A., Hulmes, L., Hulmes, S., Nuttall, P., Sparks, T.H., Critchley, C.N.R, Sherwood, A., 2006. Effectiveness of new agri-environment schemes in providing foraging resources for bumblebess in intensively farmed landscapes. *Biological Conservation* **129**, 192–206.

Rathcke, B.J., Jules, E.S., 1993. Habitat fragmentation and plant-pollinator interactions. *Current Science* **65**, 273–277.

Roberts, H.A., 1958. Studies on the weeds of vegetable crops. I. Initial effects of cropping on the weed seeds in the soil. *Journal of Ecology* **46**, 759–768.

Roberts, H.A., 1962. Studies on the weeds of vegetable crops. II. Effect of six years of cropping on the weed seeds in the soil. *Journal of Ecology* **50**, 803–813.

Robinson, R.A., Sutherland, W.J., 2002. Post-war changes in arable farming and biodiversity in Great Britain. *Journal of Applied Ecology* **39**, 157–176.

Root, R., 1973. Organisation of a plant-arthropod association in simple and diverse habitats: the fauna of collards (*Brassica oleracea*). *Ecological Monographs* **43**, 95–124.

Rowan, J.S., Black, S., Franks, S.W., 2012. Sediment fingerprinting as an environmental forensics tool explaining cyanobacteria blooms in lakes. *Applied Geography* **32**, 832–843.

Safarzoda, S., Bahlai, C.A., Fox, A.F., Landis, D.A. 2014. The role of natural enemy foraging guilds in controlling cereal aphids in michigan wheat. *PLoS ONE* **9**(12), e114230. doi:10.1371/journal.pone.0114230

Scherer-Lorenzen, M., Palmborg, C., Prinz, A., Schulze, E.-D., 2003. The role of plant diversity and composition for nitrate leaching in grasslands. *Ecology* **84**, 1539–1552.

SEERAD, 2003. *The Rural Stewardship Scheme*. Scottish Executive, Edinburgh.

Siemann, E., Tilman, D., Haarstad, J., Ritchie, M., 1998. Experimental tests of the dependence of arthropod diversity on plant diversity. *The American Naturalist* **152**(5), 738–750.

Squire, G.R., Hawes, C., Begg, G.H., Young, M.W., 2009. Cumulative impact of GM herbicide tolerant cropping on arable plants assessed through species-based and functional taxonomies. *Environmental Science and Pollution Research* **16**, 85–94.

Stutter, M.I., Chardon, W.J., Kronvang, B. 2012. Riparian buffer strips as a multifunctional management tool in agricultural landscapes: Introduction. *Journal of Environmental Quality* **41**, 297–303.

Tattersall, F.H., Macdonald, D.W., Hart, B.J., Johnson, P., Manley, W., Feber, R., 2002. Is habitat linearity important for small mammal communities on farmland? *Journal of Applied Ecology* **39**, 643–652.

Taylor, R.L., Maxwell, B.D., Boik, R.J., 2006. Indirect effects of herbicides on bird food resources and beneficial arthropods. *Agriculture Ecosystems and Environment* **116**, 157–164.

Taylor, M.E., Morecroft, M.D. 2009. Effects of agri-environment schemes in a long-term ecological time series. *Agriculture, Ecosystem and Environment* **130**, 9–15.

Tolonen, T., 1995. Importance of generalist epigeal predator species in a cereal field: predation on baits. *Journal of Applied Entomology* **119**, 113–117.

Verheijen, F.G.A., Jones, R.J.A., Rickson, R.J., 2009. Tolerable versus actual soil erosion rates in Europe. *Earth-Science Reviews* **94**, 23–38.

Vickery, J.A., Carter, N., Fuller, R.J., 2002. The potential value of managed cereal field margins as foraging habitats for farmland birds in the UK. *Agriculture, Ecosystem and Environment* **89**, 41–52.

Vickery, J.A., Bradbury, R.B., Henderson, I.G., Eaton, M.A., Grice, P.V., 2004. The role of agri-environment schemes and farm management practices in reversing the decline of farmland birds in England. *Biological Conservation* **119**, 19–39.

Webster, C.P., Shepherd, M.A., Goulding, K.W.T., Lord, E., 1993. Comparisons of methods for measuring the leaching of mineral nitrogen from arable land. *Journal of Soil Science* **44**, 49–62.

White, P.J., George, T.S., Dupuy, L.X., Karley, A.J., Valentine, T.A., Wiesel, L., Wishart, J., 2013. Root traits for infertile soils. *Frontiers in Plant Science* **4**, 193.

Williams, P.H., Colla. S., Xie, Z-H., 2009. Bumblebee vulnerability: common correlates of winners and losers across three continents. *Conservation Biology* **23**, 931–940.

Wilson, B.J., Lawson, H.M., 1992. Seedbank persistence and seedling emergence of seven weed species in autumn-sown crops following a single year's seeding. *Annals of Applied Biology* **120**, 105–116.

Chapter 10

A Framework for the Design of Agroecological Livestock Farming Systems

Bertrand Dumont*, Magali Jouven†, Thierry Bonaudo‡, Raphaëlle Botreau* and Rodolphe Sabatier‡

*INRA-VetAgro Sup — Theix, Saint-Genès-Champanelle, France
†SupAgro-CIRAD-INRA, Montpellier, France
‡INRA-AgroParis Tech, Paris, France

1. Introduction

Livestock farming systems perform numerous functions: besides supplying proteins for human nutrition, livestock provide income for rural populations and thus support a large part of the world's rural communities. Livestock produce not only meat, milk, and eggs, but also wool, fibre, and leather. Herbivore farming systems shape rural territories by supporting low population densities and associated infrastructures and public services. In temperate and Mediterranean areas, grassland-based systems keep landscapes open, preserve biodiversity, and provide various ecosystem services such as climate regulation via carbon sequestration, and prevention of erosion and running fires. They also provide socio-cultural services including tourist events such as transhumance, and products with a local image and geographic link such as Protected Designation of Origin cheeses. Livestock have been integrated with cropping systems worldwide. Small multipurpose farmers who grow crops and rear animals produce most of the food supply and promote economic development in Africa, South Asia, and Latin America. Animal traction is particularly important in these farming systems where cattle, equids, and camels directly assist crop production for ploughing, planting, weeding, and water-lifting. Livestock consume crop by-products and grass from field boundaries. Herbivore dungs make good farmyard manure for crop fertilisation, being made of 'predigested' plant organic matter easily utilised by soil microfauna; dungs are also used as slurry associated with biogas production, for vermicompost, or are dried for burning.

Despite these numerous functions, livestock farming systems are also regarded as a major cause of the world's most pressing environmental issues (e.g. Gerber *et al.* 2013). A first major criticism is that producing animal proteins requires staggering amounts of water and energy compared with plant foods. Second, the livestock sector is a major driver of land use, since most of the world's maize, soybean, and barley harvest goes for animal feed. This puts the livestock sector in direct competition with human food supply, and makes it partly responsible for deforestation. Third, livestock, mainly cattle, are responsible for a significant proportion of world greenhouse gas emissions. Fourth, the dumping of pharmaceutical residues and metabolites in the environment and the spread of antibiotic resistance threaten public health and the environment. Fifth, livestock frequently exceed the carrying capacity of grazing lands. Depending on the biogeographical area, overgrazing has effects, both direct and indirect via mineral fertilisers, on water pollution and biodiversity, and causes irreversible damage to rangelands that can lead to desertification. Finally, cattle feedlots and pig and poultry industrial production units not only cause pollution, but also raise major animal welfare issues. With their very high animal concentrations and associated economies of scale, such industrial units also compete strongly with smallholder farms, which are declining worldwide.

Unlike industrial farming systems, agroecology advocates food sovereignty, collaboration among farmers, social justice, and strong links between farmers and consumers, fostering consumer trust in food (Altieri

2002, Francis *et al.* 2003). It thereby offers a promising way to increase the sustainability of livestock farming systems (Gliessman 2007). Agroecology can provide alternative dual-benefit solutions such as producing protein-rich foods, while reducing the environmental footprint of livestock farming. It calls for stimulating natural processes for input cost reduction and income gain, and for designing self-sufficient systems by using complementarities between system components, and giving priority to pasture and by-products to feed animals, since these resources do not compete with human food supply (Röös *et al.* 2016, van Zanten *et al.* 2016). The resilience of agroecological livestock farming systems is expected to be enhanced by plant and animal diversity, animal adaptive capacities, and by the strength of interactions between system components. In this chapter, we propose a framework for the design of agroecological livestock farming systems. We first show key ecological processes that can be used in livestock farms. We then propose five principles as a guideline to implement combinations of agroecological practices in livestock farming systems, and illustrate them with examples from ruminant, pig, poultry, and integrated crop-livestock farms. Finally, we discuss the technical and organisational levers to be taken into account in implementing agroecological transition in livestock farming systems, which extends from the farm scale to the landscape/territory and food-supply chain scales.

2. Ecological processes in livestock systems

The paradigm shift underpinning agroecology involves considering the farming system and its biotic and abiotic environment as an agroecosystem whose dynamics are driven by the farmer. The focus is thus no longer on controlling environmental variability by applying predetermined management rules for animals to express their full genetic potential, but rather on keeping the system in a viable trajectory that benefits from the diversity of available resources (Bell *et al.* 2008). Diversity is a fundamental characteristic of livestock farming systems tied to their physical environment and that need to cope with uncontrolled events. Agroecology recognises this diversity and seeks to make good use of it instead of merely enduring it. Key ecological processes that are gainful for agroecological systems include animal adaptive capacities, functional diversity of animals and resources, and interactions between system components (Bonaudo *et al.* 2014).

2.1. *Adaptive capacity*

An important challenge in agroecological livestock farming systems is animal adaptation to fluctuating environmental conditions, including feed quantity and quality, host-pathogen interactions, and climate hazards. Animal adaptive capacities can be enhanced by management practices that orientate animal behaviour, but remain based on animal flexibility, plasticity (Blanc *et al.* 2006, 2010) and variability between breeds, cross-breds, or individuals. For instance, cattle breeds that have been selected in tropical environments are more resistant to trypanosomes, gastrointestinal parasites, and ticks (Mandonnet *et al.* 2011). Behavioural traits such as maternal behaviour ensure female autonomy at parturition, and breed differences have been reported in sheep and pigs (Dwyer and Lawrence 1998, Canario *et al.* 2009). Breeding programs need to balance adaptive and production traits to improve the fit of animals with their rearing conditions. For instance, interindividual variability of the capacity of growing chicks to digest low-quality wheat is partially inheritable and so could be improved by selection (Mignon-Grasteau *et al.* 2004). As the result of genotype × environment interactions on production, health, and welfare traits, breeding programmes would benefit from being sometimes conducted under limiting conditions. For example, the estimated genetic parameters for reproduction traits in organic and low-input dairy cattle herds differ from those estimated for high-input production systems (Yin *et al.* 2012). Farming practices in the fetal stage and early life affect behaviour, response to stress, immunity, and performance of adult animals. Hence, epigenetic marks reversibly or inheritably present in the genome can help animal adaptation (Dumont *et al.* 2014). A pioneer study has revealed the inheritance of acquired behaviour adaptation to unpredictable food access in chickens that could result from variations in chromatin structure (Nätt *et al.* 2009). After birth, transmission of adaptive capacities occurs through learning, mainly from the mother or social peers. Shepherds have also developed management practices enabling young animals to acquire flexibility in feeding choices, so that they exploit a wider range of rangeland resources and maintain performance better under harsh environmental conditions (Krätli 2008).

2.2. *Functional diversity of animals and resources*

Many ecological processes that can be used in agroecological systems are related to animal and plant diversity. Functional groups differ in their strategies of adaptation to environmental conditions, and thus in their structural, physical and/or metabolic characteristics. It has often been assumed that higher levels of functional diversity should result in higher productivity of agroecosystems. In grassland communities, several experiments have demonstrated higher sward productivity in multispecies swards than with monocultures (e.g. Hector *et al.* 1999). However, others have stressed that this 'diversity effect' saturates at low levels of species richness so that fertilised systems usually outperform those based on native vegetation. The advantages of species diversity for primary and secondary production result from a combination of contrasting functional types that exploit a broader range of resources with limited competition. In mixed grazing systems, the combination of livestock species with complementary ecological niches increases the overall use of pastures. In a recent meta-analysis, daily live weight gain of sheep was on average greater by +14.5 g/day when mixed with cattle than when grazed alone. Mixed grazing of cattle and sheep leads to less parasite infestation and increased meat production per hectare: average daily gain per hectare was 29% higher in mixed grazing systems than with sheep alone, and 25% higher than with cattle alone (d'Alexis *et al.* 2014).

Besides production, diversity of functional types is assumed to increase system resilience. Redundancy is the key underlying mechanism, since the collapse of any given species can then be offset by another species with similar functional characteristics. The "portfolio effect" states that communities with higher species richness are more likely to include the species that is best adapted to any given condition in fluctuating environments, so that the number of species *per se* has a positive effect on system resilience.

Within herds, there is a diversity of individual trade-offs between functions that has been quantified in dairy cows (Ollion *et al.* 2016), suckler cows (de la Torre *et al.* 2015), and dairy goats (Puillet *et al.* 2010). Diversity of trade-off profiles does not increase herd milk production, but brings stability under fluctuating environmental conditions (Blanc *et al.* 2013; Figure 1). Multispecies forage systems (Murgueitio *et al.* 2011, Altieri *et al.* 2014) and multispecies herds (Tichit *et al.* 2004) have further been shown to be more resilient to climatic events.

2.3. *Interactions between system components*

In addition to the diversity of system components, the number and strength of interactions between system components is a crucial aspect on which agroecological systems can be based (Bonaudo *et al.* 2014). This was evidenced in a network of mixed crop–beef cattle farms in the French Charolais breed area, in which farms with few interactions between herd and crops made less efficient use of inputs than farms with stronger interactions, and missed environmental performance and economies of scope (Veysset *et al.* 2014). Interactions between system components lend the

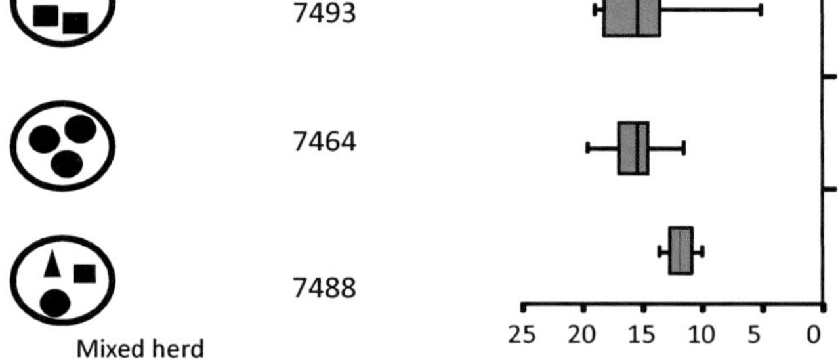

Figure 1. Simulation of annual dairy production (kg/cow) and interannual variability of dairy production (%) in homogeneous and mixed cattle herds (adapted from Blanc *et al.* 2013). Homogeneous herds were composed of cows adapted to nutritional challenges (triangles), heat waves (squares), or parasite challenges (circles). The bottom row represents a mixed herd composed of cows with contrasting trade-off profiles.

system further properties that go beyond a redundancy-based insurance effect or portfolio effect. Nutrient cycling, biomass and energy flows in agroecosystems occur through a complex set of abiotic and biotic interactions including biogeochemical processes and trophic interactions. Ecosystem resistance to disturbance involves mechanisms of population regulation and relies on proper functioning of natural pest control (Bonaudo *et al.* 2014). The structure of grassland communities results from the timing and intensity of plant–herbivore interactions, i.e. selective defoliation, plant growth and competitive interactions, and temporal stability of vegetation patches. Patch stability along the grazing season and in successive years can increase functional diversity within semi-natural grasslands (Dumont *et al.* 2012). Habitat heterogeneity under moderate grazing intensity also provides more ecological niches that benefit grassland biodiversity. Conversely, trophic interactions between plants and insects can become disrupted by a disturbance such as a high grazing intensity.

3. Management principles

Based on this set of ecological processes, and on those from Altieri (2002) for sustainable agriculture, Dumont *et al.* (2013) have proposed five principles that can be used as a guideline to implement combinations of agroecological practices in livestock farming systems. These principles are: (i) achieve integrated animal health management, (ii) decrease the external inputs needed for production, (iii) decrease pollution by optimising the metabolic functioning of farming systems, (iv) enhance functional diversity within animal production systems to strengthen their resilience and efficiency, and (v) preserve biological diversity by adapting management at farm and landscape scales.

3.1. *Achieve integrated animal health management*

Applying agroecology to the management of animal health implies focusing on the causes of animal diseases in order to reduce their occurrence and the use of chemical drugs. Major attention will be given to the choice of the animal type and/or diversity of animals (species, breeds, crossbreds, individuals) best suited to taking advantage of the farm environment (biotic environment, housing conditions, type of management). Adaptation to harsh or fluctuating environments also requires farmers to adopt management practices that make best possible use of livestock adaptations. For instance, advances have been made in feeding strategies to maintain normal rumen function in dairy cows during heatwaves, e.g. by increasing nutrient density of the diet (West 2003). As such feeding practices increase the use of cereals, and thus compete with human food, there is still a need to select animals that are less vulnerable to heat stress, and have a high feeding efficiency (Dumont *et al.* 2014). Feeding animals with bioactive plants or essential oils can limit the need for chemical drugs. Consumption of tannin-rich plants reduces the infestation of small ruminants by parasitic nematodes; these plants improve animal clinical status (diarrhoea, etc.) and reduce mortality under parasite challenge. It is also possible to adapt the frequency and seasonality of ruminant reproduction according to period and intensity of nutritional stress. Farming practices combining diet, housing and strain choice to stimulate host defences are crucial for pigs, rabbits, and poultry, usually housed at high densities. The size, composition and genetic structure of animal groups, and how they are housed, coupled with tools for early detection of diseases, can limit the use of chemical drugs. Maintaining stable dominance hierarchy is likely to increase social tolerance, and so can limit feed inputs and the use of chemical drugs in pig and bull-finishing systems (De Groot *et al.* 2001, Mounier *et al.* 2006). Adoption at age one day of renewal rabbit females by reproductive females permits early implantation of a functional, diverse gut microbiote, which increases rabbit resistance to pathogens (Gidenne *et al.* 2010). In poultry, susceptibility to dietary stress has been shown to be strain-dependent (Hangalapura *et al.* 2005), further emphasising the importance of choosing genotypes adapted to particular production conditions.

3.2. *Decrease the external inputs needed for production*

Economy of inputs is a crucial challenge, since these form a large proportion of operational costs in most livestock farming systems. As an alternative to mineral fertilisation, grass-legume mixtures achieve high forage yields due to symbiotic nitrogen fixation in legume nodules. Functional diversity of plants within a pasture or crop should be promoted, not only to increase vegetation biomass, but also to enhance resistance to weed invasion.

For the animal component of the system, economy of inputs can be achieved by changing the type of feed, the sequence of feed presentation, or by introducing strategic supplementation. Improving animal feeding efficiency is crucial for pigs and poultry, which consume cereals and leguminous plant grains in competition with human food. Diet supplementation with natural microbial phytases improved phosphorus digestibility in pigs, and thus feed conversion (Dourmad *et al.* 2009). Sequential feeding of wheat grain and protein–mineral concentrate improved feed conversion in laying hens, and can facilitate the use of local feedstuffs introduced as whole grains (Faruk *et al.* 2010). Geese grazing between tree rows in a walnut plantation (Figure 2) increased walnut production by 26% and tree growth by 6% (Dubois *et al.* 2008). There was no microbial contamination

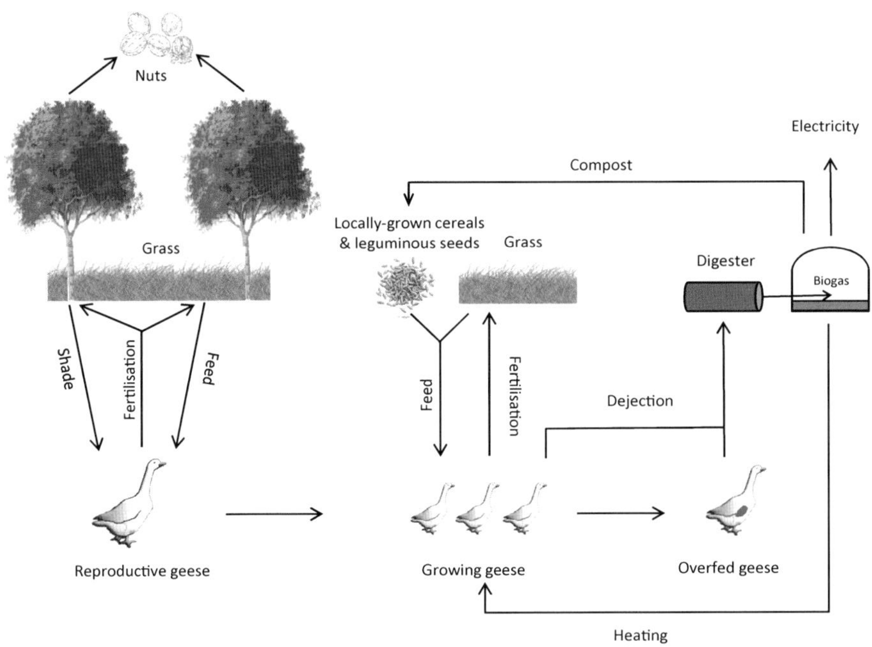

Figure 2. Complementarity between geese and walnut production on a farm in south-west France. Growing geese are fed with locally-grown cereals and leguminous plant grains. A digester produces biogas from geese manure, which is used for electricity and heat production. *Digesta* are used as compost for crop fertilisation (adapted from Dubois *et al.* 2008 and Thomas *et al.* 2014).

of the fruits by *Escherichia coli* if the geese were removed at least two months before harvesting.

Use of permanent pastures and rangelands is considered an efficient way to reduce inputs, since these are natural resources, cheap to produce, and can be used only to feed herbivores. To cope with year-to-year variability in the amount and quality of forages and improve grazing efficiency, a flexible organisation of flock and pasture management is required based on within- and between-pasture diversity; supplementation with some concentrate can improve roughage digestibility and utilisation during summer or winter (Jouven *et al.* 2010). Use of crop residues, industrial by-products associated with the production of biofuels or biomaterials, and alternative resources such as algae and insects are other non-competitive options to feed livestock while reducing wastes.

3.3. *Decrease pollution by optimising the metabolic functioning of farming systems*

An ecologically sustainable, productive system can be obtained by recycling farm wastes within the system or between neighbouring systems. Integration of cropping with livestock allows better regulation of biogeochemical cycles and environmental fluxes to the atmosphere and hydrosphere through spatial and temporal interactions among farm units. It allows the recycling of animal manures, turning a nuisance into a resource, since farmyard manures are rich in nutrients, and provide soil microorganisms with a key source of energy. Studies in France and Brazil have found that integrated systems strongly decrease the inputs required for production and thus reduce the system's environmental footprint, while increasing ecosystem services (Bonaudo *et al.* 2014). Commercial products either increased or slightly decreased, but the farms investigated always achieved a higher added value and higher net income per work unit as the result of their cost-cutting strategy (Figure 3).

Perennial soil–vegetation interactions lead to strong coupling in grasslands between the carbon, nitrogen and phosphorous cycles (Bol *et al.* 2011). These couplings operate in plants where nitrogen and phosphorous are linked to carbon during photosynthesis and in soils through the dynamics of organic matter, with the capacity of microbes to recapture

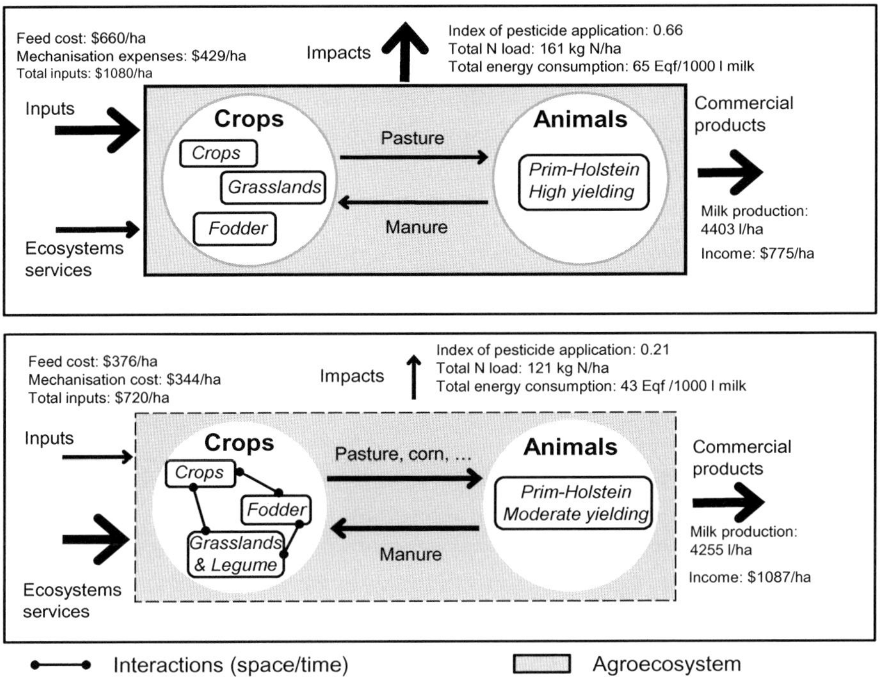

Figure 3. Conventional crop–dairy system (top) and agroecological farm from a sustainable agriculture network (bottom) in Brittany, France (Bonaudo *et al.* 2014). The example highlights the fact that cost-cutting management can be a win–win strategy combining good economic and environmental performance.

and recycle mineral nitrogen and phosphorous through mineralisation-immobilisation turnover. Ruminants tend to uncouple the carbon and nitrogen cycles by emitting CO_2 and CH_4, and by returning high concentrations of reactive nitrogen in urine patches. Biological nitrogen fixation by legumes introduces quantities of nitrogen already coupled with carbon in the system, which reduces N_2O emissions (Soussana and Lemaire 2014). A win–win option increasing plant productivity, soil carbon stocks and animal production has been reached in Portugal (www.terraprima.pt) through the use of phosphorous fertilisation and species-rich mixtures. In the tropics, nitrogen-fixing trees such as *Acacia* and *Gliricidia* species can be used to restore degraded pastures, while providing forages during seasonal droughts and shade for the animals (Murgueitio *et al.* 2011).

Intensive pig farming systems are characterised by a low nitrogen efficiency (around 50%), partly due to gaseous emissions of NH_3, N_2, and N_2O. Nitrogen efficiency can be improved at farm scale by reducing gaseous emissions from the excreta and including legume forages in crop rotations. A digester can be associated with intensive pig farming systems to produce biogas from solid and liquid pig manure. This has proved the best way to avoid environmental losses of CH_4 from liquid manure, while also reducing the biological activity of drug residues before using manure to fertilise crops (Petersen *et al.* 2007). Biogas can be used for electricity and heat production as also shown in the geese farming system depicted in Figure 2.

3.4. *Enhance functional diversity within animal production systems to strengthen their resilience*

Biological diversity can be used to improve system productivity and resilience. Functional diversity within any livestock production system can be varied by the farmer, who selects appropriate combinations of animal species, breeds, and crossbreds in order to benefit from genotype × environment interactions and adapt livestock type to available resources. Integrated crop–livestock systems are another form of diversification. Benefiting from system diversity requires appropriate management, i.e. a flexible strategy based on knowledge of the response and adaptive capacities of system components. Gaining knowledge on the interactions between system components is needed to prompt farmers to switch to these diverse farming systems and strengthen their decisional self-reliance. As previously stated, combining several livestock species on the same farm can optimise the use of feed and forage resources; it also decreases the parasite burden of each species by clearing pastures of parasites with a non-susceptible species (Dumont *et al.* 2014). In some cases however, combining several species can increase health risks, since some infectious agents such as influenza viruses adapt to different species and increase their virulence after recombination; viruses can cross the species barrier when poultry and pigs are bred together, e.g. in South-East Asia, and a virus thought to originate from ducks and wild birds is transmitted efficiently among humans (Kuiken *et al.* 2011).

At herd level, rearing different herbivore species offers a risk-spreading strategy against droughts, disease outbreaks, and market price fluctuations (Tichit *et al.* 2004). Integration of livestock and crops increases system resilience to market shocks (Bonaudo *et al.* 2014). Diversity of grassland types has been shown to improve farm self-sufficiency for forages in dairy and suckler farms as it ensures the provision of a variety of resources that livestock can use along the year (Andrieu *et al.* 2007). Recent work has emphasised that the diversity of grazing management practices (e.g. stocking rate, grazing period) can also enhance a system's ability to overcome drought events, but that a trade-off can occur between adaptive capacity and production of the system (Sabatier *et al.* 2015). In agro-pastoral systems, the feeding system is based on complementarities between cultivated grasslands, which are used to secure animal performance in crucial periods such as mating and lactation, and rangelands, which are mostly grazed at times when animals have low nutrient requirements (Jouven *et al.* 2010). The resilience of silvo-pastoral systems is also closely linked to their level of diversification in terms of tree species and management practices (Altieri *et al.* 2014); several observations in Central America have emphasised that enhancing plant diversity and structural complexity also lets silvo-pastoral systems recover more rapidly after a hurricane.

3.5. *Preserve biological diversity by adapting management at farm and landscape scales*

By considering biodiversity as both a resource and an output in livestock systems, agroecology lends food production and ecosystem integrity the same level of priority. A number of ecological functions and ecosystem services such as recycling of nutrients, forage yield, pollination, and resistance to weed invasion are closely linked to biodiversity; their persistence thus depends on maintaining biological diversity in agroecosystems. For domesticated species, the conservation of a wide genetic basis requires including males with some specific adaptive capacities in selection schemes for the most widespread breeds. In addition, local breeds are well-adapted to specific environmental and management conditions, and it is therefore useful to conserve them not only for their cultural or existence value, but also because they are a product of long-term genetic × environment adaptation.

Besides preserving livestock breed diversity, adapting management practices, e.g. stocking density, cutting frequency, and preserving land-use diversity and landscape connectivity are major levers for wild flora and fauna conservation. For instance, late grazing, legume-rich grasslands and sown margin strips at the edge of arable fields favour pollinators through positive trophic interactions with nectar-providing plants. Not grazing some plots at flowering peak can also be a win–win option, since this increases butterfly abundance and species richness, while keeping animal performance constant (Farruggia *et al.* 2012).

Jouven and Baumont (2008) modelled grassland-based beef systems and found that meat production could be maintained and biodiversity improved by deploying biodiversity-friendly practices on up to 40% of farm area. Overall, the practices resulting in the optimal production-biodiversity equilibrium depend on the farming context, i.e. type of grasslands, stocking density, and livestock species. The choice of plots managed with biodiversity-friendly practices should account for the biodiversity "potential" of any grassland, but also their location so that they can act as dispersal sources or ecological corridors (Öckinger and Smith 2007). Modelling studies have shown that both land use and the spatial arrangement of management practices affect the long-term dynamics of birds in grassland-based systems (Sabatier *et al.* 2014). For instance, converting intensive practices into more extensive ones affects production, but acting on the spatial arrangement of practices to increase landscape heterogeneity helps to reconcile production and biodiversity (Figure 4).

4. Accompanying agroecological transition in livestock farming systems

4.1. *Evaluation of agroecological systems*

Agroecology implies considering systems as a whole, so that the evaluation of any agroecological system should focus on its overall performance (based on a range of indicators) rather than on single performance indicator; systems also need to be observed long-term to assess their adaptation

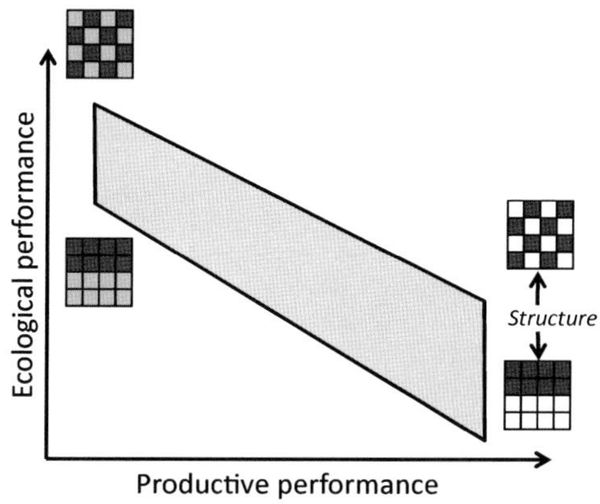

Figure 4. Effect of the proportion of different management regimes and landscape structure on the performance, both ecological (Vanellus vanellus population size) and productive (grassland production), of grassland-based systems in the Marais-Poitevin area, western France (Sabatier *et al.* 2014). This example highlights the fact that increasing landscape heterogeneity shifts the production-biodiversity trade-off line towards improved ecological performance.

and resilience (Bonaudo *et al.* 2014). Several principles apply in all agroecological livestock production systems, making their design different from the regreening of an existing system in which only some particularly detrimental practice is improved.

Murgueitio *et al.* (2011), Dumont *et al.* (2013), Bonaudo *et al.* (2014), and Thomas *et al.* (2014) have analysed a total of eight case studies covering a long gradient of production systems (ruminants, pigs, poultry, and integrated agriculture-aquaculture), intensification levels (from a pastoral sheep farming system to an integrated pig–crop system associated with a digester) and biogeographic areas (temperate, Mediterranean, tropical). Three or four of the five previously discussed management principles were applied in each system. They were, however, not mobilised equally according to production and system intensification (Table 1). Principles of the reduction of inputs and pollution emerged in nearly all the systems as the result of economic and regulatory constraints. Integrated management of animal health is based on a marked recent research effort that is gradually being transferred into practice. Here, it was only mobilised in the organic rabbit production system, and in the pig farming system associated

Table 1. The agroecological principles of livestock systems that apply in eight case studies covering a broad range of production, intensification levels, and biogeographic areas.

	Pastoral sheep Mediterra-nean France (2)	Dairy cows Brittany (2) (3)	Dairy cows Colombia (1)	Cattle ranching Brazil (3)	Integrated fish South-east Asia (2)	Geese + walnuts Dordogne (4)	Organic rabbits North-west France (2)	Pigs, crops and digester Center France (2)
Health ma-nagement							**	*
↓ inputs	**	***	***	*	***	**	***	***
↓ pollutions	*	***	*	**	**	***		***
Diversity for resilience	***	**	**	***	**	**	**	*
Biodiversity	**	*	**	*			*	

Notes: ***: key principle for system design; **: important principle; * principle of marginal use in the system or resulting from other management choices.
The geese and dairy cow systems in Brittany are presented in more detail in Figures 2 and 3, respectively.
References: (1) Murgeitio *et al.* (2011), (2) Dumont *et al.* (2013), (3) Bonaudo *et al.* (2014), (4) Thomas *et al.* (2014).

with the production of feed ingredients and with a digester. Taking advantage of system diversity was a key management principle across the whole range of systems, while preservation of biodiversity was a secondary objective in grazing-based systems in all biogeographic areas. An adapted genetic strain was used in the organic rabbit production system.

Agroecological systems thus take contrasting forms and mobilise different agroecological principles depending on the context. This calls for a multicriteria evaluation process to assess system performance and level of compliance with agroecological principles. Genericity of the five management principles makes it possible to use them for evaluating all types of livestock farming systems. A final objective could be to provide clear bases for certifications or 'eco-labels' that would inform consumers on system management and environmental impacts. Such a multicriteria evaluation is in progress for upland dairy systems (Botreau *et al.* 2014). It required subdividing each management principle into a limited set of sub-principles, e.g. increasing feed efficiency of the animals (Table 2), which were crossed with key state variables on which fulfilment of the principle should have a positive impact, e.g. concentrate feed, mineral fertilisers, water, non-renewable energy, and phytosanitary and veterinary agrochemicals for the input reduction principle. In each cell resulting from this crossing, we identified up to 3–4 practice-based criteria (Table 2) that need to be evaluated by relevant sets of indicators.

It is noteworthy that sub-principles remain generic, while state variables can be specific to a given production, e.g. it is relevant to consider mastitis in dairy cattle but not in poultry, and practice-based criteria are specific to the local context (Botreau *et al.* 2014). This diagnostic tool was developed for upland dairy systems, but the methodology proposed is transferable to other dairy systems, e.g. intensive grass-based systems in lowlands. The diagnostic tool is applied at farm scale, but it also considers flows towards the atmosphere or surface water, and the benefits of preserving agroecological structures beyond farm boundaries. The same five principles were used to characterise dairy goat systems in an upland area of central France, and their compliance with agroecological principles (Gomes *et al.* 2014). Surveys have stressed the variability of farm profiles that reflect differences in land area, integration of farm operations, and product diversification. This calls for analysing system characteristics that

Table 2. Example of practice-based criteria defined for assessing the reduction of inputs in upland dairy systems (adapted from Botreau *et al*. 2014). Only the raw of state variable "reduction of concentrate feed" is presented. We identified a total of seven criteria related to three sub-principles: increasing feed efficiency of the animals, increasing natural productivity of the environment, and increasing the use of resources that do not compete with human food supply.

P2. Reducing inputs	(A) Increasing feed efficiency of the animals	(B) Increasing natural productivity of the farm environment	(C) Increasing the use of resources that do not compete with human food supply	(D) Preserving supporting services provided by farm environment
(1) **Concentrate feed**	(1) Adapt the level and nature of concentrate feed to animal requirements (2) Split the distribution of concentrates during the day (3) Choose genotypes that meet their requirements under harsh feeding conditions (i.e. with less concentrates)	(1) Produce cereals or leguminous plant grains on farm to increase farm autonomy; these can be used as concentrates (if seeds are harvested) or as forages (if plots are grazed or mown at vegetative state) (2) Improve grazing and harvesting techniques to enhance sward biomass and feeding value	(1) Use industrial by-products rather than cereals or leguminous plant grains as source of concentrate feed (2) Adapt cow productivity to the natural productivity of the environment in order to feed animals with grass instead of concentrate feed	Not relevant

contribute to the success of agroecological systems, or limit the transition of existing systems into more agroecological forms (Dumont *et al.* 2014).

4.2. *Horizontal and vertical integration within agroecosystems*

In the definition adopted so far in this chapter, agroecology has been considered as "the application of ecological concepts and processes to the design and management of sustainable agroecosystems", and we have focused mainly on system performance and environmental impacts. Francis *et al.* (2003) proposed a wider definition of agroecology as "the integrative study of the ecology of the entire food system, encompassing ecological, economic, and social dimensions". The food system includes all processes involved in human diet, from agricultural production to processing of foods and by-products and transport. This second definition highlights the fact that agroecology is a holistic approach and that ecosystem management should be implemented at broad spatial and temporal scales, which implies shifting from an individual level (farm) to a collective one (territory).

Outputs from the model of Sabatier *et al.* (2014) given in Figure 4 are first relatively simple evidence for "horizontal integration". They show that the habitat heterogeneity hypothesis that applies at plot and farm scale (e.g. Farruggia *et al.* 2012) is also relevant to combined production and biodiversity objectives at landscape scale. Applying ecological concepts for the management of agroecosystems can also be extended to the supply chain level, which has been called "vertical integration". Socio-economic processes are largely dependent on collective organisation to structure any agroecological value-chain, increase and diversify production, develop a distribution network and strengthen links between producers and consumers. The management and development of rural areas no longer result exclusively from the agricultural sector; many stakeholders and economic activities such as tourism, industry, and urbanisation compete for access to resources, e.g. land, plant and soil resources, water, and energy. Participatory certification in Brazil is a good example of joint management that combines vertical and horizontal integration (Radomsky *et al.* 2014). In this system, the new agroecological producers are certified by a group composed of producers and consumers; collective organisation helps overcome the sectoral short-term interests that have been exacerbated by the top-down linear structuring of value-chains. In order to shift from farm to territory scale, quantitative methods such as territorial metabolism have been developed to identify and locate flows of materials and energy involved in livestock farming systems and other activities in a given territory (Chatzimpiros and Barles 2013, Bonaudo *et al.* 2015). Such an approach can help find complementary features among activities in the same territory, favouring territorial self-sufficiency while completing material cycles. The identification of flows allows prioritisation and sharing of local resources between stakeholders. An exchange network with a broad swathe of society must be rewoven to allow territorial management of resources and discuss broader questions such as ethical issues about the status of animals, acceptability of farming practices, and the environmental footprint of livestock production.

4.3. *Towards a collective organisation of agroecosystem management at territorial scale*

We assume that the scaling-up of agroecological livestock farming systems is made easier when innovations have been codesigned by various stakeholders, e.g. land-owners, legislators, farmers, consumers, and NGOs involved in nature conservation, from the start of the process (e.g. Kosgey *et al.* 2006 for small ruminant breeding programs in the tropics). Collective actions among breeders have long existed, but have usually aimed to improve system productivity or economic performance. The number and nature of stakeholders is context-dependent. For instance, in collective grazing lands a number of basic rules are generally defined and shared by the group of farmers involved. These rules include dates of entry and turn-out from the area, stocking rate, and priority access to strategic areas. In addition to these rules, tactical adjustments are based on indicators related to the amount and/or quality of resources, e.g. sward height or phenological stage. Cooperation between crop and livestock farms has been analysed in both marginal areas (46% of the nitrogen in crop residues and manure is returned to the soil of common pastures or areas of transhumance in southern Mali; Blanchard *et al.* 2013) and highly market-integrated areas (e.g. exchange of straw and manure between specialised farms in France; Moraine *et al.* 2014). Collective actions have led to a wide variety of collaborations for input supply, and access to market or services. These include access to capital, knowledge, and technology. Economically, the logic of such collective action is to limit inputs through linking crop and livestock enterprises, sharing risks, and benefiting from economies of scope (Ryschawy *et al.* 2014, Veysset *et al.* 2014). Protected Designation of Origin (PDO) areas show a more complex combination of actors. Producers and processors of a product demonstrate that the quality and specificity of this product is linked to a specific territory and local practices. PDO products thus become common goods at the territorial level with a "defence and management organisation". This collective organisation is in charge of guaranteeing compliance with production and transformation rules, and of product defence against imitation and unlawful use of its name. These examples show how stakeholders develop collaborative forms that not only target system productivity and economic performances, but also provide various environmental and social benefits.

4.4. *Knowledge management for agroecology*

Adapting management to biotic and abiotic diversity lies at the core of the agroecological transition of livestock farming systems. This implies that farmers no longer apply ready-made solutions developed at the national or international scale by research and development institutions, but instead develop their own skills and apply specific management practices adapted to the local context. Horizontal sharing of locally adapted know-how and knowledge between farmers thus becomes much more important than the vertical diffusion of information that formed the basis of post-World War II modernisation of agriculture. The success of agroecological management is often dependent on the efficient organisation of labour supply, and requires some time for learning and adaptation during the transition to a particular agroecological system (Tripp 2008). Technicians and extension services then become expert contacts that stimulate the ability of farmers to learn and strengthen their decisional self-reliance. Construction and transmission of knowledge thus need to be rebuilt with new paradigms, methods, and tools. Adaptive management is a learning-by-doing process that involves increased dedication to system observation. On that basis, existing networks between stakeholders should be revitalised to discuss practices or experiments, and innovative case studies need to be informed. This can be done through discussion forums, field days, or discussion arenas. New links between empirical and scientific knowledge should also be encouraged. Empirical knowledge is numerous and fragmented, but can provide new subjects for research and new hypotheses. Research must consider livestock farming systems both holistically, on different scales from the animal to the landscape, and in their diversity, associating animal sciences with ecology, economics, and sociology (Dumont *et al.* 2014). Evaluation models (cf. Section 4.1) can thus help stakeholders take a holistic view by providing synthetic results. In addition to evaluation, simulation models appear to be a good option for discussing and exploring new management options and making the link between local situations and generic thinking (Sterk *et al.* 2007). The use of models coupling simulation and evaluation models (e.g. relying on multiple-objective optimisation algorithms) makes it possible to understand trade-offs and synergies among agroecosystem components, while exploring the range of management options that can be used for the collective design of farming systems (Groot *et al.* 2012). All these kinds of models applied to agricultural systems can benefit from stakeholder involvement. Codesigning of models by scientists and stakeholders is thus a powerful way to facilitate knowledge integration, formalisation, transparency, and a shared view, and yield appropriate decision support tools for livestock farming systems (Voinov and Bousquet 2010, Lamanda *et al.* 2012).

5. Conclusion

Agroecological transition in the livestock sector will entail profound changes and more dynamic and integrative perceptions of each system component and their interactions, and a new relationship with society based on several forms of collective action. This opens a vast research agenda for the animal science community (Dumont *et al.* 2014), and calls for interdisciplinary research with ecology, economy and the social sciences. Bridging the gap between scientific and empirical knowledge is another requisite. Historical forms such as cooperative structures or unique leader actors that prescribe rules for managing environmental capital at landscape scale (inter-farm) have proved efficient for production management, but the new challenge set by agroecological thinking calls for more collective management, more innovation, and more widely distributed coordination networks. It must allow the construction of a common multiobjective point of view, and the allocation of individual tasks, depending on the overall system of interactions. To enable implementation, it is important to adapt individual tasks to local constraints considering both material and immaterial resources. This involves collecting and processing information at both farm and inter-farm levels to assess the state and functioning of the agroecosystem, and renewing knowledge management. The promotion of agroecological livestock systems will also demand significant investments in research, development, and education. It will need to be supported by sound agricultural policies, and by funding associating public and private sectors, such as new rural capitalisation incentive or low-risk loans. Initiatives are thus needed from local to international level to speed the uptake of innovation, and the scaling-up of agroecology.

Acknowledgements

We are grateful to Laurence Fortun-Lamothe, Marielle Thomas, and Muriel Tichit for fruitful discussions in the INRA animal think tank on agroecology, and to Fabienne Blanc and Laurence Puillet for allowing us to use model outputs in Figure 1.

References

Altieri, M.A., 2002. Agroecological principles and strategies for sustainable agriculture. In: Uphoff, N.T. (Ed.), *Agroecological Innovations: Increasing Food Production with Participatory Development*, Earthscan, London, UK, pp. 40–46.

Altieri, M.A., Koohafkan, P., Nicholls, C., 2014. *Strengthening Resilience of Modern Farming Systems: A Key Prerequisite for Sustainable Agricultural Production in an Era of Climate Change*. Third World Network Briefing Paper No. 70.

Andrieu, N., Poix, C., Josien, E., Duru, M., 2007. Simulation of forage management strategies considering farm-level land diversity: example of dairy farms in the Auvergne. *Computers and Electronics in Agriculture* **55**, 36–48.

Bell, M.M., Lyon, A., Gratton, C., Jackson, R.D., 2008. Commentary: the productivity of variability, an agroecological hypothesis. *International Journal of Agriculture Sustainability* **6**, 233–235.

Blanc, F., Bocquier, F., Agabriel, J., D'hour, P., Chillard, Y., 2006. Adaptive abilities of the females and sustainability of ruminant livestock systems. A review. *Animal Research* **55**, 489–510.

Blanc, F., Dumont, B., Brunschwig, G., Bocquier, F., Agabriel, J., 2010. Robustesse, flexibilité, plasticité: des processus adaptatifs révélés dans les systèmes d'élevage extensifs de ruminants. *INRA Productions Animales* **23**, 65–80.

Blanc, F., Ollion, E., Puillet, L., Delaby, L., Ingrand, S., Tichit, M., Friggens, N.C., 2013. Evaluation quantitative de la robustesse des animaux et du troupeau: quels principes retenir? *Rencontres autour des Recherches sur les Ruminants* **20**, 265–272.

Blanchard, M., Vayssières, J., Dugué, P., Vall, E., 2013. Local technical knowledge and efficiency of organic fertilizer production in South Mali: diversity of practices. *Agroecology and Sustainable Food Systems* **37**, 672–699.

Bol, R., Dunn, R.M., Pilgrim, E.S., 2011. Managing C and N in grassland systems: the adaptive cycle theory perspective. In: Lemaire, G., Hodgson, J., Chabbi, A. (Eds.), *Grassland Productivity and Ecosystem Services*. CAB International, Wallingford, UK, pp. 73–82.

Bonaudo, T., Burlamaqui Bendahan, A., Sabatier, R., Ryschawy, J., Bellon, S., Leger, F., Magda, D., Tichit, M., 2014. Agroecology principles for the redesign of integrated crop-livestock systems. *European Journal of Agronomy* **57**, 43–51.

Bonaudo, T., Billen, G., Garnier, J., Barataud, F., Bognon, S., Marty, P., Dupré, D., 2015. Le système agro-alimentaire. Un découplage progressif de la production et de la consommation. In: Buclet, N. (Ed.), *Essai d'ecologie territoriale. L'exemple d'Aussois en Savoie*. CNRS édition, Paris, France, pp. 161–182.

Botreau, R., Farruggia, A., Martin, B., Pomiès, D., Dumont, B., 2014. Towards an agroecological assessment of dairy systems: proposal for a set of criteria suited to mountain farming. *Animal* **8**, 1349–1360.

Canario, L., Billon, Y., Caritez, J.C., Bidanel, J.P., Laloë, D., 2009. Comparison of sow farrowing characteristics between a Chinese breed and three French breeds. *Livestock Science* **125**, 132–140.

Chatzimpiros, P., Barles, S., 2013. Nitrogen food-print: N use related to meat and dairy consumption in France. *Biogeosciences* **10**, 471–481.

d'Alexis, S., Sauvant, D., Boval, M., 2014. Mixed grazing systems of sheep and cattle to improve live weight gain: a quantitative review. *Journal of Agricultural Science* **152**, 655–666.

De Groot, J., Ruis, M.A.W., Scholten, J.W., Koolhaas, J.M., Boersma, W.J.A., 2001. Long-term effects of social stress on antiviral immunity in pigs. *Physiology and Behavior* **73**, 145–158.

de la Torre, A., Recoules, E., Blanc, F., Ortigues-Marty, I., D'Hour, P., Agabriel, J., 2015. Changes in calculated residual energy in variable nutritional environments: an indirect approach to apprehend suckling beef cows' robustness. *Livestock Science* **176**, 75–84.

Dourmad, J.Y., Rigolot, C., Jondreville, C., 2009. Influence de la nutrition sur l'excrétion d'azote, de phosphore, de cuivre et de zinc des porcs, et sur les émissions d'ammoniac, de gaz à effet de serre et d'odeurs. *INRA Productions Animales* **22**, 41–48.

Dubois, J.P., Bijja, M., Auvergne, A., Lavigne, F., Fernandez, X., Babilé, R., 2008. Qualité des parcours de palmipèdes: choix des espèces végétales, rendement et résistance au piétinement. In: *Actes des 8èmes Journées de la Recherche sur les Palmipèdes à Foie Gras*, Arcachon, France, pp. 107–110.

Dumont, B., Rossignol, N., Loucougaray, G., Carrère, P., Chadoeuf, J., Fleurance, G., Bonis, A., Farruggia, A., Gaucherand, S., Ginane, C., Louault, F., Marion, B., Mesléard, F., Yavercovski, N., 2012. When does grazing generate stable vegetation patterns in temperate pastures? *Agriculture, Ecosystems and Environment* **153**, 50–56.

Dumont, B., Fortun-Lamothe, L., Jouven, M., Thomas, M., Tichit, M., 2013. Prospects for agroecology and industrial ecology for animal production in the 21st century. *Animal* **7**, 1028–1043.

Dumont, B., González-García, E., Thomas, M., Fortun-Lamothe, L., Ducrot, C., Dourmad, J.Y., Tichit M., 2014. Forty research issues for the redesign of animal production systems in the 21st century. *Animal* **8**, 1382–1393.

Dwyer, C.M., Lawrence, A.B., 1998. Variability in the expression of maternal behaviour in primiparous sheep: Effects of genotype and litter size. *Applied Animal Behaviour Science* **58**, 311–330.

Farruggia, A., Dumont, B., Scohier, A., Leroy, T., Pradel, P., Garel J.P., 2012. An alternative rotational stocking management designed to favour butterflies in permanent grasslands. *Grass and Forage Science* **67**, 136–149.

Faruk, M.U., Bouvarel, I., Meme, N., Rideau, N., Roffidal, L., Tukur, H.M., Bastianelli, D., Nys, Y., Lescoat, P., 2010. Sequential feeding using whole wheat and a separate protein-mineral concentrate improved feed efficiency in laying hens. *Poultry Science* **89**, 785–796.

Francis, C., Lieblein, G., Gliessman, S., Breland, T.A., Creamer, N., Harwood, R., Salomonsson, L., Helenius, J., Rickerl, D., Salvador, R., Wiedenhoeft, M., Simmons, S., Allen, P., Altieri, M., Flora, C., Poincelot, R., 2003. Agroecology: The ecology of food systems. *Journal of Sustainable Agriculture* **22**, 99–118.

Gerber, P.J., Steinfeld, H., Henderson, B., Mottet, A., Opio, C., Dijkman, J., Falcucci, A., Tempio, G., 2013. *Tackling Climate Change Through Livestock — A Global Assessment of Emissions and Mitigation Opportunities*. Food and Agriculture Organization of the United Nations (FAO), Rome.

Gidenne, T., Garcia, J., Lebas, F., Licois, D., 2010. Nutrition and feeding strategy: interactions with pathology. In: De Blas, C., Wiseman, J. (Eds.), *Nutrition of the Rabbit*, CABI Publishing, pp. 179–199.

Gliessman, S.R., 2007. Animals in Agroecosystems. In: *Agroecology: The Ecology of Sustainable Food Systems*, 2nd Edition. CRC Press, Boca Raton, FL, USA, pp. 269–285.

Gomes, L.C., Pailleux, J.Y., Dedieu, B., Alcade, C.R., Cournut, S., 2014. An approach for assessing the ecological intensification of livestock systems. In: *Farming Systems Facing Global Challenges: Capacities and Strategies*, 11th European IFSA Symposium, Berlin, Germany, pp. 1352–1363.

Groot, J.C.J., Oomen, G.J.M., Rossing, W.A.H., 2012. Multi-objective optimization and design of farming systems. *Agricultural Systems* **110**, 63–77.

Hangalapura, B.N., Nieuwland, M.G.B., Reilingh, G.D., Buyse, J., Van Den Brand, H., Kemp, B., Parmentier, H.K., 2005. Severe feed restriction enhances innate immunity but suppresses cellular immunity in chicken lines divergently selected for antibody responses. *Poultry Science* **84**, 1520–1529.

Hector, A., Schmid, B., Beierkuhnlein, C., Caldeira, M.C., Diemer, M., Dimitrakopoulos, P.G., Finn, J.A., Freitas, H., Giller, P.S., Good, J., Harris, R., Högberg, P., Huss-Danell, K., Joshi, J., Jumpponen, A., Körner, C., Leadley, P.W., Loreau, M., Minns, A., Mulder, C.P.H., O'Donovan, G., Otway, S.J., Pereira, J.S., Prinz, A., Read, D.J., Scherer-Lorenzen, M., Schulze, E.D., Siamantziouras, A.S.D., Spehn, E.M., Terry, A.C., Troumbis, A.Y., Woodward, F.I., Yachi, S., Lawton, J.H., 1999. Plant diversity and productivity experiments in European grasslands. *Science* **286**, 1123–1127.

Jouven, M., Baumont, R., 2008. Simulating grassland utilization in beef suckler systems to investigate trade-offs between production and floristic diversity. *Agricultural Systems* **96**, 260–272.

Jouven, M., Lapeyronie, P., Moulin, C.H., Bocquier, F., 2010. Rangeland utilization in Mediterranean farming systems. *Animal* **4**, 1746–1757.

Kosgey, L.S., Baker, R.L., Udo, H.M.J., Van Arendonk, J.A.M., 2006. Successes and failures of small ruminant breeding programmes in the tropics: a review. *Small Ruminant Research* **61**, 13–28.

Krätli, S., 2008. Time to outbreed animal science? *A Cattle-breeding System Exploiting Structural Unpredictability: The Wodaabe Herders in Niger*. STEPS Working Paper 7. STEPS Centre, Brighton, UK.

Kuiken, T., Fouchier, R., Rimmelzwaan, G., van den Brand, J., van Riel, D., Osterhaus, A., 2011. Pigs, poultry and pandemic Influenza: how zoonotic pathogens threaten human health. *Advances in Experimental Medicine and Biology* **719**, 59–66.

Lamanda, N., Roux, S., Delmottec, S., Merot, A., Rapidela, B., Adamd, M., Werye, J., 2012. A protocol for the conceptualisation of an agro-ecosystem to guide data acquisition and analysis and expert knowledge integration. *European Journal of Agronomy* **38**, 104–116.

Mandonnet, N., Tillard, E., Faye, B., Collin, A., Gourdine, J.L., Naves, M., Bastianelli, D., Tixier-Boichard, M., Renaudeau, D., 2011. Adaptation des animaux d'élevage aux multiples contraintes de régions chaudes. *INRA Productions Animales* **24**, 41–64.

Mignon-Grasteau, S., Muley, N., Bastianelli, D., Gomez, J., Peron, A., Sellier, N., Milet, N., Besnard, J., Hallouis, J.M., Carré, B., 2004. Heritability of digestibilities and divergent selection for digestion ability in growing chicks fed a wheat diet. *Poultry Science* **83**, 860–867.

Moraine, M., Duru, M., Nicholas, P., Leterme, P., Therond, O., 2014. Farming system design for innovative crop-livestock integration in Europe. *Animal* **8**, 1204–1217.

Mounier, L., Colson, S., Roux, M., Dubroeucq, H., Boissy, A., Ingrand, S., Veissier, I., 2006. Links between specialization in the finishing of bulls, mixing, farmers' attitudes towards animals and the production of finishing bulls: a survey on French farms. *Animal Science* **82**, 561–568.

Murgueitio, E., Calle, Z., Uribe, F., Calle, A., Solorio, B., 2011. Native trees and shrubs for the productive rehabilitation of tropical cattle ranging lands. *Forest Ecology and Management* **261**, 1654–1663.

Nätt, D., Lindqvist, N., Stranneheim, H., Lundeberg, J., Torjesen, P.A., Jensen, P. 2009. Inheritance of acquired behaviour adaptations and brain gene expression in chickens. *PLoS ONE* **4**, e6405.

Öckinger, E., Smith, H.G., 2007. Semi-natural grasslands as population sources for pollinating insects in agricultural landscapes. *Journal of Applied Ecology* **44**, 50–59.

Ollion, E., Ingrand, S., Delaby, L., Trommenschlager, J.M., Colette-Leurent, S., Blanc, F., 2016. Assessing the diversity of trade-offs between life functions in early lactation dairy cows. *Livestock Science* **183**, 98–107.

Petersen, S.O., Sommer, S.G., Béline, F., Burton, C., Dach, J., Dourmad, J.Y., Leip, A., Misselbrook, T., Nicholson, F., Poulsen, H.D., Provolo, G., Sorensen, P., Vinnerås, B., Weiske, A., Bernal, M.P., Böhm, R., Juhász, C., Mihelic, R., 2007. Recycling of livestock manure in a whole-farm perspective. *Livestock Science* **112**, 180–191.

Puillet, L., Martin, O., Sauvant, D., Tichit, M., 2010. An individual-based model simulating goat response variability and long term herd performance. *Animal* **4**, 2084–2098.

Radomsky, G., Niederle, P., Schneider, S., 2014. Participatory systems of certification and alternative marketing networks: the case of the Ecovida Agroecology Network in South Brazil. In: Hebinck, P., van der Ploeg, J.D., Schneider, S. (Eds.), *Rural Development and the Construction of New Markets*, Routledge Publisher, London and New York, pp. 79–98.

Röös, E., Patel, M., Spångberg, J., Carlsson, G., Rydhmer, L., 2016. Limiting livestock production to pasture and by-products in a search for sustainable diets. *Food Policy* **58**, 1–13.

Ryschawy, J., Choisis, J.P., Joannon, A., Gibon, A., Le Gal, P.Y., 2014. Participative assessment of innovative technical scenarios for enhancing sustainability of French mixed crop-livestock farms. *Agricultural Systems* **129**, 1–8.

Sabatier, R., Doyen, L., Tichit, M., 2014. Heterogeneity and trade-off between ecological and productive functions of agro-landscapes: a model of cattle-bird interactions in a grassland agroecosystem. *Agricultural Systems* **126**, 38–49.

Sabatier, R., Oates, L.G., Jackson, R.D., 2015. Management flexibility of a grassland agroecosystem: a modelling approach based on viability theory. *Agricultural Systems* **139**, 76–81.

Soussana, J.F., Lemaire, G., 2014. Coupling carbon and nitrogen cycles for environmentally sustainable intensification of grasslands and crop-livestock systems. *Agriculture, Ecosystems and Environment* **190**, 9–17.

Sterk, B., van Ittersum, M.K., Leeuwis, C., Wijnands, F.G., 2007. Prototyping and farm system modeling — patterns on the road towards more sustainable farm systems? *European Journal of Agronomy* **26**, 401–409.

Thomas, M., Fortun-Lamothe, L., Jouven, M., Tichit, M., González-García, E., Dourmad, J.Y., Dumont, B., 2014. Agro-écologie et écologie industrielle:

deux alternatives complémentaires pour les systèmes d'élevage de demain. *INRA Productions Animales* **27**, 89–100.

Tichit, M., Hubert, B., Doyen, L., Genin, D., 2004. A viability model to assess the sustainability of mixed herds under climatic uncertainty. *Animal Research* **53**, 405–417.

Tripp, R., 2008. Agricultural change and low input technology. In: Snapp, S., Pound, B. (Eds.), *Agricultural Systems: Agroecology and Rural Innovation for Development*, Elsevier, Amsterdam, The Netherlands, pp. 129–160.

van Zanten, H.H.E., Meerburg, B.G., Bikker, P., Herrero, M., de Boer, I.J.M., 2016. Opinion paper: The role of livestock in a sustainable diet: a land-use perspective. *Animal* **10**, 547–549.

Veysset, P., Lherm, M., Bébin, D., Roulenc, M., 2014. Mixed crop-livestock farming systems: a sustainable way to produce beef? Commercial farms results, questions and perspectives. *Animal* **8**, 1218–1228.

Voinov, A., Bousquet, F., 2010. Modelling with stakeholders. *Environmental Modelling and Software* **25**, 1268–1281.

West, J.W., 2003. Effects of heat-stress on production in dairy cows. *Journal of Dairy Science* **86**, 2131–2144.

Yin, T., Bapst, B., Borstel, U.U.V., Simianer, H., König, S., 2012. Genetic parameters for Gaussian and categorical traits in organic and low input dairy cattle herds based on random regression methodology. *Livestock Science* **147**, 159–169.

Chapter 11

Agroecological Principles and Practices for Grass-based Farming Systems

Alain Peeters and Alexander Wezel

RHEA Research Centre, Corbais, Belgium
Department of Agroecology and Environment, ISARA, Lyon, France

1. Introduction

On a global scale, grasslands cover about 26% of total land area and 80% of agriculturally productive land (Boval and Dixon 2012) or about 41% of the terrestrial area excluding Greenland and Antarctica (White *et al.* 2000). Grasslands provide the feed base for grazing livestock. They are the source of important ecosystem services and fulfil several ecological, economic, and social functions. For instance, grasslands have a huge potential to mitigate climate change by absorbing CO_2 from the atmosphere and storing it in their soils (FAO 2009a). Pastures and rangelands represent a carbon sink that could be greater than forests if properly managed. They are an irreplaceable reservoir of biodiversity. Livestock is the largest land use sector on Earth (Herrero and Thornton 2013). In addition to high-quality foods, it provides products and services such as fertiliser, transport, traction, fibre, and leather. It contributes 40% of the global value of agricultural output and supports the livelihoods and food security of almost a billion people (FAO 2009b).

Although livestock systems can considerably vary from extensive to intensive systems, from grassland-based to "factory farming" or "confined animal feeding operation systems", from one breed to multiple breeds or multiple species systems, from specialised to mixed crop–livestock systems, they have the common primary objective to produce optimal quantities of meat, milk, eggs, or fibre in relation to resource inputs. Nevertheless, the underlying principles to attain this objective can be quite contrasting, e.g. feeding the animals exclusively with grass produced on the farm or with a high proportion of concentrate feeding from imported maize, soybean, or other resources.

Herbivores can be raised in very different environments and systems. In conventional systems, beef and meat sheep are often maintained in stables or feed-lots during the fattening period and are "finished" with cereals and protein-rich feed. The contribution of grasslands in the diet of dairy cows was also reduced in the last decades in conventional systems to the benefit of grain and green maize. In factory farming systems, pigs or poultry are concentrated in buildings on very small surfaces and mainly fed with grains that can have overseas origin, such as soybean imported from South America for European systems (Boerema *et al.* 2016). As a result, livestock production is very much criticised for its negative impact on the environment (Steinfeld *et al.* 2006). In contrast, in agroecological systems, ruminants tend to be exclusively fed on grass and monogastrics are at least partly fed on grass. Agroecology was often mentioned in the last decade for having the potential to develop sustainable agriculture in the future. Developing agroecological livestock systems is thus highly desirable.

Grazing/browsing livestock are large and small ruminants such as cattle, sheep and goats, and monogastrics such as horse, donkey, pig, poultry, and rabbit that also have the ability to graze. All these animals are herbivores if "herbivory" is defined as "a form of consumption in which an organism principally eats autotrophs such as plants, algae, and photosynthesising bacteria" (https://en.wikipedia.org/wiki/Herbivore).

As other agroecological systems, agroecological grazing/browsing livestock systems are based on a "system" or "holistic" approach (Peeters

2015a) considering simultaneously multiple factors, multiple scales, and having multistakeholder involvement (Wezel and David 2012). During a transition between conventional to agroecological systems, many if not all components of the system need to be modified at the same time and in the right proportion. This system optimisation procedure requires an interdisciplinary approach. This approach is fundamentally different from the "reductionist" or "analytical" approach of conventional systems that tries to solve problems one by one without looking very much to interactions.

Agroecology is inspired by the biomimicry principle (Benyus 1997). It tries to adopt the principles of nature functioning such as diversity at all scales, nutrient cycling, permanent soil cover, soil–plant–animal relationships, and self-regulation processes. In contrast with the Green Revolution and other commercial input-based systems, it does not rely directly or indirectly on massive use of fossil fuel but it is ecosystem-based. It aims to restore ecosystems and to rely on ecosystem services provided by different types of biodiversity (Peeters *et al.* 2013). Agroecological systems are low fossil energy and commercial input systems, but they are observation and knowledge intensive. This is particularly true in soil–plant–animal chains of grassland-based systems because interactions between system components are often more complex than in cropping systems.

Agroecological principles and practices were mainly defined for cropping systems, market gardening, and permanent crops, less so for livestock and grassland-based farming systems. After 20 years of research on agroecology, it is time to generalise agroecological principles and practices to grazing systems. We recently made a first attempt to do it (Wezel and Peeters 2014). This paper is a more elaborated and argued version. This chapter focuses on grazing/browsing livestock in temperate areas. It uses examples from Australia, Europe, Japan, New Zealand, and North America.

2. Agroecological principles

Principles of agroecological grass-based farming systems are rules that should be adopted concerning the functioning of agroecological processes. They can be classified into seven groups (Figure 1).

Farmers

↓ (i) Building on farmers' knowledge and culture

Individual resource and system management

↓ (ii) Conserving and managing biodiversity
(iii) Maximising resource management efficiency
(iv) Optimising agroecosystem management

Sustainability in farming and society

(v) Building a sustainable socio-economic farming system
(vi) Ensuring citizens' food and health
(vii) Restoring quality social relations in the society

Figure 1. The seven categories of agroecological principles related to grass-based farming systems.

2.1. *Building on farmers' knowledge and culture*

The farmer and the farmer's family are the decision centres of agroecological grazing/browsing systems (Figure 2). They take decisions every day on the basis of their perception of the state of the system and its ecological, sociological, scientific, technical, policy, and economic environments. Their decisions are based on their experience, skill, and knowledge. Farmers are innovating continuously. They try new solutions and learn from a trial-and-error process. After years of activities and intergenerational transmission, farmers' knowledge accumulates and adapts to local environments. It is thus very valuable and should be the basis of any attempt of system improvement by scientists or farmers' advisers. In agroecological system development, farmers' knowledge should be combined in a participatory process with the most up-to-date scientific knowledge and adequate techniques for the development of agroecological practices (Peeters 2015a). Actually, farmers combine their own knowledge with information they get from different sources, notably from advisers, neighbourhood farmers, scientists, and diverse media.

Renowned farmers can be important sources of inspiration. Some of them can be cited (Figure 3). Joel Salatin developed holistic management methods of animal husbandry on his Polyface Farm in Virginia (USA). He produces high-quality "beyond organic" meats. He associates cows and chicken in a first-last system of rotational stocking (Section 3.4.7) (see definition in Allen *et al.* (2011)). Chickens are kept in portable coops

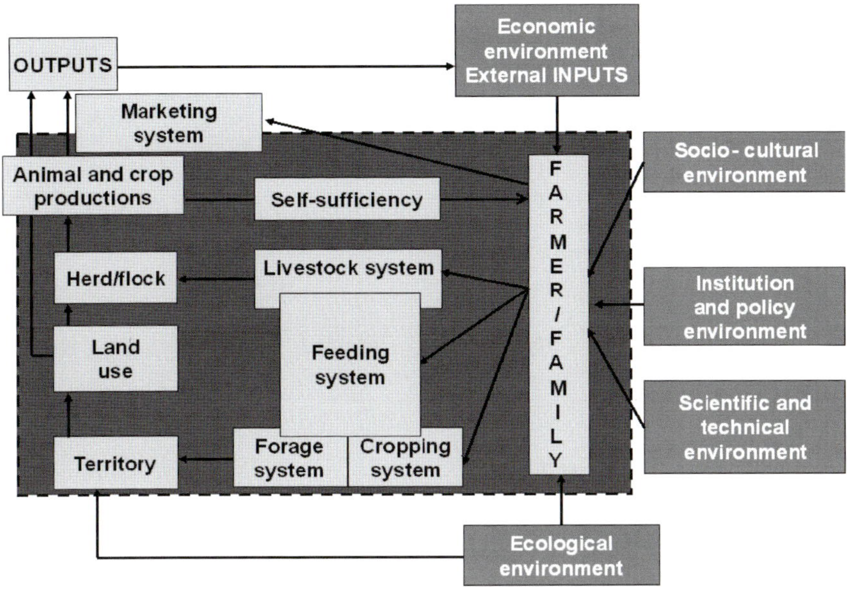

Figure 2. An agroecological mixed crop–livestock farm system (delimited by the dotted line) and its environments.

Figure 3. (Left to right) Joel Salatin © Rachel Salatin, André Pochon © André Pochon, Julius Ruechel © Julius Ruechel, and Masanobu Fukuoka (photo: Larry Korn).

moved on grassland plots. They feed on the short grass grazed some days before by cows and dig through the cow dung to eat protein-rich fly larvae while further fertilising the field with their droppings (Salatin 1996a, 1996b, http://www.polyfacefarms.com/). Julius Ruechel managed his family farm in south-western Canada. He developed a grass-based system for improving cattle health and reducing costs. He experimented winter grazing, began the transition to certified organic production, started developing a direct-marketing program, and began experimenting with grass finishing. He wrote a book on grass fed cattle on the basis of this experience (Ruechel 2006, http://www.grass-fed-solutions.com). André Pochon is a French farmer from Brittany. He is one of the defenders of peasant farming. He designed and promoted a rotational stocking system based on white clover (*Trifolium repens*) (Pochon 2002, http://www.cedapa.com/). Harry Wier, a New Zealander farmer, developed the TechnoGrazing™ concept and trained other farmers to apply it. The concept is based on the division of a farm into a grid of a high number of small grazing cells allowing very intensive and efficient pasture utilisation and a significant increase in animal production. It requires only a minimum investment and a low labour input (Charlton and Wier 2001).

A few other famous farmers used monogastric herbivores as "collaborators" of the system for achieving some specific tasks. Sepp Holzer is a pioneer of permaculture and uses specific techniques at high altitudes (1,100–1,500 m above sea level) on his farm in Austria. He experimented with different animal types. He used for instance pigs to dig grassy plots by throwing some corn and fruit on the spot for encouraging pigs to forage and disturb soil. After a couple of days, the ground was sufficiently prepared for sowing a crop (Holzer 2011, www.krameterhof.at). Bernard Poujol is also using animals as "collaborators" of his rice cropping system in Camargue, France (Figure 4). Large flocks of ducks are weeding his rice fields (http://www.canard-desrizieres.fr/). Masanobu Fukuoka was a Japanese farmer and philosopher (Figure 3). He is considered as one of the most influential personality of the organic movement. He ran a farm in southern Japan. His method is known as the "natural way of farming". Although his system was mainly based on vegetables, citrus, rice, and winter grain, he also used ducks in his rice fields as one component of

Figure 4. Bernard Poujol © Sylvie Curty.

his integrated agroecosystem approach. Ducks ate slugs and deposited nitrogen-rich fertiliser in the paddy fields, as well as provided high-protein eggs and meat (Fukuoka 2009).

Agroecological systems should be locally and culturally adapted. Practices cannot just be copied and pasted from one region to another. They should be adapted to local conditions by taking into account soil, climate, social, and economic conditions. However, systems developed in very distant areas could be comparable such as the grassland-based systems of New Zealand and Ireland, and can be a source of inspiration for more grain-based systems in different parts of the temperate world.

Agroecological systems should empower people and make them less dependent on commercial influence and input use. Local farmers should be able to control their own development process. This could be achieved by increasing farmers' participation in official research and development process through holistic and participatory approaches. Development should thus be "people-centric" and based on farmer participation (Gliessman, http://www.agroecology.org/Principles_List.html).

2.2. *Conserving and managing biodiversity*

Biodiversity should have a central place as the driver of the agroecosystem. Agroecological systems should include as much as possible multibreeds, multicultivars, and multispecies components, and a diversity of habitats. Biodiversity should be enhanced from soil to landscape. All biodiversity types — agricultural, functional, and heritage biodiversity — should be associated to the system and conserved (Peeters *et al.* 2004).

Semi-natural landscape elements should be conserved and developed on the farm and at landscape scale (e.g. semi-natural grasslands, hedges, thickets, woodland strips, traditional orchards, herbaceous field margins, ditches, ponds). These habitats should be carefully managed and benefit from a level of care comparable to crops, grasslands, or livestock.

2.3. *Maximising resource management efficiency*

Nitrogen fixing legumes and especially forage legumes should be at the core of agroecological system fertility and productivity. Continuous soil cover should be ensured for optimal nutrient cycling, soil organic matter accumulation, soil fertility, and erosion prevention. There should be minimum or ideally no use of inorganic nitrogen, other synthetic fertilisers, pesticides, and drugs. Organic fertilisation should be carefully planned with priority to on-farm produced manure or slurry. Irrigation water should be minimised. The system should include a carbon storage and greenhouse gas (GHG) emission control strategy. Energy use should be optimised and fossil fuel use minimised.

2.4. *Optimising agroecosystem management*

The main ecological relationships should be identified and managed in the context of the farm soil–plant–animal chains. Agroecological systems should be ecosystem-based and not fossil fuel-based. This requires an overall design and a set of innovative management techniques. Nutrient cycling should be optimised by nutrient loss reduction, continuous soil cover, no inversion tillage, fertility transfer from excess to deficit fertility areas, soil–plant–animal integration, mixed farming, and agroforestry. Low commercial input (e.g. feed, fertiliser, fuel) and low investment systems (e.g. machines, buildings) should be developed. High commercial input systems should only be adopted if not in contradiction with other agroecological principles.

The system should make room for a diversity of structures over time and in space, a diversity of land use and land cover types, of plant traits, and of livestock types. Management should thus integrate biodiversity from soil life to landscape, from production types to producers. This diversity should reflect the variability of local conditions and correspond to a specific management.

Agroecological systems should not be adapted to techniques and animal breeds; techniques and breeds should be chosen in relation to the system. This means that sustainable systems, adapted to local conditions, should be designed first and then techniques should be chosen and developed, cultivars and breeds should be adopted or bred in the framework of these systems. Indeed, adoption of highly productive breeds often leads to modifications of production systems that are unsustainable.

Mixed crop–livestock systems should be adopted as much as possible. Specialised grazing livestock systems could also be sustainable and particularly well adapted to marginal areas. A maximum rate of self-sufficiency in energy, protein, and fodder should be targeted at farm, landscape, and regional scale.

Animal health should be managed by (i) optimum adaptation of animals to the system by a reasoned choice of species and breed (see above) including a moderate specialisation of animal production, (ii) optimum feeding, hygiene, rotational stocking, and all other prevention methods through a holistic approach at system level (Section 3.5.4), (iii) creation of animal welfare conditions by the integration of animal ecological and social needs into the system functioning, (iv) curative instead of systematic disease and parasite treatments, and (v) when necessary, treatments based on products available in nature instead of synthetic substance.

2.5. *Building a sustainable socio-economic farming system*

Agroecological systems should be "economically viable" for farmers in order to provide them a decent income, and "liveable" for ensuring quality of life to farmer's family. Bank debts and production costs should be kept to a minimum. Systems should be "inheritable", they should be transferred to the next farmer's generation. Low bank debts and production costs are also important with regard to this criterion. Systems should be resilient to climate change and economic crisis. They should be socially equitable and environmentally sound in relation with water, soil, climate, energy, and genetic resources.

2.6. *Ensuring citizens' food and health*

In advanced agroecological systems, systems are managed in a way that provides optimum quantity of healthy and tasty food for citizens. At a regional scale, they should provide the largest possible diversity of food in a balanced way for fulfilling nutritional needs of local population. Agroecological systems should reconcile local supply and demand for food by local food processing, and short and local marketing chains.

2.7. *Restoring quality social relations in the society*

In agroecological systems, social relations seek to develop a new harmony in human societies. Human social relations should be founded on collaborations and the development of synergies instead on competition and antagonisms. Agroecological systems should link farmers to consumers as a form of empowering. It should develop alternatives to mass and prepared food distribution channels.

2.8. *Other principles defined in literature*

Very few publications were published on agroecological principles and practices in livestock system. Gliessman (2007) integrated a chapter about animals in agroecosystems in his book, but did not describe or define principles or practices for agroecological livestock systems. Recently, Dumont *et al.* (2013) did not explicitly define agroecological practices in their paper on "agroecology and industrial ecology for animal production", but they described five groups of principles: (i) adopting management practices aiming to improve animal health, (ii) decreasing the inputs needed for production, (iii) decreasing pollution by optimising the metabolic functioning of farming systems, (iv) enhancing diversity within animal production systems to strengthen their resilience, and (v) preserving biological diversity in agroecosystems by adapting management practices. These principles remain mostly technical. They do not mention a principle related with the recognition that farmers are the decision-makers and the managers of the agroecosystem and that their knowledge is thus essential. Also principles that are related with the future and the social role of agriculture in society and food and health are not explicitly developed. Within the framework of the farming and food system approach in agroecology (Francis *et al.* 2003, Wezel and David 2012, Wezel *et al.* 2015), these groups of principles are also important to be considered. In contrast, Dumont *et al.* (2013) have defined one specific principle on animal health, which we consider under the principle of system management optimisation. Bonaudo *et al.* (2014) also dealt with agroecological principles, but they focussed specifically on integrated crop–livestock systems. They provided some principles which are more related to cropping systems within cropping–livestock systems, but did not explicitly define the principles of the livestock system. Their major principles are also based on diversity, maximisation of ecological interactions, closing nutrient and energy cycles, optimising nutrient availability, and collective management at the landscape scale. Baumont *et al.* (2014) defined their view on sustainable intensification of grass-based ruminant production and discussed definition and challenges of sustainable intensification. They addressed also different aspects of grassland and livestock management.

3. Agroecological practices

If it is important to define a list of principles, the main concern for farmers is to translate them into concrete measures, practices, and management. Therefore, agroecological practices are described with some details in this paper (Figure 5). Seven categories are proposed. The individual components of these practices should be regarded as tools that can be adopted or not according to the structure and the characteristics of specific livestock systems.

Land and resource management:
(i) Diversification of land use, land cover, livestock types and productions
(ii) Specific resource management in mixed cropping–livestock systems
(iii) Biodiversity conservation

↓

Technical production aspects:
(iv) Grassland and rangeland management
(v) Livestock types and management

↓

Socio-economic aspects:
(vi) Diversification and increase of income sources
(vii) Optimisation of farming and food system

Figure 5. The seven categories of agroecological practices related to herbivore farming systems.

Three categories of practices are related with **land and resource management**: (i) diversification of land use, land cover, livestock types, and productions, (ii) specific resource management in mixed crop–livestock systems, and (iii) biodiversity conservation.

Two categories of practices are linked with **technical production aspects**: (iv) grassland and rangeland management and (v) livestock types and management.

The last two categories of practices are (vi) diversification and increase of income sources and (vii) optimisation of farming and food system. They are related with **socio-economic aspects** through the strong synergies and partnerships that exist in agroecological systems between different types of stakeholders and especially between farmers and consumers.

3.1. *Diversification of land use, land cover, livestock types, and productions*

In mixed crop–livestock farms, high crop diversity and long crop rotations including the integration of legume-based temporary grasslands can improve fodder quality, reduce nutrient inputs, increase soil fertility, control weeds, pests, and diseases. Arable plots should be sufficiently narrow in size for increasing spatial heterogeneity by alternating crop species and crop and grassland species.

Instead of trying to standardise permanent grassland quality and production potential on the whole farm territory, it may be more interesting to combine the use of different grassland types (Allen *et al.* 2011, Peeters *et al.* 2014) for different types of animals. Even in a single herd or flock, different animal categories most often coexist (Table 3). This diversity of animal types is even higher when different livestock species or breeds are raised on the same farm. These animals have different feeding needs, for instance a highly productive lactating dairy cow and a dry beef cow have very different protein and energy requirements. In this context, agricultural productive or less productive permanent grasslands and semi-natural grasslands can be associated to improve resource use, maintain biodiversity, improve product quality, and decrease production costs. For instance, it would not be economically beneficial to invest large amounts of commercial inputs for improving quality and yield of extensive grasslands for feeding low requirement animals. Some geographical reasons such as the necessary proximity of grazed grasslands with the milking parlour or the difficulty to graze remote and isolated plots even with non-milked animals justify the combination of different management types such as only mown, only grazed, and grazed/mown grasslands. The combination of permanent and temporary grasslands may also provide different quality forage, usually higher in temporary grasslands for winter-feeding of high performance animals. Mainly cut temporary grasslands can also complement mainly grazed permanent grasslands for extending the grazing season in early spring and late autumn since temporary grasslands may start growing earlier in spring and may exhibit a prolonged growth in autumn.

Diversity of livestock species and breeds may increase resilience of the system. This diversity makes mixed stocking possible (Section 3.4.7) which may improve grassland use and animal health, and increase farm income sources. Diversity of animal productions may increase economic resilience in case of price instability and price decrease of some products compared to others (e.g. meat versus milk or beef meat versus sheep meat for instance).

The integration of trees into silvo-pastoral or agro-silvo-pastoral systems can present several assets. Animal and fruit productions can be combined in traditional meadow orchards or even in high-density low-size orchards and vineyards grazed by Shropshire sheep for instance, a breed with a well-established reputation for being "tree safe" (Figure 6). This is an application of the "Two Crops From One Acre" principle. Even more than two crops can be harvested from a single plot. Fruits from traditional meadow orchards can be used for preparing fruit juice or cider. The energy rich by-product of this beverage production, fruit pulp, can be used for feeding pigs. Dairy cows grazing under fruit trees produce meat and milk. A part of the milk can be processed into cheese and butter. The protein rich by-products of these dairy products, whey and buttermilk, can also be incorporated in pig feeding. Laying hens and broilers can feed on grass, worms, and arthropods they find in grassland under the protection of fruit tree canopy. During the spectacular flowering of fruit trees, nectar can be collected by honeybees for honey production. The combination of permanent grassland and fruit trees constitutes thus a very productive example of a silvo-pastoral agro-ecosystem. Trees for timber or firewood can be planted in grassland plots, or in hedges or woodland strips on the margin of grassland plots. Hedges can be used for fencing and providing shelter to livestock (Figure 7). Forage woody species can provide nutritious forage not only for goats in shrubby vegetation but also for cattle and sheep in hedges. Although this practice was widespread in traditional systems, it has been forgotten in intensive conventional systems. Recently, a knowledge-sharing network and a database have been created in the Netherlands "to promote the multifunctional use of trees and shrubs in modern farming systems, and to create agroforestry and silvo-pastoral systems" (http://www.voederbomen.nl/). Different livestock types can ingest acorns from oaks in forests or forest grasslands. All these woody elements increase land productivity, diversify productions, provide shelter and forage to livestock, fix carbon, and produce tannins for controlling livestock internal parasites.

3.2. *Specific resource management in mixed crop–livestock systems*

The application of agroecological principles is much easier in mixed crop–livestock systems than in specialised arable systems. A sustainable nutrient and carbon cycle can indeed more easily be implemented in

Figure 6. Shropshire ewes in an orchard (photo: Pierre Moran Mouchard).

Figure 7. A silvo-pastoral system in a "bocage" region where hedges surround grassland plots (photo: Serge Rouxhet).

mixed crop–livestock systems. One of the reasons for this is the possibility to introduce legume-based temporary grasslands in crop rotations since livestock is present for using the forage produced. Although this is also sometimes possible in specialised arable systems when there are possibilities to sell grass hay or silage to surrounding livestock farms, horse owners or riding schools for instance, this is not possible everywhere. Other reasons for improved nutrient and carbon cycle in mixed crop–livestock systems include the possibility to produce and spread farmyard manure and other livestock effluents on arable land. Forage plants can also be cropped and harvested in intercropping; they protect soil, store carbon in the soil, fix nitrogen, and produce quality forage.

All temporary grassland (ley) types are important components of grazing livestock systems but especially legume-based mixtures. Perennial legumes such as lucerne (*Medicago sativa*) and red clover (*Trifolium pratense*) can fix annually up to 300–400 kg N/ha (Deprez *et al.* 2004). A part of this nitrogen is distributed to animals as a source of protein, for instance as hay or silage for the winter period. The main proportion of this ingested nitrogen, about 80–92% in beef cattle (Marini and van Amburgh 2005, Yan *et al.* 2007, Waldrip *et al.* 2013) and about 65–80% in lactating dairy cows (Peyraud *et al.* 1995, Huhtanen and Hristov 2009, Mills *et al.* 2009, Weiss *et al.* 2009), is excreted in dungs and urine and can be partly spread as manure on annual crops. Another part of the total nitrogen fixation by legumes is directly available for the following crops of the crop rotation through soil plant residues, root excreta, and dead rhizobium nodules. Not surprisingly, legume-based temporary grasslands are the main source of nitrogen for large-scale agroecological systems. They fix much more nitrogen than the two other important nitrogen sources: annual pulses and legume-based intercrops (Peeters *et al.* 2006; Jensen *et al.* 2010 in Murphy-Bokern *et al.* 2016; Baddeley *et al.* 2013; Gebhard *et al.* 2013). Large use of nitrogen fixing legumes in temporary grassland mixtures and in fodder crops such as cereal–pea mixtures harvested green at grain milky stage for silage or for dry grain may reduce significantly commercial nitrogen inputs in crop rotations. The inclusion of leys in crop rotations is also an efficient mean for controlling weeds, crop pests and diseases.

Optimum management of organic matter and transfer between livestock and arable land (farmyard manure or slurry) and between arable land and livestock (cereal straw as litter or fodder, green forage, crop by-products such as straw or bran) create an ecological cycle of energy, matter, and nutrients that is one of the main bases of agroecological systems. In some cases, this virtuous cycle can be implemented at regional or landscape level by a cooperation between arable and livestock farmers for straw-manure exchanges, use of crop residues by sheep grazing cereal stubbles for instance (Figure 8), forage hay and silage sales. Compared to commercial input-based systems, this farmers' cooperation also reduces energy for transport, optimises nutrient use and matter cycling. It is notably a good way to improve soil organic matter content in arable farms.

Mixed crop–livestock farms have the possibility to maximise forage/feed self-sufficiency by giving priority to grass-based systems in combination with local arable forage production for reducing fodder imports and production costs.

3.3. *Biodiversity conservation*

Semi-natural grasslands are among the most species-rich ecosystems in temperate areas (Pärtel *et al.* 2005). Traditional extensive systems produced these wildlife habitats and those have sometimes survived in current more intensive systems. They can be integrated in agroecological systems by a careful management: the use of adapted stocking rates on semi-natural pastures and late or low frequency mowing on semi-natural meadows (Janssens and Peeters 2000, Plantureux *et al.* 2005, Gaujour *et al.* 2012). These techniques may conserve or increase species richness and improve product quality (Figure 9).

The maintenance of high nature value grasslands often requires the use and conservation of rare or rustic breeds that have the potential to produce animal products from low quality forage. Management of species-rich grasslands can thus be combined to the conservation of livestock genetic resources and to the production of quality meat and dairy products, possibly promoted by protected designation of origin labels.

Still existing or abandoned pastoral areas can be used or re-used by adapted stocking pressures on marginal land (e.g. for sheep, goats, and horses), for example in dry grasslands, open woodlands, and other types or rangelands. This would conserve biodiversity of these systems,

Figure 8. Sheep grazing clover and cereal stubbles (photo: Pierre Moran Mouchard).

Figure 9. A species-rich, semi-natural grassland (photo: Serge Rouxhet).

conserve specific wild species and domestic livestock breeds, and decrease production costs.

Agroecological systems can maintain or establish semi-natural landscape elements on the farm or in the landscape (e.g. semi-natural grasslands, hedges, ponds) for increasing species richness and ecosystem diversity.

3.4. *Grassland and rangeland management*

In contrast with conventional livestock systems where, increasing amounts of grain were used in ruminant and monogastric livestock in the last decades, agroecological systems give priority to grasslands as the main source of quality feed for grazing animals. Grasslands are the cheapest source of nutrients and can thus decrease feeding costs. In a large number of temperate countries, Dillon *et al.* (2005) showed a strong negative relationship between total production costs and the grazed-grass proportion in the diet of dairy cows, when the proportion of grass in their diet is higher than 50% (Figure 10).

Agroecological grassland management is understood in this paper as a system without or with little inorganic fertiliser, no maize silage and low or no supplementation with concentrate feed. This management relies on several components.

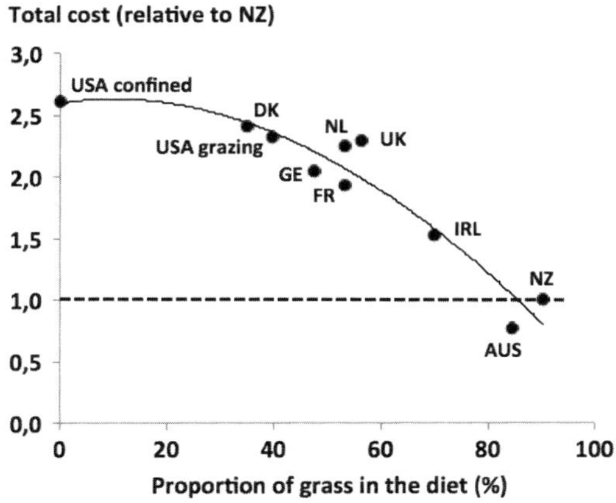

Figure 10. Total cost (expressed relative to New Zealand costs) of milk production according to the proportion of grass in the annual diets of cows (adapted from Dillon *et al.* 2005).

Note: AUS, Australia; DK, Denmark; NL, The Netherlands; GE, Germany; FR, France; UK, United Kingdom; IR, Ireland.

3.4.1. Using complex mixtures in sown grasslands and favouring species-rich swards in permanent grasslands

During the 20th century, grassland mixtures were increasingly simplified ending by the use of single species sowing in grazed or mown temporary grasslands. In many areas of Europe, perennial ryegrass (*Lolium perenne*) was often the only component of regularly sown swards. In permanent grasslands, most if not all dicotyledon species were seen as weeds and destroyed by selective herbicides. However, favouring multiple forage species swards in grasslands may increase yield, resilience, and improve feeding quality. Legumes and several other dicotyledon species may indeed exhibit higher forage digestibility and increase intake of grazing animals compared to grass species (Wilkins *et al.* 1995, Schubiger *et al.* 2001, Champion *et al.* 2004).

The combination of early with later growing species can sustain production during all the growing season. Many legumes such as white and red clovers (*Trifolium repens* and *T. pratense*) and lucerne (*Medicago sativa*) continue growing during summer drought when many grass species such as perennial ryegrass stop producing biomass (Figure 11). Even the combination between grasses can induce a more regular sward growth by combining for instance perennial ryegrass with drought resistant grasses such as cocksfoot (*Dactylis glomerata*) and tall fescue (*Festuca arundinacea*) in mown grasslands. Complex mixtures can not only improve yield but also resilience in case of severe drought and extreme weather conditions in general.

Roca-Fernández *et al.* (2016) demonstrated that increasing sward botanical complexity from one (perennial ryegrass) to five (perennial ryegrass, white and red clovers, chicory (*Cichorium intybus*) and tall fescue) species has positive effects on animal performance. Inclusion of both clovers and chicory in a pure ryegrass sward improved pasture chemical composition (e.g. protein, energy, and minerals) and enhanced milk production and milk fat and protein productions on a per cow basis (Figure 12). Benefits of grazing ryegrass/clover/chicory mixture over ryegrass/clover mixture were also noted. They are linked to the

Figure 11. Red clover (front) and birdsfoot trefoil (back) based mixtures (photo: Alain Peeters).

Figure 12. Chicory, plantain, white and red clover mixture (photo: Kevin Morel).

cumulative effects of improved sward quality and increased pasture dry meter (DM) intake in all seasons, due to the positive effect of chicory.

In permanent grasslands, tannin-rich forbs and legumes can be introduced by clod planting or over-sowing. Some species can be introduced even in productive grasslands such as yarrow (*Achillea millefolium*) and narrowleaf plantain (*Plantago lanceolata*). Grazing plantain mixtures is particularly attractive (Judson and Moorhead 2014). Dandelion (*Taraxacum* spp.) usually spread spontaneously. Birdsfoot trefoil (*Lotus* spp.) can be used in dry and/or low fertility soils. Woody species can be sometimes conserved or planted in hedges as a source of tannins too. Some species are a source of essential oil. They can decrease methane emissions from the rumen (Benchaar and Greathead 2011) and improve animal health (Section 3.5.4).

3.4.2. Giving priority to legumes in grasslands

Perennial forage legumes can fix significant amounts of nitrogen (Table 1). They can thus reduce nitrogen external input use in temporary and permanent grasslands. For instance, the amount of nitrogen fixed by lucerne and red clover and assessed by different methods varied from 251 to 546 kg N/ha*year for red clover and from 292 to 502 kg N/ha*year for lucerne

Table 1. Examples of amounts of nitrogen fixed annually by a selection of perennial temperate legumes (after Peeters *et al.* 2006).

Species	Country	Amount (kg N ha^{-1} $year^{-1}$)	References
White clover	UK	100–200	Whitehead (1995)
White clover	Switzerland	270–370	Boller and Nosberger (1987)
White clover	New Zealand	100–350	Hoglund *et al.* (1979), Caradus (1990), Ledgard *et al.* (1990)
Birdsfoot trefoil	New Zealand	Maximum: 140	Laidlaw and Teuber (2001)
Red clover	Belgium	300–400	Deprez *et al.* (2004)
Red clover	North America	100–250	Smith *et al.* (1985)

in Belgian temporary grasslands (Deprez *et al.* 2004). Dry matter and N yields of non-fertilised grass-legume mixtures were close to those of a pure perennial ryegrass fertilized with 400 kg N/ha*year. Average yields of legume and legume-grass mixtures can reach for instance 11–17 t DM/ha in year 1 and 8–15 t DM/ha in year 2 (Peeters *et al.* 2006). Nyfeler *et al.* (2009) obtained similar results in Switzerland. Grass/clover mixtures fertilised with 50 or 150 kg/ha*year attained similar annual DM yields as the heavily fertilised (450 kg N/ha*year) monocultures of the highly productive grass species.

Several studies have shown that yields of legume/grass mixtures are more resilient than pure grass swards in case of drought (Lüscher *et al.* 2005, Nyfeler *et al.* 2009).

Peyraud *et al.* (2009) wrote a synthesis on the potential of forage legume-based systems in Europe. They concluded that animal intake of legumes is higher and the rate of decline of legume nutritional quality with advancing maturity is slower than for grasses, especially in the case of white clover. This makes grass/clover pastures easier to manage than pure grass swards. Grass-legume mixtures can for instance be grazed at a slower rhythm after a longer rest period than pure grass mixtures in rotational stocking since legume forage nutritional quality declines more slowly (Frame and Newbould 1986, Vertès and Simon 1992). The nutritional advantage and voluntary intake of white clover over grasses is well established (Thomson 1984, INRA 2007). Grazing animal performances are similar or higher on grass/clover pastures compared to pure nitrogen fertilised grass swards. Forage legumes increase also the concentration of beneficial α-linolenic acid in meat and dairy products.

System energy efficiency, calculated as herbage net energy lactation (UFL) produced per MJ of energy consumed for input production, is three times higher for white clover/grass pastures compared to nitrogen fertilised grass pastures (2.5 versus 0.8 UFL/MJ, Besnard *et al.* 2006). Similarly, Le Gall *et al.* (2009) have shown that energy consumption decreases from 5.0 MJ/kg milk for intensive dairy farm in The Netherlands to 4.0 MJ/kg for French farms using maize silage and nitrogen fertilised grasses and to 3.1 and 1.4 MJ/kg for systems based on grazing in Ireland and New Zealand, respectively.

3.4.3. Including perennial forage species in crop rotations

Temporary grasslands (leys) are important in crop rotations in agroecological mixed farms. They should be legume-based with a high proportion of productive legumes. The most important temperate legume species for mainly cut temporary grasslands are lucerne and red clover. White clover, and on wet soils crimson clover (*Trifolium hybridum*), can be added to the mixture. Lucerne is usually mixed with cocksfoot. Red clover, possibly mixed with Ladino type white clover and crimson clover, can be mixed with Italian ryegrass (*Lolium multiflorum*), perennial ryegrass, cocksfoot, and other grasses. On dry, calcareous and other marginal soils, sainfoin (*Onobrychis viciifolia*) and birdsfoot trefoil can play an important role (Frame and Laidlaw 2011).

Legume-based temporary grasslands are the main tool for increasing soil organic matter, improving soil life, and soil fertility. They are often the main source of nitrogen by biological nitrogen fixation for the whole system and the following crops of the crop rotation. Temporary grasslands improves usually a lot water infiltration, thanks to the increase of earthworm populations and other soil life species that improve soil structure (van Eekeren *et al.* 2008). Earthworms, in particular, are very active in grasslands by digging horizontal and vertical galleries that accelerate water infiltration in soils, reduce soil erosion and flooding. Temporary grasslands play an important role in weed control. Lucerne dominated mixtures can for instance totally eradicate thistle (*Cirsium* spp.), after two or three years with a cutting frequency of four cuts per year (Schreiber 1967). They also decrease soil borne disease of arable crops.

3.4.4. Finding the right balance between permanent and temporary grasslands

In North-west Europe, there is a tradition to renovate mainly grazed grasslands every four to five years. After sowing with cultivars, the new sward looks "nicer" and is more productive. This practice implicates though high destruction and sowing costs and has negative impacts on the environment because of CO_2 emissions, herbicide pollution (if used), and nitrate leaching risks. Hopkins *et al.* (1990) demonstrated that this practice is not justified and that most permanent spontaneous swards are more productive than regularly resown swards with cultivars except the first production year after sowing. Agroecological systems should thus use spontaneous permanent swards in priority. In such swards, good stocking management improves sward botanical quality over time. This does not exclude the over-sowing of some species such as white clover in existing swards (Section 3.4.7).

Over time, increasing amounts of carbon and nutrients are stored in permanent grassland soils which is favourable for these grassland types but, as a consequence, these amounts of carbon and nutrients are not available for crops on arable land in mixed crop–livestock systems. Permanent grassland is a fertility sink in these farming systems. If a part of the permanent grassland area is converted to arable land and replaced by temporary grasslands, more nitrogen becomes available for annual crops which increases their yield and more carbon is stored in the arable soil organic matter which improves soil fertility, life, and structure. The right balance between both grassland types depends on each farm characteristics. An indicator could be an approximate 50% share of temporary grasslands on arable land. The permanent grassland area is then deduced for a given stocking rate on the whole grassland area calculated at farm level.

3.4.5. Using annual forage crops as intercrops

Annual forage crops can be sown in the interval between two main crops, and thus function also as a cover crop. This technique was more widespread in the past but has progressively disappeared in conventional farming. These forage crops are a source of high quality feed for livestock although they produce much less than temporary grasslands. They are also a secondary tool for increasing arable soil organic matter, improving soil life and soil fertility, and nitrogen fixation if the mixture contains legume. They can significantly reduce nitrate leaching after crop harvest and before the following crop is able to absorb nutrients. They play also an important role in weed control.

3.4.6. Devoting great attention to grassland management

In the wild, large herds such as those of bisons (*Bison bison*) in North America or mixed herds of wildebeests (*Connochaetes* spp.), zebras (*Equus* subgenus *Dolichohippus* and *Hippotigris*) and other large herbivores in Africa, migrate and exert a high stocking pressure on natural grasslands by very high instantaneous stocking rates. This kind of system experiences daily pasture moves and long rest periods. In agroecology, a stocking rotation should mimic this process and might improve its productivity by a careful and precise management based on up-to-date knowledge and technology. Improved stocking management techniques were described by pioneers such as Voisin (1988, first published in 1957) and Murphy (1998, first edition in 1987) for rotational stocking. The Voisin system was very much developed in the last 60 years.

In a modern and agroecological rotational stocking system, optimum stocking management should rely on the following five main principles:

(1) Finding a trade-off between forage quantity and quality.
(2) Adoption of optimum stocking rate according to fluctuating seasonal grass productions and grassland plot potential.
(3) Combining requirements of grassland plants with those of grazing livestock.
(4) Offering grass quantity and quality according to different livestock group requirements.
(5) Adoption of a multifunctional stocking system.

These principles are detailed below.

Trade-off between forage quantity and quality

In grasslands, yield increases and quality decreases over time during a period of growth (Figures 13(a) and 13(b)). Forage should be grazed or mechanically harvested at a time that is a compromise between quantity and quality. This optimum defoliation time is earlier for grazing than for cutting for silage, haylage, or hay, respectively. In all cases, plants should have enough time to achieve their growth surge (the vertical part of the

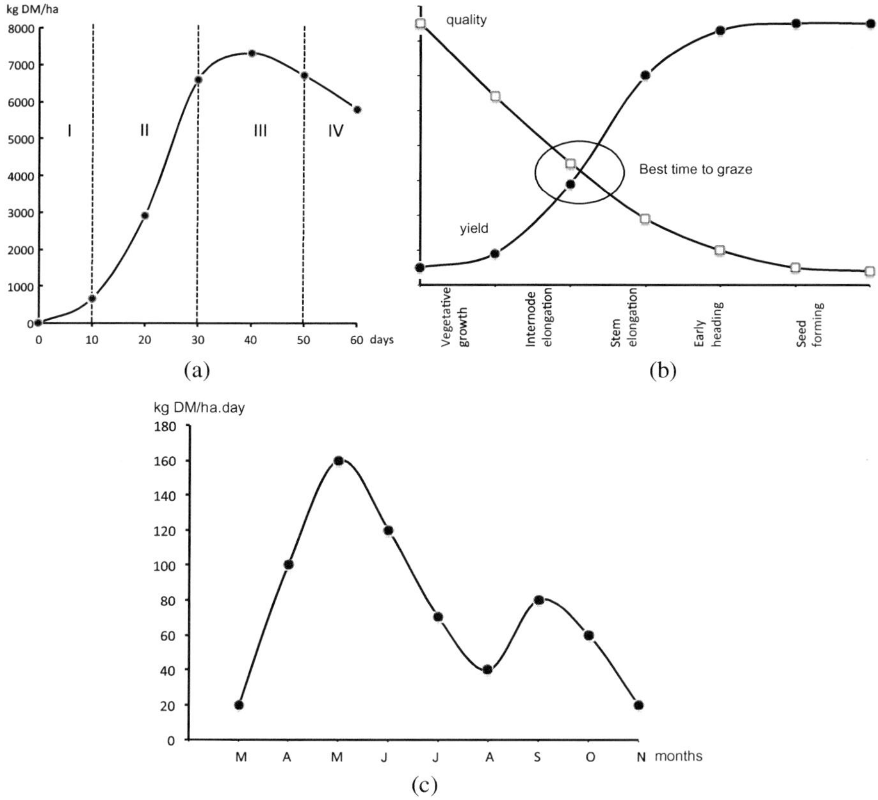

Figure 13. (a) Evolution of forage yield (kg DM/ha) during an uninterrupted growth period (Phases I–IV) (after Voisin 1988); (b) evolution of forage yield (kg DM/ha) (black circles) and quality (white squares) during an uninterrupted growth period (after Undersander *et al.* 2002). Quality can be quantified as energy or crude protein contents or dry matter or organic matter digestibility; (c) evolution of grassland growth rate (kg DM/ha.day) from March to November (Northern hemisphere) (after Corral and Fenlon 1978).

exponential growth curve; Phase II in Figure 13(a)) and reconstitute their stocks in stubble and roots.

Optimum stocking rate

Since the grass growth rate is highly variable over the growing period with a high peak in spring and a second smaller peak in autumn (Figure 13(c)), stocking rate should follow these grass growth variations. Optimum stocking rate has thus to be defined locally and season-by-season. This can mainly be done by reducing the surface grazed in spring, by harvesting the excess grass growth and conserve it as hay or silage for winter-feeding. This surface is put back in the grazing area later in spring or summer. In this case, stocking rate is adapted to the low summer growth period. Other strategies are possible with a higher average stocking rate, lower amount of grass conservation, summer forage supplementation or insertion of new grazing areas in summer in the grazing circuit. Optimum stocking rate maintains or improves sward quality. Avoiding overstocking limits sward degradation, animal diseases and parasites, nitrate and phosphate pollutions, and soil erosion. Understocking not only reduces animal production per ha, but may also lead to sward degradation by the decrease of good grasses and white clover proportions. These species require indeed frequent defoliations for persisting in the sward.

When stocking rate increases on a grazed area, daily liveweight gain (LWG) or daily milk production per animal decreases almost linearly. LWG or milk production per hectare increases first, reaches a maximum and then decreases (Figure 14). A strategic choice has thus to be done, between adapting stocking rate for maximising performances per animal

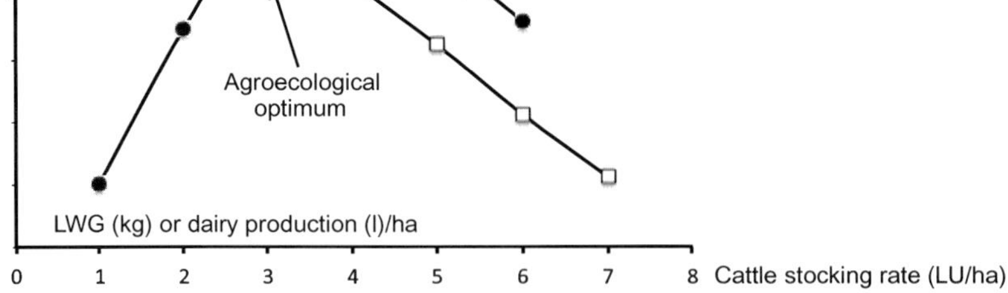

Figure 14. Influence of stocking rate on live weight gain (kg LWG) or dairy production (l) per animal (white squares) and per hectare (black circles) during a 200-day grazing period (after e.g. Derner *et al.* 2008).

Note: No units are given for the *Y*-axis since units depend on grassland productivity and cattle breeds.

or per hectare. Conventional systems maximise production per animal and then often complement them with forages and concentrates produced on other surfaces. Agroecological systems tend to find a compromise between productions per animal and per hectare of the grazed area as Irish and New Zealander systems do (Le Gall *et al.* 2001).

Combining requirements of grassland plants with those of grazing livestock

Herbivores coevolved with grassland species and habitats. Grasses are adapted to frequent defoliations by their tillering system where most plant buds are located close to the soil surface out of reach of animal teeth or tongues, or can be easily replaced when destroyed. On the other hand, animals are adapted to the nutritional properties of grassland plants. On a fertile soil, a short grass sward covers most feeding requirements of productive animals. It offers high net energy, protein, and mineral contents. This kind of grassland forage is even too high in protein and potassium for animal needs. It may be though too low in some elements such as sodium and selenium for high animal performances (Frame and Laidlaw 2011). This justifies complementation in some macro and microelements.

In a managed stocking system, requirements of grassland plants should be combined with those of grazing livestock in an optimal way. That is the reason why Voisin (1988) defined grazing as "the meeting of cow and grass!" Grass requires that growth or regrowth periods should be long enough for achieving its growth surge. Relatively long growing or rest periods associated with quite severe defoliations produces the highest annual productions per hectare. On the other hand, grass growth or regrowth period should be short enough for offering grazing animals high quality grass. Moreover, the sward should not be grazed too close to the soil surface since forage quality decreases with canopy depth; the biomass close to the soil surface has the lowest quality. A trade-off is achieved in the rotational stocking system.

Figure 15 shows the antagonism between maximum cow intake and maximum grass use. Offered grass amount (recorded at soil level) ranging between 25 and 40 kg DM per hectare is a good trade-off. When the different requirements of grass and animals are combined, the grass growth

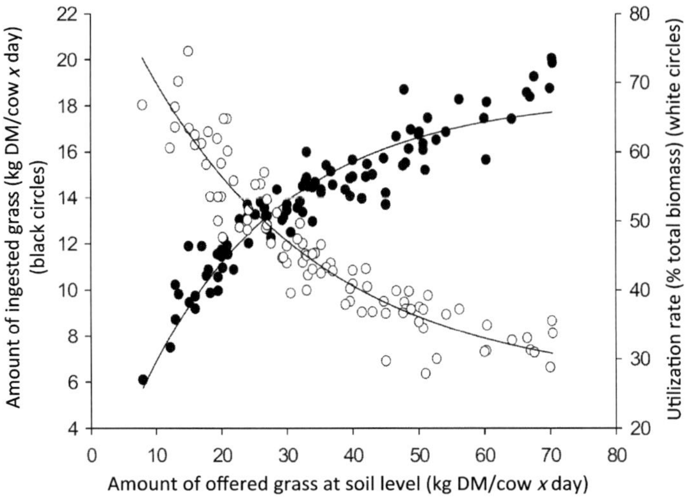

Figure 15. Influence of offered grass amount per day at soil level on grass ingested by dairy cows and grass utilisation in rotational stocking (adapted from Delagarde *et al.* 2001 in Peyraud 2010).

period should usually be included between 30 and 45 days, but it can be as short as 12 days in spring when grass growth and stem elongation are fast. Sward height at the start of a stocking episode (pregrazing height) should be about 10–14 cm for cattle either for pure grass sward and grass–white clover mixtures (Table 2).

Optimum sward height at the animal exit from the plot (post-grazing height) is a controversial topic. After severe defoliations, grass regrowth is slower compared with more lenient stocking. This is due to a strong reduction of grass root biomass and a low remaining leaf area. However, after severe defoliation, tissue senescence is reduced because most biomass has been harvested. Trade-off between the two processes is reached for an optimum height of about 5–7 cm for cattle or about the half of pregrazing height (Mayne *et al.* 1987, Dequin *et al.* 1998, Leconte 2002).

These values vary also according to sward type (e.g. clover–grass mixtures can be grazed at a bit longer intervals than pure grass mixtures), animal type (e.g. dairy cows should graze shorter grass than beef cows since they have higher nutritional requirements) and season (in the beginning of spring, in the first grazing round, grass should be grazed shorter

Table 2. Typical values of key indicators of optimum rotational stocking management.

	Spring	Summer		Autumn
Rest period (days)	12–30	30–45		24–30
Number of grazed paddocks for one-day occupation	13–21		25–46	
Number of grazed paddocks for five-day occupation	7		6–10	
	Dairy cows	Suckler cows		Sheep, goats
Grazing height (entrance) (cm)	10–12	13–14		8–12
Grazing height (exit) (cm)	5–6	7		5
Occupation period (days)		Ideally: 0.5–3 (maximum 5)		
Stocking rate (LU/ha)		About 2.0–2.5		
Stocking density (LU/ha)		One-day occupation paddock: 110–150		
Grazing area per LU in spring (ha)		About 0.25		

Note: For the definitions of terms in the text and in Table 2, see Allen *et al.* (2011).

for controlling stem elongation; after summer drought, grass should be grazed taller for grass reconstituting its stocks).

Grass intake is maximum for half a day or a one-day stay in each paddock. The period of occupation (the sum of the periods of stay of different groups grazing consecutively the same paddock; period of occupation = period of stay if only one group) should ideally not be longer than 3–5 days, period after which a grass plant grazed the first day of stay starts regrowing and could be grazed again. This would lead to plant stock exhaustion. The number of paddocks can be deduced from the longest summer rest period, the period of stay of each animal group in paddocks in each rotation, and the number of animal groups (Table 2). In spring, about one third to one half of the number of plots can be excluded from the grazing circuit, harvested and conserved for periods of grass growth deficit.

Hoden *et al.* (1991) defined a simplified rotational stocking for high and regular rainfall areas. The number of paddocks is restricted to two in spring and four during the rest of the grazing period. The occupation periods varies between 10 and 14 days which implies that grass ingested by cattle is partly produced during the rest period and partly during the occupation period. With dairy cows, milk production per day is more variable compared to one-day rotational stocking but total amounts are comparable at least in nitrogen fertilised paddocks. (see comment on multifunctionality below).

TechnoGrazing™ is a technique that derives from rotational stocking (Charlton and Wier 2001) and pushes its productivity to an optimum. The stocking area is divided into a grid of stocking cells, each usually about 0.1 ha (40 × 25 m). It is particularly adapted to high requirement animals (Table 3). Long narrow lanes are established by permanent fences. Temporary fences are placed across these lanes to form the one-day stocking cells delimited by front and back electric wires. The cross fences can be erected or dismantled easily and quickly. The system is characterised by long rest periods and low post-stocking residuals. It boosts grassland and animal productivity, because: (i) small animal groups are less stress-prone than larger groups, (ii) daily moves on fresh pasture changes animal psychology, animals focus on grazing instead of walking, trampling, and wasting part of the available forage such as in longer stocking periods. As a result, intake and overall forage quality are higher. Further points are (iii) animal disorders such as bloat and internal parasites are reduced, (iv) dungs and urine are more evenly distributed on the surface and more efficiently used for plant growth, and (v) energy spent for movements is reduced (water and forage are more accessible).

Table 3. Grazing groups based on nutritional requirements (modified after Undersander *et al.* 2002).

Livestock type	High requirement	Medium requirement	Low requirement (maintenance)
Dairy cattle	Lactating cows	Heifers	Dry cows
Beef cattle	Fattening steers and heifers	Suckler cows with calves, heifers	Dry cows
Goats	Milked goats	Goats with kids	Dry goats
Sheep	Weaned lambs	Ewes with lambs	Dry ewes

On the other hand, continuous stocking includes only one paddock whose size varies between spring and summer. It has been promoted in rainy climates and high inorganic nitrogen fertilisation rates by fertiliser companies (see comment on multifunctionality in the following that).

Ration or strip stocking, first-last stocking, and forward creep stocking are variants of rotational stocking (Allen *et al.* 2011).

White clover leaves should always have enough light which implies that they should not be shaded by grass leaves and stems. Aggressive Hollandicum type of white clover cultivars are thus useful for a well-balanced productive sward. They produce large leaves that can follow the growth in height of grass canopy. These clover types can be introduced in a permanent grassland sward by oversowing. This can be easily done by broadcasting seeds in the sward after the period of maximum spring growth rate during a rainy period and after a severe defoliation by stocking or, even better, after silage or hay cut.

Management has also to control weeds and grass stem production in order to minimise refused areas. After one or two fast stocking rotations of 10–15 days in spring, close grazing and short periods of stay of one or two days per plot are the most efficient techniques for doing so.

Grassland management is complex because it has to integrate a large number of parameters and decisions must be taken in a highly variable and unpredictable environment.

One of the first parameters that has to be continuously evaluated is grass availability. In a production context, grass availability has to be assessed by fast and easy methods. This measure is important for evaluating grass stocks at a given time, to decide if a paddock can be grazed and when animals have to leave it. Grass availability can be measured by simple tools such as "Cut and Weigh" system, Pasture Probes, the HFRO sward stick (Barthram 1986) or Wellington boots with white-painted horizontal stripes, but the (electronic) "Plate Pasture Meter" or "Rising Plate Meter" that combines measurement of sward density and height is the most common (Figure 16). The most widespread method is through eyeballing that can be effective if correctly calibrated for instance with the Rising Plate Meter (Frame and Laidlaw 2011). In the future, drones could be used for assessing available biomass in a grassland area.

(a) (b)

Figure 16. (a) Wellington boots in a ryegrass sward (photo: Kevin Morel), and (b) Rising Plate Meter (photo: GAEC l'Epillet dans l'herbe (Goueltas Le Berrigaud) (right)).

Decision-support tools have been developed for grazing planning and management. One of oldest is the grazing calendar. This tool is very useful especially for learning rotational stocking management. Electronic spreadsheets (Mosimann *et al.* 2008) or softwares are now available, as for instance the Grass Wedge (Hennessy 2012, O'Donnovan and Delaby 2016) (Figure 17).

Offering grass quantity and quality according to livestock group requirements

All animals from a farm do not have the same feeding requirements (Table 3). In dairy farms for instance, lactating cows should have the highest quality forage while dry dairy cows and heifers could graze lower quality plots and receive lower quality conserved forage. In mixed first-last stocking system (Allen *et al.* 2011), meat sheep (last group) can graze the lowest quality grass left by dairy cows (first group). In agroecology, these differences have implications on the use of different grassland types of variable forage yield and quality, instead of standardising the whole grassland area by investing in fertilisers, drainage system and other intensification means.

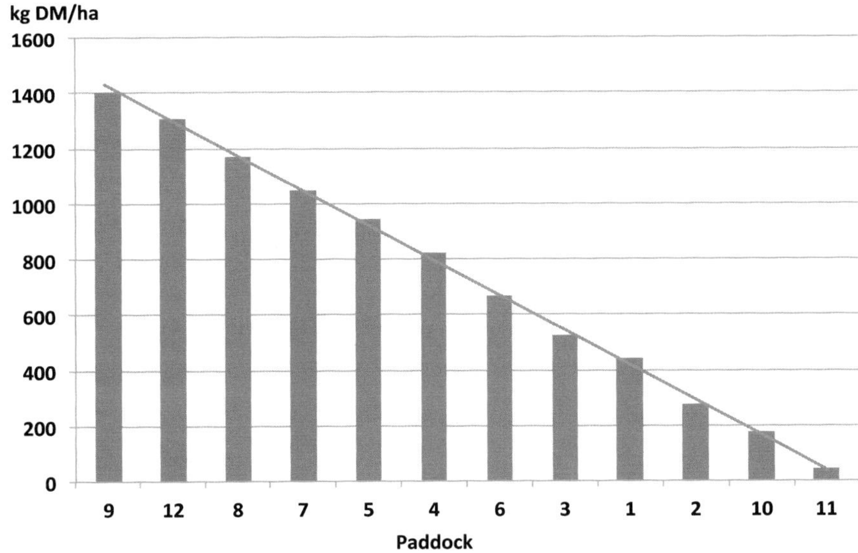

Figure 17. An ideal wedge showing the availability of grass (kg DM/ha) in 12 paddocks of a farm (Hennessy 2012).

Adoption of a multifunctional stocking system

The stocking system should be designed in such a way that it is multifunctional. It should fulfil several functions. Rotational stocking satisfies this best. Simplified rotational stocking is less effective and continuous stocking is characterised by a low number of functions. For instance, the latter system does not control internal parasites. Rotational stocking allows transforming a fluctuating grass growth into a regular quantity and quality of grass. It controls weeds and increases forage quality. It enhances the proportion of white clover in the sward. In legume/other dicotyledons/grass mixtures, long rest periods can be adopted, since this sward maintains its quality over relatively long periods (45 days resting period for instance). This induces a combination of good grassland growth pattern and feeding quality and the control of livestock internal parasites. Parasite eggs excreted by livestock on the grass are usually largely destroyed by UV radiations after a period of 40–45 days in rainy climates. In more sunny climates, this destruction is faster (Neeson and Love 2014).

Figure 18. Mixed stocking, associating cattle and sheep grazing together (photo: George Goodfellow).

Whenever possible, adoption of mixed stocking of herds and flocks (e.g. cattle and sheep grazing together) can be organised in two ways: combination of several grazing livestock species grazing together (Figure 18) or in sequence (e.g. the Salatin system where poultry follow cattle on the same plot, poultry eating grass and insects attracted by cow dungs (Salatin 1996b) in a first-last stocking system). Mixed stocking allows better grassland resource use. It is an efficient way of controlling livestock internal parasites because the parasites of one livestock species can be destroyed by another species in its digestive tract. Mixed stocking may also contribute to controlling grassland weeds by associating different animal types (Section 3.5.5). Cuts taken in periods of excess forage prevents weeds such as docks (*Rumex* spp.), thistle (*Cirsium* spp.) or nettle (*Urtica* spp.) to produce seeds.

Rotational stocking of grassland plots can be combined with a permanent access to a woodland area or a woodland strip where cattle can rest, shelter, or calve. A surface of a few tens of meter square per animal is sufficient. Woodland plot rotation allows annual or multiannual restoration of tree roots. Shade can also be provided by hedges, isolated trees and "portable shade" equipment. Access to shade can improve grass intake (Palacio *et al.* 2015).

Pastoralism is also an agroecological system where livestock moves according to forage availability. Pastoral systems can be transhuman or sedentary in temperate areas. Transhumance often involves movements between lowlands in winter and mountains in summer. Sedentary extensive systems are mainly located on marginal lands such as the highlands of Scotland or the dry areas of the USA. Stocking rates and productivity are usually low. Grazing livestock have the opportunity to select their diet. Internal parasite pressure is normally low because of low stocking rate and long intervals between two grazing episodes.

3.5. *Livestock types and management*

3.5.1. Being concerned about animal welfare

Respect of animal welfare should consist in giving the opportunity to each livestock species to express its natural behaviour in feeding, reproduction, social needs, and preferences, and to fulfil its ecological requirements. It is thus reasonable to consider livestock farming conditions that built as much as possible on the natural conditions of their wild ancestors.

It is inspiring to analyse the behaviour of aurochs (*Bos primigenius*) for instance to better understand the needs of modern cattle (see synthesis by Van Vuure 2002). The European auroch subspecies (*B. p. primigenius*), now extinct, lived in a distribution area from Europe to the Middle East. Its habitat was linked with woodland alternating with marshes and other open areas. It was certainly searching for open habitats that could have been tall sedge vegetation along riversides. In spring and summertime, its diet consisted mostly in grasses and grass-like plants, as well as herbaceous dicotyledon species and leaves of woody species. In autumn, they ate somewhat less grass and more from trees and bushes, including acorns. In winter, aurochs still grazed green and dried leaves of herbaceous species but they increased the proportion of branches and even included some bark in their diet. Auroch populations organised themselves in different groups. Cows lived in mixed herds with calves and young bulls. Older bulls lived in separate small herds. Some old bulls, probably the most powerful ones, lived solitary. At the beginning of the mating periods, old bulls fought, sometimes heavily. The winner had the privilege to mate with females of the large group. After two months of breeding, reproductive males resumed their previous living. Cows lived thus separate from reproductive males for most of the year. In Poland, were aurochs survived the longest, the breeding period was concentrated in August and September and consequently the calving period was in May–June. In West Europe, it is likely that these two periods started one month earlier (breeding in July–August and calving in April–May). Calving started thus with grass growth. Social links between cows were certainly strong for all their lives and especially between mothers and their daughters.

Several things can be learned or deduced from this information: (i) the diet of modern cattle should not be restricted to grass. It should include fibrous and woody material at all seasons. Cattle would benefit from an access to hedges or woodland all year long for browsing, calving in a protected place and finding a shelter during warm, very cold, windy, rainy, or snowy periods. (ii) Aurochs lived, of course, in full outdoor conditions but they found shelter in woodland. They were actually forest animals. Modern cattle as aurochs does not need buildings during the cold period provided that they are adequately fed; they can overwinter outdoors, for instance on outwintering pads protected from cold winds by hedges or woodlands (French and Boyle 2008). In very cold areas, where soil is frozen in winter and where grassland sward degradation is not feared, such as in continental areas of North America, cattle can overwinter and even graze under the snow (Ruechel 2006). (iii) Calving time should be synchronised with the start of grass growth. Reproductive males and females should live most of the time separately. The weaning could be natural and progressive in autumn. (iv) Cattle are highly social animals, individuals, especially females, develop strong bonds with other members of the herd. When these bonds are disturbed, animals become stressed, growth and dairy production can be affected, chronic diseases can even be induced. The same approach can be applied to all domestic herbivores.

3.5.2. Choosing the right livestock types and genetics

Highly productive dairy or beef breeds such as "Holstein–Friesian" and "Belgian Blue" were not bred for grazing. It is thus not surprising that these animals are not well suited to grazing systems (Delaby *et al.* 2010). In case of restricted feed availability, during a period of slow grass growth for instance, the organisms of animals adapted to grazing maintain adequate body reserves because they do not give absolute priority to milk production such as "Holstein–Friesian" cows do. This physiological feature prevents them from reproduction and health problems. In contrast, in these feed restriction conditions, hyper-specialisation for meat and milk production induces health and fertility problems, reduces animal welfare and lifetime expectation.

Animals adapted to grazing are able to graze effectively and ingest large amounts of green fodder. They can move on relatively long distances. They are robust and autonomous for calving for instance. They are more flexible, have improved health and fertility, and higher milk and beef value per litre or kilogram compared to highly specialised breeds (Coleman *et al.* 2009, Prendiville *et al.* 2009, Delaby *et al.* 2014). They can be dual-purpose breeds or crossbreeds. It has been shown that the "Normande" breed or "Jersey × Holstein–Friesian" crosses for instance are better adapted to grazing than purebred "Holstein–Friesian". "Jersey × Holstein–Friesian" cows achieve improved reproductive performance, larger intake per kilogram of body weight at grazing and require significantly less grass to produce 1 kg of milk fat plus protein compared to pure "Holstein–Friesian" (Prendiville *et al.* 2009). Locally adapted breeds that exhibit these characters of rusticity have usually declining populations and some are threatened by extinction. They know an interest renewal in the framework of agroecological systems.

Moreover, two double-goal cows can roughly produce the same amount of milk and meat than one specialised dairy cow and one specialised beef cow, with lower nitrogen, phosphorus and methane emissions in the environment (after Rosenberger and Rutzmozer 2002 in Kampschulte 2009).

3.5.3. Reducing livestock feeding costs

With adequate breeds and genetics, it is possible to target maximum feed/forage self-sufficiency and low cost feeding at farm level. Giving priority to feed (e.g. fresh grass, grass hay or silage) compared to food (e.g. cereal grain, pulses) reduces costs. Grazed grass is the cheapest feed compared to all others (Deprez *et al.* 2005, Finneran *et al.* 2012). Grass and maize silage costs are usually similar. Concentrate feed such as cereals and soybean are the most expensive. These concentrate feed are actually food since they are also consumable by humans. Giving priority to grass for herbivores prevents thus, feeding competition between grazing livestock and people. Giving priority to perennial (permanent and temporary grasslands) compared to annual forage (e.g. green maize, sorghum, immature cereals, fodder beat, turnips) also decreases dramatically costs for soil tillage, sowing, fertilisation, weeding, and pesticides.

Extending the grazing season in spring and autumn reduces high winter-feeding costs. Temporary grasslands, in particular Italian ryegrass/red clover mixture starts growing faster than permanent grasslands in spring. It can be lightly grazed for about two weeks before permanent grasslands start growing. In autumn, temporary grasslands such as lucerne/cocksfoot or Italian ryegrass/red clover mixtures can be grazed until November and even December in the Northern Hemisphere, when permanent grasslands have stopped growing. In this case, grazing on usually wet soils must be done with extreme care to not degrade soil structure, by very short grazing periods on small fresh grass plots. These techniques combined may save 1–3 months of expensive conserved winter-feeding thanks to cheap grazed grass. This can also be achieved by starting grazing permanent grassland early in March compared to a traditional start in mid-April. An early part-time start of grazing in March during 3–4 hours per day for instance, not only reduces conserved forage feeding, it facilitates the subsequent management of grassland plots and produces a more leafy and digestible grass for the following stocking cycles (Peyraud 2010).

Producing the best possible quality of conserved and grazed forages reduces the need to buy expensive protein-rich concentrate feed. Quality of forages can be optimised by grazing or cutting at adequate physiological stage, by the adoption of adequate stocking (Section 3.4.7), wilting and conservation techniques. These techniques are extensively described in the literature (e.g. Frame and Laidlaw 2011). Barn hay drying by solar energy is a particularly attractive solution for producing high quality hay even in rainy climates.

Alignment of calving period with the beginning of the grass growth period allows to produce more milk and meat from grazed grass, reduce the use of conserved or concentrate feed and thus reduce production costs (Dillon *et al.* 1995). This requires controlling the start of the mating period by introducing a reproductive bull in the herd at the right period. It necessitates also reaching a high pregnancy rate in a short period just afterwards. This last criterion can be improved by optimal feeding and by selection.

3.5.4. Developing the concept of agroecological animal health

The agroecological basis of animal health care is not yet well defined, but several practices and methods are though already available. An integrated health and metabolic disorders and parasite control can be applied by:

- general use of prevention methods: e.g. rotational stocking including optimum stocking rate and long rest periods of grasslands, mixed stocking of different livestock species, balanced feeding, adequate mineral supplementation, continuous provision of straw or late cut hay at grazing especially on high quality swards and during rainy periods, use of tannin-rich forage species for parasite control, well-designed housing, hygiene, use of rustic breeds, isolation of sick animals in case of infectious disease;
- curative instead of systematic treatments of animals;
- when necessary, disease treatment with plant extracts, essential oils (phytotherapy) or other natural means such as clay, vinegar, $MgCl_2$ to replace synthesis chemical treatments when possible (Wynn and Fougere 2007).

Winter feeding is relatively high in fibre, low in water and moderate in protein contents. In early spring, short grass is high in water, protein and potash contents and low in fibre and magnesium. It is quiet unbalanced for a short period. These two diet types are thus very different. A progressive transition between winter and spring feeding must be a matter of attention for avoiding metabolic troubles such as grass tetany. This can be achieved by giving access to winter feeding in the morning and by restricting grazing to some hours in the afternoon during about 7–10 days for instance. During that period, the duration of the grazing time should be progressively increased. Magnesium supplementation can also help during this critical period.

During the full grazing period, transition of a livestock group (e.g. dairy cows) from a low nutritive value grass sward to a high nutritive value legume-dominated sward should be avoided for preventing bloat, in cattle mainly, or enterotoxemia, in sheep mainly. A permanent access to straw or low quality hay is recommended during the whole grazing season, especially if animals cannot regularly browse woody species leaves and shoots. This allows animals to regulate water and protein contents in their diet and avoid diarrhoea and other metabolic troubles. It also improves welfare.

Regular observations of animals are essential for diagnosing abnormal symptoms or behaviours. Some methods such as OBSALIM (http://www.obsalim.com/en/index.htm) are available for interpreting these signs, relating them with nutritional disorders and proposing feed adjustment for cattle, sheep and goats.

3.5.5. Considering domestic animals as “collaborators” of the system

Herbivory livestock species can provide different services and can fulfil different tasks in agroecosystems. In permanent grassland and rangeland, donkeys can control thistle (*Cirsium* spp.), goats, docks (*Rumex* spp.), goats and horses, shrubs. Shropshire sheep can graze among fruit trees for maintaining sward short, weeding and controlling the spread of fungal diseases, such as apple scab (*Venturia inaequalis*). Sheep eat fallen leaves and so can help to prevent the spread of fungal diseases (Geddes and Kohl 2009) (http://www.shropshire-sheep.co.uk/sheepintrees.html). Cattle can provide similar services in traditional orchards by ingesting fallen fruits infested by codling moth (*Cydia pomonella*) larvae and leaves attacked by apple scab. Ducks can control slugs or snails in market gardening (Hervé-Gruyer 2014) and paddy fields (Fukuoka 2009), they can also weed rice fields (Fukuoka 2009, Bernard Poujol, rice fields of Camargue (France): http://www.canard-desrizieres.fr/histoire.html).

3.6. *Diversification and increase of income sources*

The economic strategy of agroecological systems includes three main components for increasing income:

- The up-stream strategy consists in making farms less dependent from banks and input producers and suppliers such as companies active in the seed, fertiliser, pesticide, feed, and animal drugs sectors. This is achieved by a systematic quest for limiting and reducing production costs. There are also huge possibilities to reduce investment costs in building and large machinery and equipment. Investment and production costs reduction can be achieved by all techniques that can replace inputs by ecosystem services delivered by biodiversity and by simple infrastructures (e.g. a simple outwintering pad on wood chips instead of a large concrete cow shed). This implies a higher use of local resources instead of imported resources.
- The down-stream strategy includes quality production, product processing, and short, local or regional marketing chains.
- A diversification of activities.

Decreasing costs is firstly achieved by increasing fodder self-sufficiency by producing high quality forage at low costs on the farm (Sections 3.1–3.5). Inorganic fertilisers can be replaced by biological nitrogen fixation, best use of organic fertilisers, nutrient loss reduction, better soil profile exploration by plant roots including tree roots, better soil life for solubilisation and plant uptake of low available nutrients (e.g. solubilisation of P by fungi mycorrhiza). Veterinary costs can be reduced by the use of rustic breeds, priority given to productions per hectare and not per animal, optimum feeding, continuous follow-up of the animals, and the implementation of an animal welfare-based system.

In dairy farms, grassland-based systems that have lower variable costs (e.g. concentrates) and are more forage self-sufficient, have similar income per agricultural working unit and are sometimes more profitable whatever the economic context and always more profitable in a low milk price context than "more intensive" systems (Oostindie *et al.* 2013, Peyraud *et al.* 2014). These results show that paths alternative to spurred intensification are feasible.

Another agroecological strategy consists in increasing product value by high quality and tasty productions, product processing, short, local, and regional marketing chains (Section 3.7) to reduce farm income dependence on global and national market prices.

Diversification of products and activities, including agri-tourism if possible, is also a way to make the farm more resilient to price fluctuations. This is particularly important in an increasingly open and global market that considerably increased price volatility in the last decade compared to a more protected and regulated market that included measures such as milk quotas, subsidies, and technical trade barriers. The combination of dairy and meat productions and different animal products is a way to achieve this resilience. "Minor" niche productions should not be neglected; they can be locally important even if they cannot be generalised. They can also be alternatives to large-scale productions integrated in global value chains.

New alternative indicators to simple production indicators such as volume of milk per cow or weight of sold beef cattle should be developed. The ultimate goal of an agroecological system is not the maximisation of production volume or yields, it is much more increasing farmer's family income. Beyond income, the real objective is increasing farmers' living standards and well-being.

3.7. *Optimisation of farming and food system*

Francis *et al.* (2003), Gliessman (2007), and Wezel *et al.* (2015) have shown the way for new marketing chain organisation and new consumption models with the food systems approach in agroecology. This concept should be extended to meat and dairy products.

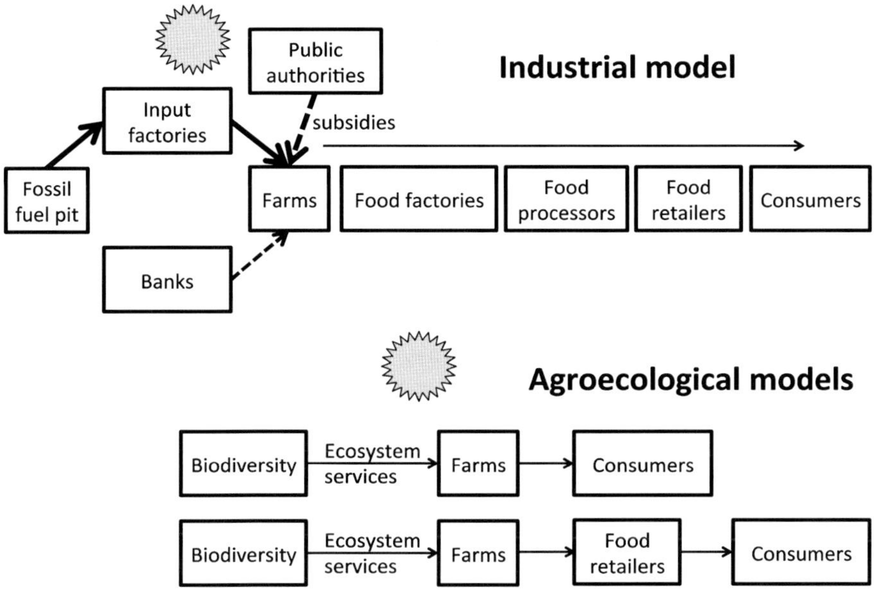

Figure 19. Food chains in industrial and agroecological models.

Food value chains can be quite different from one livestock production system to another. In general, food value chains are designed to be shorter in agroecology than in industrial farming (Figure 19). Industrial farming such as "Intensive conventional agriculture", "Green Revolution" (Gaud 1968), "Doubly Green Revolution" (Conway 1997), and "Sound Farming" ("Agriculture raisonnée" in French) (Paillotin 2000) are very fossil fuel-based. Their value chain starts from fossil fuel pits and finish in the plate of consumers. Upstream and downstream industries are important parts of the system. Farmers devote a large part of their revenue for buying inputs and paying loans to banks. Farmers' margins are limited because of low prices paid by strong dealers. A large part of total profit is captured by food processors and retailers. In agroecology, shorter value chains are favoured. They are ecosystem-based and biodiversity fulfils several functions that drive the system (Section 2.2). A proportion of the food produced is processed in farms and sold in short, local, or regional marketing chains to consumers. Production costs are thus reduced and a large proportion of increased profit is kept in farms. Since producers (farmers) and consumers are in closer contacts, the food chain is more transparent than in long industrial value chains. This induces higher consumer trust in product quality. Although short value chains are often favoured in agroecology, also midscale value chains may provide many positive aspects to farmers (Lev and Stevenson 2011, Stevenson *et al.* 2011). These midscale food value chains are often based on strategic alliances among small and mid-sized farms and other agri-food enterprises that operate at regional levels. Slaughterhouses, leather, or wool processors are also examples of industries that can be intermediate between farmers and citizens/consumers in agroecological marketing chains.

Meat processing is an opportunity to cut a carcass, to prepare sausages, salami, pates and other specialties, for preparing meat parcels adapted to consumers' needs. Milk can be processed into cheese, yogurt, ice cream, and butter for instance. These processes usually create jobs within farmer's family and in the local community. Product marketing in local and short chains is the last step of the farming and food process of the farm and the most profitable one.

The management of cooperatives can influence production systems and their sustainability. In contrast with the situation prevailing on the European continent, Irish cooperative dairy enterprises belong to farmers and farmers have still a word to say in their management. In Ireland, farmers decided to give priority to spring milk production because production costs are lower at that period of the year. Most of Irish milk is thus produced on the basis of grazed grass. Irish dairy enterprises stop their activity in winter. This decision maximises the efficiency of the biological process. The result of the Irish strategy is a higher profitability and a more resilient system for farmers (Pflimlin 2010).

Agroecological livestock systems should target the production of quality products of high nutritional and taste values. In livestock farming, this is achieved by grass-fed meat and dairy products, the use of rustic breed, and species-rich or at least diverse vegetation to improve milk quality, polyphenol content, and livestock and human health.

Compared to grass, grains have negative influence on meat and milk quality. This is particularly the case with the combination of green maize with soybean meal and cereals. Compared with grain-fed beef or milk, grass-fed beef or milk are for example lower in total fat, lower in saturated fatty acids, higher in total omega-3 and in the ratio omega3/omega6 fatty acids, and higher in conjugated linolenic acids (CLA) (Dhiman *et al.* 1999, Couvreur *et al.* 2006, Dewhurst *et al.* 2006, Duckett *et al.* 2009). Saturated fatty acids are linked with coronary heart diseases, CLA are recognised as anticancer. Modern human feeding is typically too low in omega-3 compared to omega-6 fatty acids. Grass-fed meat and grass-based dairy products can be sold at higher prices by highlighting their nutritional values notably their fatty acid profile, their taste, and their better impact on the environment.

Development of quality labels (e.g. geographical indications) and trademarks that focuses on quality is an efficient way for increasing selling price and income. The Protected Designation of Origin of the "Comté" cheese (East of France) is an outstanding example of a holistic system of high quality and tasty cheese production, mainly based on grass. The specification system includes criteria on cow breed type that should only be local "Montbéliarde" or French "Simmental" and their crosses. Temporary grasslands are limited to 15% of the farm forage area, annual fertilisation is limited to 50 kg/ha of inorganic nitrogen and 120 kg/ha of total nitrogen. Stocking rate on the whole farm forage area must be below one dairy cow per hectare, milk production is limited to 4,600 l/yr and per hectare of forage area, total stocking rate must be lower than 1.3 livestock unit per hectare of forage area, and forage must be conserved as hay only (INAO and CIGC 2015). All these measures limit intensification in the use of concentrate feed and fertilisers, aim at a high forage utilisation especially from permanent grasslands and target the maintenance of a diverse flora in permanent grasslands. The American Grassfed Association (AGA) (http://www.americangrassfed.org) is another striking example. AGA defines grassfed animals as those that have eaten nothing but grass and forage from weaning to harvest, have not been raised in confinement, and have never been fed antibiotics or growth hormones. It established a legal definition for the grassfed livestock and implemented a labelling program.

Cooperation and development of synergies between farmers, between consumers/citizens and farmers (e.g. by signed agreements like those of Community-Supported Agriculture) may contribute to increasing and stabilising farmer's income, improving access to quality food and decreasing food prices for consumers, increasing contacts between cities and rural land, and improving contacts of citizens with nature and farming.

4. Discussion

The successive crisis and the lack of perspective in conventional farming made clear that a fundamental model change is necessary. New models should be designed and developed. Self-sufficient and thrifty economic systems centred on local markets should be alternatives to global markets that are disconnected from energy and ecological realities and does not create long-term jobs at local scale. High quality grass-fed livestock systems should replace environmentally devastating grain-based systems. Grass-fed meat and dairy products can be sold at higher price by farmers which is acceptable for consumers if they do not pay intermediaries and if the proportion of these products in their diet is reduced. Grass-fed beef or milk products are usually higher quality and can be sold at higher price than conventional products.

Intensive dairy and beef systems are in a rationale of high investments, high use of concentrates, high production per animal and have to increase herd and farm size to be able to survive. That makes them very susceptible to sudden fall of milk and meat price and increase of input prices. In contrast, agroecological farmers reduce expenses linked to banks and agro-industries (e.g. fertiliser, feed) and often increase also their economic margin by food processing and short marketing chains. They try to improve the use of local resources which necessitates creativity and innovation (Peeters *et al.* 2015). Maximising grass, reducing green maize and concentrates, using the right animal, reducing investments and loans are key-parameters to achieve better income.

In a quest for more productivity and competitiveness in the sector, a large part of dairy farms disappeared since the 1960s. While farmer population decreased, farm size and the use of external inputs (e.g. fertiliser, concentrate feed, fossil energy) increased and many other parameters changed. The price of land became more and more important, dairy production required huge investments in terms of buildings, machinery, land, and cattle, while production costs for fertilisers, animal feed and veterinary costs increased a lot. The milk quota ensured a stable market in Europe between 1984 and 2004 but since then milk prices became much more unstable and do not always allow profitability of dairy farms. The farmer population is ageing since young people hesitate to enter into the sector because of the necessity to contract important loans and the lack of good economic perspectives. The system is thus questioned and many farmers started to develop alternative paths, towards more 'autonomous' farming systems (Peeters *et al.* 2015). The race for farm size enlargement and productivity per animal is abandoned in these new approaches. Workload can be reduced by replacing crops by grasslands, lengthening the grazing duration, the use of outwintering platforms, the reduction of herd size, the reduction of milking frequency or a well-reasoned adoption of automatic milking systems (AMS). Since workload for farming is reduced, more time is available for food processing and marketing.

Ruminants and other herbivores should not only be considered as "production factors" for making money with meat or dairy products, they should be treated with respect all along the food chain, including in the farm, during transport and in the slaughterhouse. In the farm, they should thus find welfare conditions that should be close to the ecological conditions that knew their wild ancestors and where they could express their feeding, reproductive, and social behaviours. On-farm slaughter of cattle via gunshot method can reduce stress and pain. It is now authorised in Switzerland and it is developing in other countries (Schiffer 2015).

Farmers should also consider the multiple functions that ruminants can fulfill by using grassland cellulosic products produced at low cost and often on soils that cannot be used for crops. Grassland-based ruminant production systems are most often much more efficient than concentrate-based systems for procuring proteins. They reduce the competition between feed for animal production and food for humans (Peyraud and Peeters 2016). Ruminants can use forages of legume-based temporary grasslands that play a strategic role on arable land in soil fertility, nitrogen fixation, crop weeds and disease control. All animals can contribute to soil fertilisation. They can also provide work by fulfilling some specialised functions (Section 3.5.5).

The transition towards agroecology requires a change of paradigm, a mind change that is not always easy. It requires trusting natural processes, observing and understanding them before acting in an inconvenient way. It requires more knowledge and thus more intellectual efforts but it can provide also more satisfaction in the farming job.

A fast transition from conventional to agroecological systems would require a strong political will, corresponding budgets and a strong involvement of many types of stakeholders: farmers at the first place, and also scientists, teachers of technical schools, farmer's advisers, and traders of the food sector. It should require in priority (Peeters *et al.* 2013):

- new agricultural policies or their substitution by a larger feeding policy;
- a reform of the training of future farmers, technicians and farmer's advisers in technical schools;
- a reform of the specialisation training (master) of agricultural scientists in higher education institutions;
- a restructuration of agricultural research that should define new priorities. The dominance of reductionist research should be drastically reduced at the benefit of holistic and participatory research.
- a technological revolution for designing and developing agroecological methods and systems, adapting them locally, disseminating them, supporting farmers in their transition period;
- the design of a biodiversity friendly agriculture: the integration of biodiversity enhancement in methods and practices;
- giving priority to agroecological products in school and administration cafeterias (canteens).

Besides changes in policy, budget allocation, and involvement of diverse stakeholders in the development of agroecological herbivore farming systems, also research has an important role to play. Major implications for research would be (Mendez *et al.* 2013, Peeters *et al.* 2013, Peeters 2015b):

- to have a more holistic, interdisciplinary or transdisciplinary, approach in taking into account the whole livestock farming system and its location in a socio-technical network of stakeholders and policy frame;
- to develop participatory approaches (see above);

- to develop a set of indicators that should help farmers to manage their system in all their dimensions including soil fertility, grassland quality and production, biodiversity associated to the system, efficiency of resource use, animal welfare, product quality, work load, etc.
- to reinforce research on pasture and rangeland management and their integration into herbivore farming systems;
- to broaden research for diagnostic, systematic use of prevention methods and non-toxic curing methods for integrated animal disease and parasite control;
- to monitor biodiversity when developing new practices and adapting farming systems.

5. Conclusions

In this chapter, we defined seven groups of agroecological principles related to herbivore farming systems and seven groups of agroecological practices.

Regarding principles, any development process of agroecological systems should recognise the central role of farmers and should associate farmers in holistic and participatory approaches. Agroecological systems should be locally and culturally adapted and should empower farmers by making them less dependent on commercial influence and input use. Agroecological systems are based on all biodiversity types, at all scales. They carefully manage natural resources such as soil, water, atmosphere, and energy. Biodiversity makes use of these resources for providing ecosystem services that may replace fossil fuel use to a large extent, for instance by legume symbiotic nitrogen fixation. Livestock farming can ideally be incorporated into mixed crop–livestock systems but specialised grazing livestock systems could also be sustainable. Animal health management is also an important component. Agroecological systems should be "economically viable", "liveable", "inheritable", "socially equitable", and environmentally sound. They should also be resilient to environmental and economic crisis. Moreover, these systems should provide optimum quantity of healthy and tasty food for citizens and a large diversity of food to local population. And finally, they should develop a new harmony in human societies by linking closely farmers to consumers.

Regarding practices, legume-based grasslands and multiple species swards should be the pillars of agroecological herbivore systems. Their management should be adapted to grassland plant and animal needs. Grazed grasslands should be preferably managed in rotational stocking. Stocking density should be high and plot sward height well controlled at the entrance and the exit of animals. Rest periods can be long thanks to the high legume proportion in swards. Different grassland types can be managed according to the needs of different animal types.

The structure of livestock systems should be designed by a holistic approach where animal breeds are chosen and adapted to the system and not the other way round. Double-goal and rustic breeds should be preferred to highly specialised animals. Moreover, The management of domestic livestock farming should be inspired by the conditions prevailing for their wild ancestors.

Agroecological systems should reduce feeding costs by giving priority to feed, to grass mainly, compared to food, by extending the grazing season, producing high quality grazed and conserved grass and aligning young animal birth period with the beginning of the grazing period. The agroecological animal health strategy should be based on prevention methods, curative instead of systematic treatments, and disease treatments based on natural means.

The economic strategy of agroecological systems for increasing income includes (i) the reduction of investments and production costs (up-stream strategy), (ii) an increase of the value of products by the production of quality products, their processing, and their marketing in short, local or regional chains (down-stream strategy), and (iii) a diversification of productions and activities.

In general, food value chains are designed to be shorter in agroecological than in industrial farming. They are ecosystem-based and not fossil fuel-based. Production costs are reduced and a large proportion of profit is kept in farms. Since farmers and consumers are in close contacts, consumer trust is higher in product quality, and thus consumers are ready to pay adequate prices. Product processing usually creates jobs and local marketing reduce transport costs and energy.

The development of synergies between farmers and consumers/citizens (e.g. within the framework of Community Supported Agriculture) often contributes to increasing and stabilising farmer's income, and to providing quality food to consumers at a reasonable price.

References

Allen, V.G., Batello, C., Berretta, E.J., Hodgson, J., Kothmann, M., Li, X., McIvor, J., Milne, J., Morris, C., Peeters, A., Sanderson, M., 2011. An international terminology for grazing lands and grazing animals. *Grass and Forage Science* **66**, 2–28.

Baddeley, J.A., Jones, S., Topp, C.F.E., Watson, C.A., Helming, J., Stoddard, F.L., 2013. *Biological nitrogen fixation (BNF) by legume crops in Europe*. Legume Futures Report 1.5.

Barthram, G.T., 1986. *Experimental Techniques: the HFRO Sward Stick*. In: HFRO (Ed.), Hill farming research. Organisation Biennial Report 1984–85. Edinburgh, Scotland, UK, pp. 29–30.

Baumont, R., Lewis, E., Delaby, L., Prache, S., Horan, B., 2014. Sustainable intensification of grass-based ruminant production. *Grassland Science in Europe* **19**, 521–532.

Benchaar, C., Greathead, H., 2011. Essential oils and opportunities to mitigate enteric methane emissions from ruminants. *Animal Feed Science and Technology* **166–167**, 338–355.

Benyus, J.M., 1997. *Biomimicry: Innovation Inspired by Nature*. William Morrow & Co., New York, USA, 308 p.

Besnard, A., Montarges–Lellahi, A., Hardy, A., 2006. Systèmes de culture et nutrition azotée. Effets sur les émissions de GES et le bilan énergétique. *Fourrages* **187**, 311–320.

Boerema, A., Peeters, A., Swolfs, S., Vandevenne, F., Jacobs, S., Staes, J., Patrick, M., 2016. Soybean trade: Balancing environmental and socio-economic impacts of an intercontinental Market. *PLoS ONE* **11**(5), e0155222.

Boller, B.C., Nosberger, J., 1987. Symbiotically fixed nitrogen from field grown white and red clover mixed with ryegrass at low levels of 15N fertilisation. *Plant and Soil* **104**, 219–226.

Bonaudo, T., Bendahan, A.B., Sabatier, R., Ryschawy, J., Bellon, S., Leger, F., Magda, D., Tichit, M., 2014. Agroecological principles for the redesign of integrated crop-livestock systems. *European Journal of Agronomy* **57**, 43–51.

Boval, M., Dixon, R.M., 2012. The importance of grasslands for animal production and other functions: a review on management and methodological progress in the tropics. *Animal* **6**(5), 748–762.

Caradus, J.R., 1990. The structure and function of white clover root systems. *Advances in Agronomy* **43**, 1–46.

Champion, R.A., Orr, R.J., Penning, P.D., Rutter, S.M., 2004. The effect of the spatial scale of heterogeneity of two herbage species on the grazing behaviour of lactating sheep. *Applied Animal Behaviour Science* **88**, 61–76.

Charlton, J.F.L., Wier, J.H., 2001. TechnoGrazing™, a new grazing concept. *Proceedings of the New Zealand Grassland Association* **63**, 33–36.

Coleman, J., Pierce, K.M., Berry, D.P., Brennan, A., Horan, B., 2009. The influence of genetic selection and feed system on the reproductive performance of spring-calving dairy cows within future pasture-based production systems. *Journal of Dairy Science* **92**, 5258–5269.

Conway, G., 1997. The double green revolution: Food for all in the twenty-first century. Comstock Pub. Associates, Ithaca, NY, USA, 334 p.

Corral, A.J., Fenlon, J.S., 1978. A comparative method for describing the seasonal distribution of production from grasses. *Journal of Agricultural Science* **91**, 61–67.

Couvreur, S., Hurtaud, C., Lopez, C., Delaby, L., Peyraud, J.L., 2006. The linear relationship between the proportion of fresh grass in the cow diet, milk fatty acid composition, and butter properties. *Journal of Dairy Science* **89**, 1956–1969.

Delaby, L., Hennessy, D., Buckley, F., 2014. Animal choice for grass-based systems. *Grassland Science in Europe* **19**, 795–797.

Delaby, L., Horan, B., O'Donovan, M., Gallard, Y., Peyraud, J.L., 2010. Are high genetic merit dairy cows compatible with low input grazing systems? *Grassland Science in Europe* **15**, 928–930.

Delagarde, R., Prache, S., D'Hour, P., Petit, M., 2001. Ingestion de l'herbe par les ruminants au pâturage. *Fourrages* **166**, 189–212.

Deprez, B., Lambert, R., Decamps, C., Peeters, A., 2004. Nitrogen fixation by red clover (*Trifolium pratense*) and lucerne (*Medicago sativa*) in Belgium leys. *Grassland Science in Europe* **9**, 469–471.

Deprez, B., Parmentier, R., Lambert, R., Peeters, A., 2005. *Les prairies temporaires : une culture durable pour les exploitations mixtes de Moyenne Belgique*. Unité d'Ecologie des Prairies (UCL)/Ministère de la Région Wallonne, Louvain-la-Neuve, Belgium, 72 p.

Dequin, A., Follet, D., Grasset, M., Le Gall, A., Roger, P., Thébault, M., 1998. *Cinq menus pour vaches laitières au pâturage*. Document EDE Chambres d'Agriculture de Bretagne. 12 p.+ fiches.

Derner, J.D., Hart, R.H., Smith, M.A., Waggoner, J.W. Jr., 2008. Long-term cattle gain responses to stocking rate and grazing systems in northern mixed-grass prairie. *Livestock Science* **117**, 60–69.

Dewhurst, R.J., Shingfield, K.J., Lee, M.R.F., Scollan, N.D. 2006. Increasing the concentration of beneficial polyunsaturated fatty acids in milk produced by dairy cows in high-forage systems. *Animal Feed Science and Technology* **13**, 168–206.

Dhiman, T.R., Anand, G.R., Satter, L.D., Pariza, M.W., 1999. Conjugated linoleic acid content of milk from cows fed different diets. *Journal of Dairy Science* **82**, 2146–2156.

Dillon, P., Crosse, S., Stakelum, G., Flynn, F., 1995. The effect of calving date and stocking rate on the performance of spring-calving dairy cows. *Grass and Forage Science* **50**, 286–299.

Dillon, P., Roche, J.R., Shalloo, L., Horan, B., 2005. Optimising financial return from grazing in temperate pastures. In: *Utilization of Grazed Grass in Temperate Animal Systems*. Proceedings of a satellite workshop of the XXth International Grassland Congress, Cork, Ireland (ed. J.J. Murphy). Wageningen Academic Publishers, The Netherlands, pp. 131–148.

Duckett, S.K., Neel, J.P.S., Fontenot, J.P., Clapham, W.M., 2009. Effects of winter stocker growth rate and finishing system on: III. Tissue proximate, fatty acid, vitamin, and cholesterol content. *Journal of Animal Science*, published online.

Dumont, B., Fortun-Lamothe, L., Jouven, M., Thomas, M., Tichit, M., 2013. Prospects from agroecology and industrial ecology for animal production in the 21st century. *Animal* **7**(6), 1028–1043.

FAO, 2009a. *Review of Evidence on Drylands Pastoral Systems and Climate Change. Implications and Opportunities for Mitigation and Adaptation.* FAO, Rome, Italy, Land and Water Discussion Paper 8, 38 p.

FAO, 2009b. The state of food and agriculture 2009. Livestock in the balance. FAO, Rome, Italy, 166 p.

Finneran, E., Crosson, P., O'Kiely, P., Shalloo, L., Forristal, D., Wallace, M., 2012. Stochastic modelling of the yield and input price risk affecting home produced ruminant feed cost. *Journal of Agricultural Science* **150**, 123–139.

Frame, J., Laidlaw, S., 2011. Improved grassland management. The Crowood Press Ltd, Ramsbury, UK, 288 p.

Frame, J., Newbould, P., 1986. Agronomy of white clover. *Advances in Agronomy* **40**, 1–88.

Francis, C., Lieblein, G., Gliessman, S., Breland, T.A., Creamer, N., Harwood, R., Salomonsson, L., Helenius, J., Rickerl, D., Salvador, R., Wiedenhoeft, M., Simmons, S., Allen, P., Altieri, M., Flora, C., Poincelot, R., 2003. Agroecology: The Ecology of Food Systems. *Journal of Sustainable Agriculture* **22**(3), 99–118.

French, P., Boyle, L., 2008. *Development of Sustainable Low Cost Animal Accommodation Outwintering Pads (OWP's).* TEAGASC, Moorepark Research Centre, Ireland, Final Report, Project Number 5232, 46 p.

Fukuoka, M., 2009. *The One-straw Revolution: An Introduction to Natural Farming.* New York Review Books Classics, New York, USA, 200 p.

Gaud, W.S., 1968. The green revolution: Accomplishments and apprehensions. Conference at Shorehan Hotel, Washington, DC. 8 Mar 1968. http://www.agbioworld.org/biotech-info/topics/borlaug/borlaug-green.html (accessed January 2014).

Gaujour, E., Amiaud, B., Mignolet, C., Plantureux, S., 2012. Factors and processes affecting plant biodiversity in permanent grasslands. A review. *Agronomy for Sustainable Development* **32**(1), 133–160.

Gebhard, C.-A., Büchi, L., Liebisch, F., Sinaj, S., Ramseier, H., Charles, R., 2013. Screening de légumineuses pour couverts végétaux: azote et adventices. *Recherche Agronomique Suisse* **4**(9), 384–393.

Geddes, P., Kohl, R., 2009. How Shropshire sheep are being used to control weeds in orchards. *Pesticides news* **86**, 21–23.

Gliessman, S.R., 2007. *Agroecology: the ecology of sustainable food systems.* CRC Press, Taylor & Francis, New York, USA, 384 p.

Hennessy, D., 2012. *Guidelines and Tools to Get the Most From Grazing in Ireland.* Teagasc, Animal and Grassland Research and Innovation Centre, Moorepark, Fermoy, Co. Cork, Ireland.

Herrero, M., Thornton, P.K., 2013. Livestock and global change: Emerging issues for sustainable food systems. *PNAS* **110**(52), 20878–20881.

Hervé-Gruyer, Ch., 2014. *Permaculture: Guérir la terre, nourrir les hommes.* ACTES SUD, Arles, France, 272 p.

Hoden, A., Peyraud, J.-L., Muller, A., Delaby, L., Faverdin, P., 1991. Simplified rotational grazing management of dairy cows: Effects of rates of stocking and concentrate. *Journal of Agricultural Science* **116**, 417–428.

Hoglund, J.H., Crush, J.R., Brock, J.L., Ball, R., Carran, R.A., 1979. Nitrogen fixation in pasture. XII. General discussion. *New Zealand Journal of Experimental Agriculture* **7**, 45–51.

Holzer, S., 2011. *Sepp Holzer's Permaculture: A Practical Guide to Small-scale, Integrative Farming and Gardening.* Chelsea Green Publishing, Hartford, VT, USA, 256 p.

Hopkins, A., Gilbey, J., Dibb, C., Bowling, P.J., Murray, P.J., 1990. Response of permanent and reseeded grassland to fertilizer nitrogen. 1. Herbage production and herbage quality. *Grass and Forage Science* **45**(1), 43–55.

Huhtanen, P., Hristov, A.N., 2009. A meta-analysis of the effects of dietary protein concentration and degradability on milk protein yield and milk N efficiency in dairy cows. *Journal of Dairy Science* **92**(7), 3222–3232.

INAO (Institut national de l'origine et de la qualité), CIGC (Comité Interprofessionnel du Comté), 2015. *Cahier des charges de l'appellation d'origine "Comté"* (Specifications of the Designation of Origin Label 'Comté'). Ministère de l'agriculture, de l'agroalimentaire et de la forêt, Paris, France, Bulletin officiel 11, 22 p.

INRA (Institut National de la Recherche Agronomique), 2007. *Alimentation des bovins, ovins et caprins* (Cattle, sheep and goat feeding). Quae Eds, Versailles, France, 330 p.

Janssens, F., Peeters, A., 2000. Restoration of species-rich grasslands in Belgium. *Grassland Science in Europe* **5**, 259–280.

Jensen, E.S., Peoples, M.B., Hauggaard-Nielsen, H., 2010. Faba bean in cropping systems. Review. *Field Crops Research* **115**, 203–216.

Judson, H.G., Moorhead, A.J.E., 2014. *Benefits and Uses of Ceres Tonic Plantain (Plantago lanceolata) in Sheep and Beef Systems*. Agricom.

Kampschulte, J., 2009. *Doppelnutzung statt Hochleistung. Der kritische Agrarbericht*, 136–141.

Laidlaw, A.S., Teuber, N., 2001. *Temperate Forage Grass-legume Mixtures: Advances and Perspectives*. Proceedings of the XIX International Grassland Congress, Sao Pedro, Brazil, pp. 85–92.

Le Gall, A., Beguin, E., Dollé, J.-B., Manneville, V., Pflimlin, A., 2009. Nouveaux compromise techniques pour concilier efficacité économique et environnementale en élevage herbivore. *Fourrages* **198**, 131–151.

Le Gall, A., Faverdin, P., Thomet, P., Vérité, R., 2001. Le pâturage en Nouvelle Zélande: des idées pour les régions arrosées d'Europe. *Fourrages* **166**, 137–164.

Leconte, D., 2002. Importance du rythme et de la hauteur de défoliation du ray-grass anglais sur la composition morphologique des repousses. *Fourrages* **169**, 47–63.

Ledgard, S.F., Brier, G.H., Upsdell, M.P., 1990. Effect of clover cultivar on production and nitrogen fixation in clover-ryegrass swards under dairy cow grazing. *New Zealand Journal of Agricultural Research* **33**, 243–249.

Lev, L., Stevenson, G.W., 2011. Acting collectively to develop midscale food value chains. *Journal of Agriculture, Food Systems, and Community Development* **1**(4), 119–128.

Lüscher, A., Fuhrer, J., Newton, P.C.D., 2005. Global atmospheric change and its effect on managed grassland systems. In: *Grassland: A Global Resource*

(Ed. D.A. McGilloway), Wageningen Academic Press, Wageningen, The Netherlands, pp. 251–264.

Marini, J.C., Van Amburgh, M.E., 2005. Partition of nitrogen excretion in urine and the faeces of Holstein replacement heifers. *Journal of Dairy Science* **88**(5), 1778–1784.

Mayne, C.S., Newberry, N.D., Woodcock, S.C.F., Wilkins, R.J., 1987. Effect of grazing severity on grass utilization and milk production of rotationally grazed dairy cows. *Grass Forage Science* **42**, 59–72.

Mendez, V.E., Bakon, C.M., Cohen, R., 2013. Agroecology as a transdisciplinary, participatory, and action-oriented approach. *Agroecology and Sustainable Food Systems* **37**, 3–18.

Mills, J.A.N., Crompton, L.A., Bannink, A., Tamminga, S., Moorby, J., Reynolds, C.K., 2009. Predicting methane emissions and nitrogen excretion from cattle. *J. Agric. Sci.* **147**, 741.

Mosimann, E., Muenger, A., Schori, F., Pitt, J., 2008. Pâturages pour vaches laitières 1. Modèle d'aide à la gestion du pâturage. *Revue suisse Agric.* **40**(1), 33–40.

Murphy-Bokern, D., Peeters, A., Westhoek, H., 2016. The role of legumes in bringing protein to the table. In Murphy-Boken, D., Stoddard, F., Watson, Ch. (Ed.) *Legumes in Cropping Systems* 364 p. CABI (Centre for Agricultural Bioscience International), Wallingford, UK, 49–76.

Murphy, B., 1998. *Greener Pasture on Your Side of the Fence: Better Farming Voisin Management-intensive Grazing*. Arriba Publishing, Open Library, 388 p.

Neeson, R., Love, S., 2014. *Managing Internal Parasites in Organic Livestock Production Systems*. NSW Department of Primary Industries, Australia, Primefact 1341, 16 p.

Nyfeler, D., Huguenin-Elie, O., Suter, M., Frossard, E., Connolly, J., Lüscher, A., 2009. Strong mixture effects among four species in fertilised agricultural grassland led to persistent and consistent transgressive overyielding. *Journal of Applied Ecology* **46**, 683–691.

O'Donnovan, M., Delaby, L., 2016. Grazed grass in the dairy cow diet – how this can be achieved better! *Grassland Science in Europe* **21**, 350–365.

Oostindie, H., van der Ploeg, J.D., van Broekhuizen, R., 2013. *Buffercapaciteit, bedrijfsstijlen in de melkveehouderij, volatiele markten en kengetallen*, WUR/ALFA, Wageningen, The Netherlands, 50 p.

Paillotin, G., 2000. L'agriculture raisonnée. Rapport au Ministre de l'Agriculture et de la Pêche, Paris, France, 41 p.

Palacio, S., Bergeron, R., Lachance, S., Vasseur, E., 2015. The effects of providing portable shade at pasture on dairy cow behavior and physiology. *Journal of Dairy Science* **98**, 6085–6093.

Pärtel, M., Bruun, H.H., Sammul, M., 2005. Biodiversity in temperate European grasslands: origin and conservation. *Grassland Science in Europe* **11**, 1–13.

Peeters, A., 2015a. Participatory and holistic approaches with grassland farmers and development of policies. In: D. Vijay, M.K. Srivastava, C.K. Gupta, D.R. Malaviya, M.M. Roy, S.K. Mahanta, J.B. Singh, A. Maity, P.K. Ghosh (Eds.) *Sustainable use of Grassland Resources for Forage Production, Biodiversity and Environmental Protection.* Army Printing Press, Lucknow, India, 290–305.

Peeters, A., 2015b. Environmental impacts and future challenges of grasslands and grassland-based livestock production systems in Europe. In: Ghosh, P.K., Mahanta, S.K., Singh, J.B., Pathak, P.S. (Eds.), *Grasslands: A Global Resource Perspective*. Army Printing Press, Lucknow, India, 365–390.

Peeters, A., Beaufoy, G., Canals, R.M., De Vliegher, A., Huyghe, C., Isselstein, J., Jones, G., Kessler, W., Kirilov, A., Mosquera-Losada, M., Nilsdottir-Linde, N., Parente, G., Peyraud, J.L., Pickert, J., Plantureux, S., Porqueddu, C., Rataj, D., Stypinski, P., Tonn, B., van den Pol-van Dasselaer, A., Vintu, V., Wilkins, R.J., 2014. Grassland term definitions and classifications adapted to the diversity of European grassland-based systems. *Grassland Science in Europe* **19**, 743–750.

Peeters, A., Delobel, V., Van der Ploeg, J.D., 2015. Are there alternative paths towards more self-sufficient and resilient systems in dairy farms? *Grassland Science in Europe* **20**, 475–477.

Peeters, A., Dendoncker, N., Jacobs, S., 2013. Enhancing ecosystem services in Belgian agriculture: a vision for a farming with a future. In: ‚Jacobs, S., Dendoncker, N., Keune, H. (Eds.), *Ecosystem Services. Global Issues, Local Practices*. Elsevier Science Publishing Co Inc, Atlanta, GA 30303, USA, pp. 285–304.

Peeters, A., Maljean, J.F., Biala, K., Brouckaert, V., 2004. Les indicateurs de biodiversité pour les prairies: un outil d'évaluation de la durabilité des systèmes d'élevage. *Fourrages* **178**, 217–232.

Peeters, A., Parente, G., Le Gall, A., 2006. Temperate legumes: key-species for sustainable temperate mixtures. *Grassland Science in Europe* **11**, 205–220.

Peyraud, J.-L., 2010. *Le pâturage des vaches laitières: comment tirer parti d'un fourrage équilibré pour assurer de bonnes performances dans des systèmes économes en intrants*. Centre de reference en agriculture et agroalimentaire du Québec (CRAAQ), 34ème Symposium sur les bovins laitiers, 21 p.

Peyraud, J.L., Delaby, L., Delagarde, R., Pavie, J., 2014. Les atouts sociétaux et agricoles de la prairie. *Fourrages* **218**, 115–124.

Peyraud, J.L., Le Gall, A., Lüscher, A., 2009. Potential food production from forage legume-based-systems in Europe: an overview. *Irish Journal of Agricultural and Food Research* **48**, 115–135.

Peyraud, J.L., Peeters, A., 2016. The role of grassland based production system in the protein security. *Grassland Science in Europe* **21**, 29–43.

Peyraud, J.L., Vérité, R., Delaby, L., 1995. Rejets azotés chez la vache laitière: effets du type d'alimentation et du niveau de production des animaux. *Fourrages* **142**, 131–144.

Pflimlin, A., 2010. *L'Europe laitière*. Editions France Agricole, Paris, France, 314 p.

Plantureux, S., Peeters, A., McCracken, D., 2005. Biodiversity in intensive grasslands: Effect of management, improvement and challenges. *Agronomy Research* **3**(2), 153–164.

Pochon, A., 2002. La prairie temporaire à base de trèfle blanc. CEDAPA, Plérin, France, 222 p.

Prendiville, R., Pierce, K.M., Buckley, F., 2009. An evaluation of production efficiencies among lactating Holstein-Friesian, Jersey and Jersey × Holstein-Friesian cows at pasture. *Journal of Dairy Science* **92**, 6176–6185.

Roca-Fernández, A.I., Peyraud, J.L., Delaby, L., Delagarde, R., 2016. Pasture intake and milk production of dairy cows rotationally grazing on multi-species swards. *Animal* **10**(9), 1448–1456.

Rosenberger, E., Rutzmoser, K., 2002. Ökologische Folgewirkungen der Zucht auf höhere Milchleistungen. *Gruber Info* **4**, 26–31.

Ruechel, J., 2006. *Grass-fed Cattle: How to Produce and Market Natural Beef.* Storey Publishing, North Adams, MA, USA, 384 p.

Salatin, J., 1996a. *Salad Bar Beef.* Polyface, Swoope, VA, USA, 384 p.

Salatin, J., 1996b. *Pastured Poultry Profit$.* Polyface, Swoope, VA, USA, 334 p.

Schiffer, K.J., 2015. *On Farm Slaughter of Cattle via Gunshot Method.* Dissertation, University of Kassel, Shaker Verlag Aachen.

Schreiber, M.M., 1967. Effect of density and control of Canada Thistle on production and utilization of Alfalfa pasture. *Weeds* **15**(2), 138–142.

Schubiger, F.X., Lehmann, J., Daccord, R., Arrigo, Y., Jeangros, B., Scehovic, J., 2001. Valeur nutritive des plantes des prairies. 5. Digestiblité de la matière organique. *Revue Suisse d'Agriculture* **33**(6), 275–279.

Smith, R.R., Taylor, N.L., Bowley, S.R., 1985. Red clover. In: Taylor N.L. (ed) *Clover Sciences and Technology, American Society of Agronomy*, Madison, Wisconsin, USA, pp. 457–470.

Steinfeld, H., Gerber, P., Wassenaar, T., Castel, V., Rosales, M., De Haan, C., 2006. *Livestock Long shadow, Environmental Issues and Options*. FAO, Rome, Italy, 390 p.

Stevenson, G.W., Clancy, K., King, R., Lev, L., Ostrom, M., Smith, S., 2011. Midscale food value chains: An introduction. *Journal of Agriculture, Food Systems, and Community Development* **1**(4), 27–34.

Thomson, D., 1984. The nutritive value of white clover. In: Thomson, D.J. (Ed.), Forage legumes. *Occasional Symposium of the British Grassland Society*. No. 16, Hurley, British Grassland Society, UK, pp. 78–92.

Undersander, D., Albert, B., Cosgrove, D., Johnson, D., Peterson, P., 2002. *Pastures for Profit: A Guide to Rotational Grazing*. University of Wisconsin, Madison, WI, USA, Extension, Publication A3529, 38 p.

van Eekeren, N., Bommele, L., Bloem, J., Schouten, T., Rutgers, M., de Goede, R., Reheul, D., Brussaard, L., 2008. Soil biological quality after 36 years of ley-arable cropping, permanent grassland and permanent arable cropping. *Applied Soil Ecology* **40**, 432–446.

Van Vuure, C.T., 2002. History, morphology and ecology of the Aurochs (*Bos primigenius*). *Lutra* **45**, 1–16.

Vertès, F., Simon, J.C., 1992. Extensification : quel rôle pour le trèfle blanc ? In: *Compte-rendu des Journées 1992 de l'Association française de Production fourragère, L'extensification en production fourragère*, pp. 102–103.

Voisin, A., 1988. Grass productivity. Conservation Classics, Island Press, Washington, DC, USA, 384 p.

Waldrip, H.M., Todd, R.W., Cole, N.A., 2013. Prediction of nitrogen excretion by beef cattle: a meta-analysis. *Journal of Animal Science* **9**(9), 4290–302.

Weiss, W.P., Willett, L.B., St-Pierre, N.R., Borger, D.C., McKelvey, T.R., Wyatt, D.J., 2009. Varying forage type, metabolizable protein concentration, and carbohydrate source affects manure excretion, manure ammonia, and nitrogen metabolism of dairy cows. *Journal of Dairy Science* **92**(11), 5607–5619.

Wezel, A., David, C., 2012. Agroecology and the food system. In: Lichtfouse E. (Ed.), *Agroecology and Strategies for Climate Change*. Sustainable Agriculture Reviews 8, Springer, Dordrecht, The Netherlands, pp. 17–34.

Wezel, A., Fleury, Ph., David, C., Mundler, P., 2015. The food system approach in agroecology supported by natural and social sciences: topics, concepts, applications. In: Benkeblia, N. (Ed.), *Agroecology, Ecosystems and Sustainability*. CRC Press, Boca Raton (FL). USA, pp. 181–199.

Wezel, A., Peeters, A., 2014. Agroecology and herbivore farming systems — principles and practices. *Options Méditerranéennes A* **109**, 753–767.

White, R.P., Murray, S., Rohweder, M., 2000. *Pilot analysis of global ecosystems. Grassland ecosystems.* World Resources Institute, Washington, DC, USA, 69 p.

Whitehead, D.C., 1995. *Grassland Nitrogen.* CAB International, Wallingford, UK, 397 p.

Wilkins, R.J., Gibb, M.J., Huckle, C.A., 1995. Lactation performance of spring-calving dairy cows grazing mixed perennial ryegrass/white clover swards of differing composition and height. *Grass and Forage Science* **50**, 199–208.

Wynn, S.G., Fougere, B., 2007. *Veterinary Herbal Medicine.* MOSBY, Elsevier, Amsterdam, The Netherlands, 714 p.

Yan, T., Frost, J.P., Keady, T.W.J., Agnew, R.E., Mayne, C.S., 2007. Prediction of nitrogen excretion in feces and urine of beef cattle offered diets containing grass silage. *Journal of Animal Science* **85**(8), 1982–1989.

Chapter 12

Agroecological Management in Fish Pond Systems

Joël Aubin*, Joël Robin†, Alexander Wezel† and Marielle Thomas‡

*INRA, Rennes, France
†Department of Agroecology and Environment, ISARA, Lyon, France
‡URAFPA, University of Lorraine, Nancy, France

1. Introduction

Aquatic products play an increasing role in human nutrition. Depending on the type of production and location in the world, they can be the basis for daily protein requirements or a luxury food. World demand for edible fish (including finfish, crustaceans, molluscs, and amphibians) increased from 9.9 kg per capita in the 1960s to 19.2 kg per capita in 2012 (FAO 2014). This growth is no longer sustained by fisheries, as fishery catches have remained stable for more than 10 years (at around 90 million tonnes). As a consequence, aquaculture production has increased an average of 6.2% per year (2000–2012) (FAO 2014). This sectorial growth is currently occurring in Asian countries, which produced 88% of world aquaculture production in 2012 (FAO 2014). Conversely, fish farming in Europe has not developed greatly due to economic and environmental constraints (environmental regulation, access to production sites, etc.).

Fish farms and fish pond systems harbour a wide variety of species with different feeding behaviours: filter feeding, detritivory, herbivory, planktivory, omnivory, piscivory. The FAO (2014) recorded 354 species of reared finfish in 2012. Nevertheless, only 18 species represent 75% of the world finfish production (FAO 2014).

Among the 44.2 million tonnes of finfish produced in 2012, 87% are from inland aquaculture (Table 1). In freshwater finfish aquaculture, filter-feeding carps, considered as non-fed species, represent 18.4%. These values show the predominance of freshwater pond production in world aquaculture. In past decades, fed aquaculture has grown faster than non-fed aquaculture, demonstrating the development of more intensive production systems generally based on finfish monoproduction and formulated feed. The older systems were often based on polyculture and exploited the natural biomass developed directly from nutrients entering the ponds with water from catchments, or by using fertiliser inputs (minerals or manure). The polyculture system was empirically designed to use multiple levels of the food chain through a balanced choice of species. This system, originally created in China using several cyprinid species, was developed and adapted in different countries using a larger panel of species. This production system is used mostly in large ponds (several hectares). This system has been slowly abandoned in favour of more intensive practices in

Table 1. World production of farmed species groups from inland aquaculture and mariculture in 2012 (FAO 2014).

	Inland aquaculture (million tonnes)	Mariculture (million tonnes)	Quantity subtotal (million tonnes)	(% by volume)	Value subtotal (USA$ billion)	(% by value)
Finfish	38.6	5.6	44.2	66.3	87.5	63.5
Crustaceans	2.5	3.9	6.4	9.7	30.9	22.4
Molluscs	0.3	14.9	15.2	22.8	15.9	11.5
Other species	0.5	0.3	0.9	1.3	3.5	2.5
Total	41.9	24.7	66.6	100.0	137.7	100.0

smaller surface areas that are less likely to vary in yield and fish size and that are more adapted to market demand. Nevertheless, the objective of having species interact in aquatic production systems became a new interest in the development of a new family of systems called Integrated Multitrophic Aquaculture (Neori *et al.* 2004, Troell *et al.* 2009). Here, finfish species and plants (algae, phytoplankton, and macrophytes), molluscs, and other invertebrates are combined to make the best use of the nutrients provided by fish feed and to diversify the species produced. This concept is associated with different production practices, using different species in separate compartments (tanks or small ponds) in open or recirculating aquaculture systems, or combining all species in the same pond or the open sea.

Another benefit of fish pond systems is their potential to recycle waste, as in integrated livestock and pond systems. These systems generally combine monogastric animal production (duck, poultry, pig) with a pond (generally a fish polyculture) to recycle the animal waste into commercial fish production. This practice, developed in Asia, Africa and to lesser extent in South America, shows the potential benefit of introducing a pond into a more complex agricultural system at the farm scale, such as occurs in Vietnam (Nhan *et al.* 2007).

In this chapter, we focus on traditional fish pond systems, with examples from France, Brazil, and Indonesia. The objective is to evaluate to what extent fish pond systems comply with the principles of agroecology, how development of pond management practices can be consistent with these principles, and how the interaction of pond and terrestrial systems may enhance the agroecological character of agriculture. More specifically, we define specific agroecological principles for fish pond systems and determine different agroecological practices for them.

2. Major characteristics of fish pond systems

Fish ponds are agroecosystems that contribute to the identity of rural areas. Hydromorphologically speaking, they can be defined as contained water bodies whose topography and shallow depth often do not allow stable thermal stratification of water, e.g. due to wind, while fostering colonisation of the shoreline by diverse flora (Bernard, 2008). This shallowness, with a typical maximum depth less than 3 m, is certainly the most distinctive morphological characteristic of ponds. Surface area is probably not the most relevant defining characteristic of a pond, since area can vary greatly depending on the site, ranging from less than 5 ha to more than 100 ha (Korinek *et al.* 1987, Céréghino *et al.* 2008). A fish pond differs from a natural pond in the way that facilities are installed to allow the management of ponds for fish production. For example, dams are built to retain water, and full gravity drainage is ensured by a discharge structure incorporated into the dams.

From a hydrological viewpoint, all ponds are fed by rain water coupled with other, sometimes temporary, water sources. On this basis, four types of fish ponds can be distinguished (Figure 1). The first type is barrage ponds. They are built on a valley floor by constructing a dam across a stream. They are usually located at the head of the catchment, thereby preventing the dam from being washed out by heavy flooding. The second type comprises diversion ponds. They are fed by diverting a water resource through a canal or a ditch, thus enabling adequate control of the water supply. Ponds can be arranged in chains of diversion, fish ponds in series or in parallel. The third type of fish pond is fed only by rain and runoff waters from the catchment or from ponds located upstream connected via ditches or channels. Lastly, a fourth type is represented by fish ponds supplied by a spring or by groundwater.

Fish pond systems are also distinguished by the quantities and species of fish produced. They are often traditional low-intensity polyculture systems which exploit the various trophic levels in the agroecosystems and often have no or low input of supplementary feed. These extensive systems, without the need for specific production equipment, produce fish yields usually between 100 and 1,000 kg/ha (Szumiec 1996). This characteristic distinguishes them significantly from semi-intensive production (1,000–2,000 kg/ha) and intensive production (up to 6,000 kg/ha or more). These systems also exploit the trophic complementarity of different fish species, but their production levels are only possible with supplementary feed or fertilisation and aeration.

Pond types and how they are perceived can vary considerably, but in all cases a pond is an agroecosystem open to its surrounding environment.

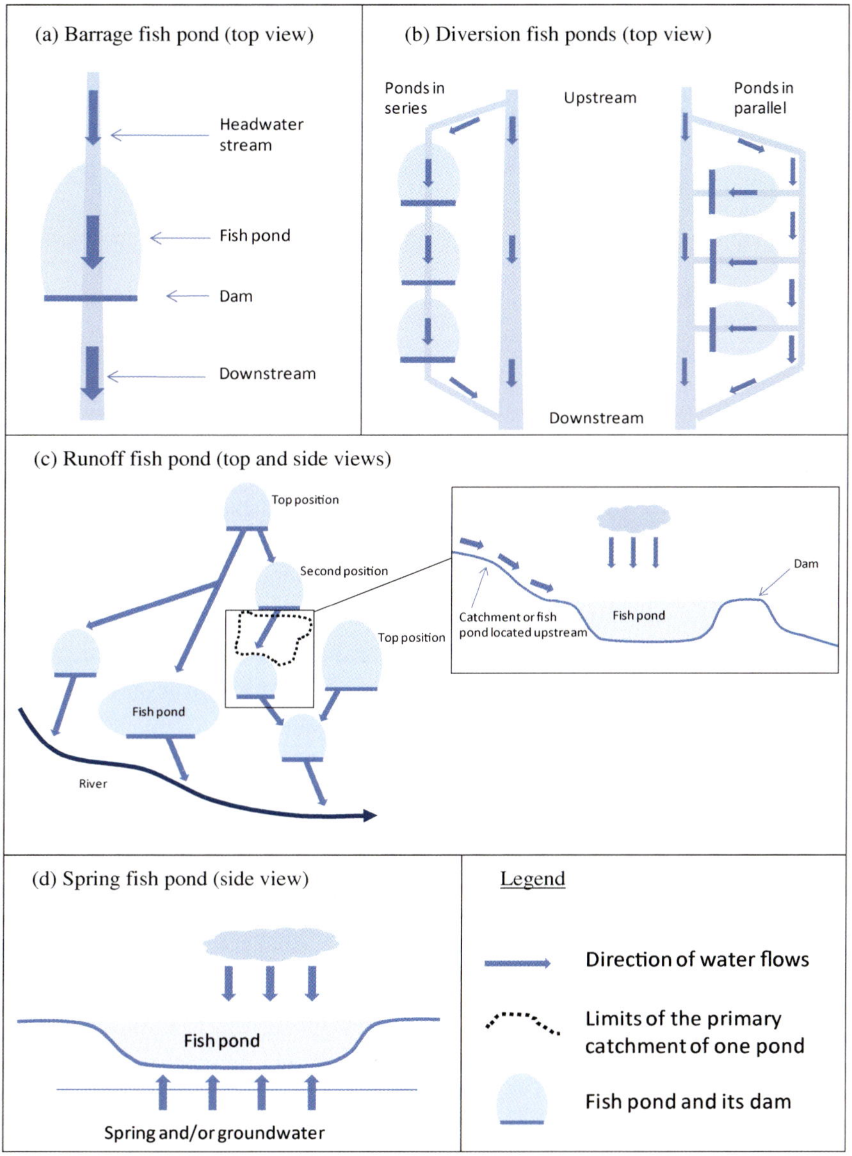

Figure 1. Four types of fish ponds according to the main water supply sources: (a) barrage ponds, (b) diversion ponds, (c) runoff pond and (d) spring (and/or groundwater) fed ponds.

Several biotic, abiotic, environmental, and anthropogenic factors influence the characteristics and functioning of ponds. Likewise, the pond affects its environment. Although ponds provide benefits beyond fish production, standing water was often considered unhealthy in the past, leading some people to be wary of ponds. Even though perception of this agroecosystem has changed today, such mistrust persists and has even grown in response to water quality targets identified in the Water Framework Directive in Europe (European Union 2000, 2008, 2013).

Fish ponds are complex systems organised into multiple compartments (open water, mudflats, vegetation belt, aquatic grass beds, and sediment). They may have high species diversity (Wezel *et al.* 2014). Traditionally (especially in Europe), pond management usually uses an ecosystem approach to exploit natural resources and ecological functions. This often contrasts with management of the surrounding land, which usually is based on practices such as tillage, chemical inputs, and pesticides. Therefore, attempting to fully understand the relationships that exist between a pond and its environment requires considering both management methods, which vary according to the desired objectives, and the multiple constraints related to the socio-economic and environmental contexts. Thus, one can distinguish between characteristics which are inherited and those which may be modified, e.g. the water matrix, species diversity in ponds, habitat diversity, and the surrounding landscape. Several characteristics must be considered when illustrating the relationships between a pond agroecosystem, its environment, and management (Table 2).

The water matrix can be used to provide a simple illustration of how the catchment affects a pond. The amount and quality of water in ponds are driven not only by the hydrogeomorphological characteristics of their catchments, but also by land use and agricultural practices (Lazartigues *et al.* 2012, 2013). Likewise, ponds affect their surrounding environment. The water coming out of ponds varies in amount and quality according to several factors, such as fish farmers' practices and processes (retention, transformation, matter transfer) occurring within the ponds (Banas *et al.* 2002). Consequently, the ratio of the surface area of ponds to the area of their catchment is important when analysing spatial–temporal changes in the water matrix.

Table 2. Site-specific or local stakeholder-driven characteristics to consider when studying pond–environment relationships.

	Characteristics to consider	
Scale studied	Site-specific features	Characteristics under the control of operators/managers
Fish pond	— Size — Depth — Bank morphology (length, slope) — Position in the catchment (single pond, series of ponds) — Time variation in hydraulic regime	— Fish stocking — Water fertilisation (organic, mineral) — Aquatic vegetation cover — Control of aquatic vegetation (herbicides, weed cutting) — Draining/filling of ponds (alternating dry periods, fluctuating water levels)
Catchment	— Total area — Slope in catchment — Soil types	— Land use — Crop diversity and area concerned — Crop rotations — Tillage, cover crops — Soil fertilisation — Pesticide application — Land drainage, irrigation

3. Fish pond landscapes in Europe and fish pond systems in the world

Many types of fish pond systems exist in the world. In Europe, several systems based on carp (*Cyprinus carpio*) production are important in terms of quantities produced and of their cultural heritage. Often, carp are mixed in ponds with pike (*Esox lucius*), pike-perch (*Sander lucioperca*), European catfish (*Ictalurus melas*), perch (*Perca fluviatilis*), tench (*Tinca tinca*), roach (*Rutilus rutilus*) and rudd (*Scardinius erythropthalmus*). The ponds have an average cycle of 1–3 years, after which the pond is emptied and harvested. Water management practices such as fertilisation and liming, as well as fish feeding and intensity of fish production, can vary considerably among systems (Horváth *et al.* 2002).

Many of these carp-based systems are more common in the fish pond cultural landscapes of several countries (Figure 2). Most cultural landscapes with a large number of fish ponds were created during the Middle Ages, and their ponds are still used, at various intensities, for fish production. Usually, these ponds are emptied every year for fish harvest and then refilled. France, for example, has several regions with a high fish-pond density: Dombes, Lorraine, Brenne, Forez, and Bresse-Jura (Table 3), as well as Sundgau. The Midden–Limburg region in Belgium is an important fish pond area (Lemmens 2014). In Poland, the highest density of fish ponds is found in the south-west (Bukacińska *et al.* 1995). In the Czech Republic, about 40,000 ha of fish ponds exist (Pokorný and Hauser, 2002). An important area is located in the Trebon basin, south Bohemia (Table 3). They also exist in the Lausitz region (Lusatia) in eastern

Figure 2. Fish pond cultural landscapes in Europe where high numbers of fish ponds are a characteristic component of the landscape.

Table 3. Fish pond cultural landscapes in Europe with pond numbers and total pond areas.

Fish pond region	Country	Number of ponds	Total area (ha)	References
Midden–Limburg	Belgium	1000	700	Lemmens (2014)
Dombes	France	1100	11,500	Bernard and Lebreton (2007)
Lorraine	France	—	7000	Le Bihan and Font (2008)
Brenne	France	1300	8500	PNR de la Brenne (1998)
Forez	France	300	1500	MEDD 2007, Vallod and Sarrazin (2010)
Bresse-Jura	France	250	—	Conservatoire des Espaces Naturels de Franche-Comté (2007)
Trebon	Czech Republic	1000	7500	Lhotský (2010)
Lausitz	Germany	335	3000	Touristische Gebietsgemeinschaft Heide und Teich im Bautzener Land (2014), UNESCO (2015)
Aischgrund	Germany	6859	2252	Bätzing (2013)
Oberpfalz	Germany	14,000	10,000	Bezirk Oberpfalz (2015)
Waldviertel	Austria	1800	1700	Bauer (2014)

Germany. In southern Germany, the Aischgrund and the Oberpfalz regions are well-known (Table 3), as well as Oberschwaben, which has more than 2,000 natural shallow lakes and fish ponds (Strehle and Trautmann 2011). In eastern Austria, the Waldviertel region is important (Table 3), and other fish ponds systems are found in southern Styria (Steiermark) (BMLFUW 2014, Matzinger 2014). Other carp pond areas are found in Bulgaria, Hungary, the former Yugoslavia and the former western part of the Soviet Union (Korinek *et al.* 1987).

Outside Europe, many fish pond systems have also been developed, and some have existed for many centuries. We will mention just a few of them, although many more exist throughout the world. A typically traditional system well known in China has endured more than 1,000 years and is based on polyculture of Chinese carps, chosen because their trophic complementarity use the many ecological compartments of the pond more effectively (Milstein 2005). The concept of complementarity has even been extended to integrating livestock as a nutrient source; addition of livestock waste to ponds creates a synergistic effect between livestock and the pond system via integrated management of resources (Little and Edwards 2005). A wide range of systems exist, especially those developed in Asia which directly integrate livestock and ponds, either directly, by using the waste of animals reared over the pond or nearby, or indirectly, by using by-products of livestock systems or meat production as inputs for ponds. The livestock species considered are often chicken, ducks or pigs, but also cattle, sheep or goats (Little and Muir 1987), and sometimes even humans.

Pond systems are also integrated with crop systems, with an exchange of nutrients and water among system compartments. One example, from upland northern Vietnam, are fish ponds which are embedded between paddy fields in valley bottoms. Paddy fields and ponds are closely linked to each other to exchange water (Steinbronn 2009). Also, on the slopes of the hills where slash and burn agriculture is practised (Wezel *et al.* 2002), small fish ponds are located between maize, cassava and upland rain-fed rice fields. The typical pond system is a polyculture containing mainly herbivorous and omnivorous cyprinid species which is integrated into the overall farming system (Steinbronn 2009). Weeds from the paddy fields, leaves from the upland fields, crop residues such as cassava peels and rice bran, as well as manure from large ruminants and pigs, serve as inputs for the ponds. In southern Vietnam, different integrated aquaculture-agriculture systems exist which integrate fish production with rice cropping and fruit production (Nhan *et al.* 2007).

Conversely, the success of striped catfish (*Pangasianodon hypophthalmus*) production in Vietnam shows the opposite approach of a highly intensive pond system. Vietnam is now the third largest aquaculture producer, producing 3.09 million tonnes in 2012 (FAO 2014), mainly supported by striped catfish. The dramatic increase in striped catfish production in Vietnam, which was less than 1,000 tonnes at the beginning of the 1990s, reached 1.14 million tonnes in 2010 (Kluts *et al.* 2012).

The yields in ponds are extremely high (427 tonnes/ha/year) due to their depth (sometimes over 5 m) and the fishes' ability to use atmospheric oxygen and low sensitivity to water quality degradation. Entirely dependent on formulated feed, this system is quite efficient and relies mainly on crop by-products. Nevertheless, the relatively high pollutant emissions of the system cause debates, due more to the number of ponds in the production regions than to the feed conversion ability (Anh *et al.* 2010) or the occurrence of disease outbreaks.

4. Agroecological principles

4.1. *General principles of agroecology*

In the general context of agricultural intensification supported by the green revolution, agroecology practices appear an alternative to using a high level of inputs. Different definitions of "agroecology" exist (Wezel *et al.* 2009), as it can be considered a set of practices, a science, or a social and political movement. Nevertheless, there is consensus around the same basic concepts, leading to five general principles popularised by Altieri (Altieri and Nicoll 2005, based on Reinjntjes *et al.* 1992): (1) recycling biomass and balancing nutrient flows and availability; (2) securing soil conditions favourable for plant growth through enhanced organic matter and soil biotic activity; (3) minimising losses of solar radiation, air, water and nutrients via microclimate management, water harvesting and soil cover; (4) enhancing species and genetic diversity of the agroecosystem in time and space; and (5) enhancing beneficial biological interactions and synergisms among agrobiodiversity components which result in promotion of key ecological processes and services.

These principles combine general guidelines (e.g. promotion of key ecological processes and services; securing soil conditions favourable for plant growth) and practices or objectives (e.g. enhancing organic matter and soil biotic activity; enhancing species and genetic diversity of the agroecosystem in time and space). Two main ideas drive these principles. One is the diversity of biotic components in soils and cultivated species, with an expected benefit for regulation and interactions with natural and domesticated biodiversity. The second is the recycling of nutrients and energy within the production system. These two ideas are not independent because diversity is considered an important factor in recycling. Agroecological principles were mainly conceived by agronomists and are almost exclusively applied to plant production. They generally consider livestock as a component of the production system with a role in nutrient cycling but not as a production objective in itself. This seems a possible reason why animal farming systems have so far been mostly ignored in agroecology when defining principles and practices. Recently, Dumont *et al.* (2013) extended agroecological thinking to animal production by reformulating the five principles of Altieri and Nicolls (2005, based on Reinjntjes *et al.* 1992) so they can be optimised to maintain yields while minimising negative environmental impacts of animal production (see Chapter 10 by Dumont *et al.* 2013). Application of agroecological principles generally varies greatly among animal production systems (Dumont *et al.* 2013). Therefore, an animal farming system may be considered an "agroecological inspiration" if it applies only one, or several, of the above-mentioned principles.

4.2. *Specific agroecological principles of fish pond systems*

Based on the general definitions and aspects of agroecology adapted to animal production, we adapted the framework of agroecological principles to the fish farming sector, specifically to fish production systems in ponds. These principles refer to general attributes of systems (López-Ridaura *et al.* 2002). They must be considered together for achieving agroecological status. Nevertheless, it is possible to define an agroecological orientation for systems which do not comply with all of the principles. From our viewpoint, an agroecological fish pond can be defined (Figure 3) as productive (Principle 1), robust and resilient (Principle 2), efficient in using local and natural resources (Principle 3), environmentally friendly (Principle 4), and having a natural and cultural value (Principle 5). These principles are described further, along with their underlying concepts and associated agroecological practices, in the following paragraphs.

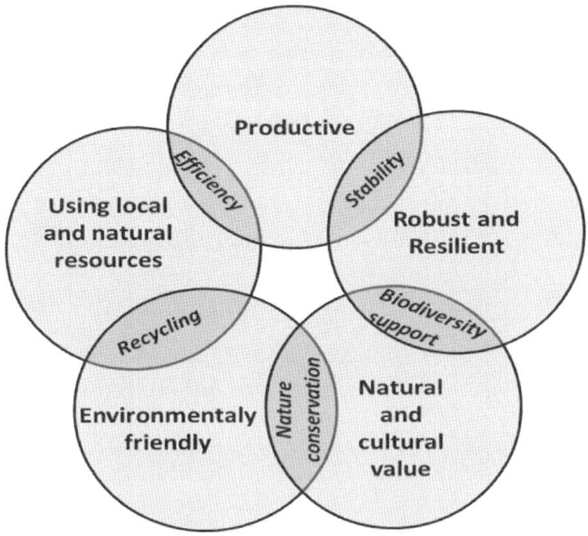

Figure 3. Agroecological principles, relations between them and interface concepts.

Principle 1: An agroecological fish pond is productive

Even though large extensive ponds are considered natural ecosystems with great influence on ecological cycles and landscape structure, they are production systems. It is because of the production objective that they were designed, built and managed. Abandoning the production activity means abandoning the management practices (e.g. reed-bed cutting, draining and filling, dam maintenance) which maintain certain ecosystem functions, which may lead to a decrease in biodiversity. From this perspective, maintaining production so that pond owners or fish farmers can earn an acceptable income is an important factor in agroecological pond management. Therefore, providing moderate and accurate amounts of active nutrients to support trophic chains and avoid malfunctions in the food web can be considered agroecological. Pond systems supply two markets: aquatic products as food, as well as fish to restock ponds for angling. In Europe, the marketed products are essentially fish, but other regions, such as Asia, produce more aquatic plants (e.g. water lilies) for food markets. Diversification of the target production species is another objective of agroecological management. It mainly concerns edible fish but can include other fish or plant species for feeding or ornamental purposes. The pond system can support other production activities, such as terrestrial crop systems, by recycling drained water and associated nutrient transfers for irrigation or for orchards, vegetables or paddy fields. This nutritional function is reinforced by the use of pond sediments with an agronomic value as fertiliser for terrestrial crop systems.

Moreover, pond systems may provide different types of ecosystem services. The provisioning services (described above) are supplemented by cultural services which can be market-oriented. One main example of the latter in Europe, is the use of ponds to hunt waterfowl. This idea extends the concept of productivity, which may include a wide range of services, and contributes to diversify the incomes of producers.

Principle 2: An agroecological fish pond is robust and resilient

Two ideas lie behind this principle. Robustness is associated with the ability of a system's productivity to remain stable when encountering potential disturbances in its context (e.g. environmental, economic, management). Resilience means that a system recovers its previous status and performance after an extrinsic disturbance. Different factors support the resilience and robustness of pond systems, but the main factor is diversity within the system, especially a high level of (agro-) biodiversity. Generally, complexity is considered a factor of stability. More diverse cultured species allow production of a wider variety of products to reach different markets, which can mitigate impacts of a market failure. The variety of biophysical or ecological preferences of the different species may help to maintain the overall productivity of the pond in the face of environmental changes, such as an increase in temperature or decrease in oxygen availability. Moreover, interactions between species (e.g. food web, behaviour) may be important for regulations of diseases, pests, algae blooms, and plants, and expansion of invasive species. Overall biodiversity (cultured and uncultured) is a complex factor; higher biodiversity increases flows of nutrients and energy, which prevent pathogen expansion, for example. In agriculture, Altieri (2002) calls this strengthening the "immune system", which includes agricultural operations such as promoting natural pest control and, in animal-production systems, adopting management practices which improve animal health (Dumont *et al.* 2013).

Therefore, pond polyculture systems favouring cultured and uncultured biodiversity have great adaptive capacity. It is possible to support multiple habitats within the same pond system. The number of habitats in pond systems depend on pond depth and distances to the benthic zone and the shore. It is possible to use the functions associated with these habitats. For example, intensive fed-fish production can be developed in small spaces (cages, happas), while the surrounding habitats support all biodiversity and potentially improve water quality. Maintenance of reed-bed habitats is an important factor for improving water quality and increasing carrying capacity.

Principle 3: An agroecological pond uses local and natural resources efficiently

Decreasing dependence on external inputs is an objective generally proposed for sustainable agriculture systems. In this sense, Dumont *et al.* (2013) describe the need to decrease inputs in animal production. This principle aims to decrease dependence on non-renewable energy sources and manufactured inputs. It decreases economic risks due to variations in market prices of inputs and is based on using natural inputs or production and/or recycling elements inside the system. It does not necessarily seek autonomy, but calls for input efficiency and a priority on using renewable resources. Therefore, practices that maximise solar radiation capture via production of aquatic plant and phytoplankton biomass, as a basis for the trophic chain, are key in the agroecological management of ponds. The often large areas of fish ponds relative to their volume are beneficial because larger areas capture more solar radiation and thus intensify photosynthesis. Similarly, nutrients provided by water from the catchment or naturally present in pond sediments are essential in non-fed pond systems to support the food web. Optimising nutrient and energy recycling by adapting practices is a factor of autonomy, as it enables producing more with the same quantity of inputs. This directly relates to the objective of balancing regulatory systems (nutrient cycles, water, energy flows, and population regulation) defined by Altieri (2002).

In pond systems, promoting a well-balanced food web of complementary species (i.e. primary producers, herbivores, detritus feeders, filter feeders, carnivores) favours use of nutrients at the system scale. Indirectly, the variety of nutrients and feed provided by well-managed fish pond ecosystems may also compensate for the decrease in fish oil and meal in a formulated fish diet.

Principle 4: An agroecological fish pond is environmentally friendly

One criticism often raised about fish pond systems is their potential emission of suspended matter and pollutants during draining. The drained water contains sediments, organic matter, nutrients, and other dissolved elements (Banas *et al.* 2002, 2008). Pond sediment is the main source of these elements; it consists of material accumulated from the catchment and from the living pond ecosystem. Decreasing this organic and mineral dead-end is a major objective of agroecological pond management. It is based on intense recycling of organic matter to increase the amount of mineral nutrients captured by the cultured species. Recycling of sediment nutrients decreases the loss of potential pollutants in drainage water. Dumont *et al.* (2013) also suggest using this process for animal-production systems. Increasing system efficiency by recycling nutrients in the food web is a major way to reduce pollution in pond systems. Among the recycling processes, denitrification of aquatic ecosystems is also important for reducing N_2O emissions, which have a potential greenhouse effect.

Ponds also remove pollution by sequestering carbon (Boyd *et al.* 2010) and transforming and metabolising nutrients (especially phosphorus), heavy metals and pesticides (Gaillard *et al.* 2016a, 2016b). These processes occur in different compartments in the pond, especially at the air-water and water-sediment interfaces, due in part to activities of living organisms, such as aquatic plants and bacterial biofilms. This function must be supported with appropriate practices to avoid re-emissions of pollutants during draining.

Principle 5: An agroecological fish pond has natural and cultural values

Around the world, pond systems have multiple objectives, such as supplying protein or other nutrients, hosting recreational activities (e.g. angling, hunting, bird watching, hiking), and serving as locations of cultural or religious importance. In Europe, for example, many pond systems were created in the Middle Age to supply fish during Easter or other diet prescriptions for monks. Up to the present, they have supported traditional knowledge, customs, and culinary traditions, and are mentioned in many folk tales. Ancient ponds can have a real cultural influence in rural regions. Even new ponds are appreciated as a structural part of the landscape and can contribute to the general scenic beauty and to the overall attractiveness of landscapes. In general, pond systems support a variety of cultural ecosystem services (Blayac *et al.* 2014).

Besides cultural heritage value, fish ponds also possess a natural value. Pond systems support specific types of biodiversity by creating or conserving habitats at the junction of aquatic and terrestrial ecosystems. In particular, they host endangered or rare species and provide habitat for endemic and migratory birds, amphibians, and insects. Maintaining these pond habitats requires particular ecological management practices based on a compromise between production efficiency and nature conservation. Habitat quality is directly influenced by alternating dry periods, when ponds are emptied for fish harvest, with water-filled periods, and by controlling the expansion of reed-beds and certain aquatic plants, controlling riparian woodlands, and following different fish stocking practices. In addition, several other ecosystem services are associated with the cultural and heritage values of ponds. Ponds have influence fresh water reservoirs and hydrological regulation (e.g. protection against floods). They are associated with ancestral knowledge about catchment management in the sharing of water among multiple users: fish farmers, farmers, municipalities, nature conservation associations managing different types of ecosystems, and in some cases hunter associations.

5. Agroecological management in selected fish production systems

In this section, we present fish pond systems in France, Brazil, and Indonesia and then define agroecological practices for fish pond systems. We then illustrate which of these practices are applied in the studied systems and evaluate to what degree these systems correspond with the principles of agroecological fish pond systems presented in this chapter.

5.1. *The fish pond systems studied*

5.1.1. Fish pond systems in the French Dombes and Lorraine regions

In France, fish ponds have a surface area of more than 112,000 ha, the largest of any European country. The mean surface area of single fish ponds in France ranges from 1.7 ha to 10.1 ha (Le Quéré and Marcel 1999). Generally, fish ponds are extensively managed for polyculture, including species with different diets and behaviour: carp, roach, rudd, tench, perch, pike and/or pike-perch. Two regions are of specific focus: Dombes and Lorraine. These regions differ in their fish pond characteristics (runoff fish ponds in Dombes versus barrage fish ponds in Lorraine) and management practices, as described below (Table 4).

The Dombes region was formed by glacial activity during the Quaternary period (Avocat, 1975). It is a plateau of about 1,000 km^2, slightly inclined to the north-west, with long, fan-shaped morainic mounds (drumlins). The average elevation is about 280 m. In the east, south and west, the plateau is flanked by fluvial valleys of the Ain, Rhône and Saône Rivers, respectively; the three valleys are located about 50–100 m below the plateau. The northern portion is defined by the Bresse landscape, which has soil types (sandy and gravelly) that differ from those in the Dombes region. Consequently, the plateau's water supply comes exclusively from precipitation. Annual precipitation ranges from 900 mm to 1,300 mm, and the average yearly temperature is approximately 11°C (Table 4).

The history of the shallow pond systems of the Dombes region began in the 13th century (Guichenon 1650 cited in Sceau 1980). The ponds were created for two reasons. First, farmers needed to drain the dominant clay

Table 4. Main characteristics of the studied fish pond regions in France.

Fish pond characteristics	Region in France	
	Dombes	Lorraine
Type	A network of ponds connected to each other by a system of ditches	Mainly dam ponds fed by surface water runoff and small streams
Total area (ha)	12,000 ha (about 1,100 fish ponds)	7,000 ha
Mean area (ha)	10.0 ha	10.1 ha
Traditional practices	Alternating extensive fish farming (for 3–4 years) and grain farming (oats, maize, sorghum) for 1 year	Extensive polyculture with few added inputs
Annual precipitation	900–1,300 mm	800 mm
Average yearly temperature	11.0°C	10.7°C
Fish production	700–1,000 tonnes per year	860–1,100 tonnes per year
Species produced (% of total fish produced):		
Carp	56%	54%
Roach and rudd	28%	25%
Tench	6%	13%
Pike, pike-perch, and Eurasian perch	10%	8%
Use of fish products	Supply for restocking and angling; Fish processing	Supply of juvenile fish for restocking and recreational activities (fishing trips)
Human resources	330 owners, 6 traders and 3 processors	350 owner-operators and 10 trader-wholesalers
References	Avocat (1975), Billard (1979), Bernard and Lebreton (2007), and own data	Le Quéré and Marcel (1999), Regional Chamber of Agriculture Lorraine (2005), Fontaine and Banas (2012), Blayac *et al.* (2014), and own data

Note: *Scientific names*: Carp (*Cyprinus carpio*), Roach (*Rutilus rutilus*), rudd (*Scardinius erythrophthalmus*), Tench (*Tinca tinca*), Pike (*Esox lucius*), pike-perch (*Sander lucioperca*), Eurasian perch (*Perca fluviatilis*).

soils to be able to grow cereals. Because it was difficult to empty the water into smaller rivers or creeks, they usually stored drainage water behind dykes in selected areas or sometimes collected it at the lowest point of the area. Second, ponds were also created to produce fish (Guichenon 1650 cited in Sceau 1980). This activity expanded greatly during the medieval period due to a need for fish in response to strict food prescriptions. More than 1,000 ponds were created from the 12th to 18th centuries. During the 19th century, the local government drained many ponds because of health risks due to malaria. Economic benefits from the ponds and new activities such as hunting and nature conservation brought a new period of pond creation at the beginning of the 20th century. Today, the Dombes region has about 1,100 shallow ponds covering about 12,000 ha located in an agricultural area with pastures, fields and forests (Bernard and Lebreton 2007) (Figure 4). Pond size varies considerably, from less than 1 ha up to more than 100 ha. Net fish yield varies from 100 kg/ha to 700 kg/ha, with an average of 250 kg/ha, corresponding to a very low density of 45–70 g m^{-3} (Wezel *et al.* 2013, Robin *et al.* 2014).

Use of these shallow pond systems is based on a particular traditional practice which is rare in Europe: alternating extensive fish farming and grain farming on the same plot of land. Ponds are emptied every year to harvest fish and are then refilled. After 3–4 years, the ponds are left to dry out to cultivate mainly oats, maize or sorghum for one year (Figure 4). The water that fills the ponds during the wet phase comes from either an upstream pond or a system of ditches which collects rainwater from the

(a) (b)

Figure 4. Typical landscape of the Dombes region with a mosaic of cropping systems, ponds and forests (a), and a fish pond during a drying period (b).

catchment. A dense network of parallel or slightly diverging ditches is used to fill and empty ponds during the dry phase (Avocat 1975, Billard 1979). Thus, most ponds have a high degree of connectivity.

Currently, fish farmers are modifying their practices because fewer fish from the area are consumed and because of strong competition with fish from supermarkets. Fish farmers also must consider environmental regulations, the local importance of hunting, and the increasingly intensified agricultural land use around the ponds. Therefore, since the 2000s, the district council has developed several programmes to enhance agroecological practices adapted to fish farming. Since 1990, successive programmes have been developed to conserve the biodiversity of the fish ponds; they focus on reed-belt conservation or protection of endangered bird species. The district council funds and manages a large programme, with the main goal of providing support to local fish production at all stages. This programme currently directly supports fish production based on agroecological practices. It also promotes research and development of new prepared products based on fish from the fish ponds.

The Lorraine region is located in the upstream Rhine river basin. It has a high density of fish ponds covering around 7,000 ha and yearly produces 1,100 tonnes of fish (Le Quéré and Marcel 1999, Lorraine Chamber of Agriculture 2005) (Table 4). Fish ponds were created in lowland zones, mainly in two zones: *Le pays des étangs* ("The land of fish ponds") and the Woëvre River plain. These two zones are vast clayey plains belonging to the Parisian Basin. Originally, they were swampy and forested areas favourable for establishment of natural and artificial ponds. The elevation ranges from 220 m to 250 m. Annual precipitation is about 800 mm, and average yearly temperature is 10.7°C (Table 4).

On these hydric soils, many ponds were built as early as the 10th century at the instigation of the monasteries (Bernard 2008). Their primary function was to supply fish on Good Friday by managing the wetlands (Figure 5). Some ponds were also established to supply water to canals. Most of these agroecosystems are barrage ponds generally fed by surface water runoff and small streams, in which extensive polyculture is practised, with common carp as the emblematic species (Billard 2010). Management operations include three steps. The first is drainage in late autumn or early spring. During this step, all of the water is discharged

(a) (b)

Figure 5. (a) Typical vegetation belt in the tributaries upstream from a barrage fish pond in the Lorraine Region (photo: Yannick Jouan, Filière Lorraine d'Aquaculture Continentale) and (b) production of fish in extensive ponds in Lorraine.

downstream to allow fish harvest every 1–3 years. Subsequently, pond level increases in a filling step due to water inputs from tributaries and precipitation. Finally, the third step corresponds to a pseudo-balance phase during which fish are stocked and grown in ponds whose water level depends on the water balance (Chary 2013). The hydrological cycles of fish ponds are therefore largely controlled by the fish farmers. Additional feeding and direct use of fertilisers or pesticides during dry year in ponds are not common practices in the Lorraine region.

Currently, fish farmers are adapting their practices due to growing global concerns about environmental issues, as well as recreational development. Thus, they produce fingerlings intended for restocking or recreational fishing (sale of living fish). They also offer nature-discovery activities mainly based on hiking and fishing. Thus, the work of fish farmers is changing along with perceptions in Lorraine of the services provided by their ponds (Blayac *et al.* 2014). Several studies in this region show that barrage fish ponds can efficiently trap pesticides transported in runoff from agricultural fields, thus helping to reduce peak pesticide concentrations and potential risks of adverse effects in downstream ecosystems (Gaillard *et al.* 2016b). Today, many recognise the importance of regulating and supporting services supplied by fish ponds, which strengthen the territorial integration of fish farmers' practices (Bekefi and Varadi, 2007). The Lorraine region also produces the most organic fish in

France: 2 tonnes per year (Fontaine and Banas 2012). Similarly, some aqua-environmental measures have been developed with fish farmers to adopt environmentally friendly practices (reed-protection program).

5.1.2. Brazil

An interesting example of agroecological development of pond systems in Brazil is provided by two models created in Santa Catarina state in the 1990s. Aquaculture in Brazil is generally growing rapidly. For example, from 2008 to 2010, production increased by 31% (MPA 2012, cited by Oliveira Santos *et al.* 2015), mainly due to tilapia production in fresh water aquaculture systems. In Santa Catarina state at the beginning of the 20th century, limited and extensive production of common carp existed due to a large local population which descended from German and Italian immigrants. In the 1980s, it developed modestly with the creation of hill reservoirs for irrigation. Some of the ponds added a few pigs as a nutrient source due to the lack of available formulated feed. In the 1990s, non-governmental organisations (NGOs) claimed that ponds were polluting water resources. This prompted the Santa Catarina state to search for efficient and profitable systems by associating scientists and advisory agents to promote adapted or new pond production systems.

The systems developed today consist of a polyculture which mimics the Chinese model. It involves common carp and/or Nile tilapia (*Oreochromis niloticus*) as the main species, combined with herbivorous carp (*Ctenopharyngodon idella*) which feeds on aquatic plants, silver carp (*Hypophthalmichthys molitrix*) which feeds on phytoplankton, and big-head carp (*Hypophthalmichthys nobilis*) which feeds on zooplankton. These species are associated to best exploit the natural biomass produced by the pond system, which is supported by the addition of pig production. The pigs are reared in small wooden houses over the ponds, allowing pig manure and pig feed losses to fall into the ponds.

Two models were developed from this basic system in two different areas. The "Modelo do Oeste de Santa Catarina de Peixes Integrado" (MOCAPI) model is based on a moderate-input system (Figure 6). Pig production is around 60–80 pigs/ha of pond/year. The carp polyculture is supplemented by the use of local fish species which have a complementary

(a) (b)

Figure 6. Two Brazilian systems of fish polyculture integrated with pig production in Santa Catarina, Brazil: (a) MAVIPI and (b) MOCAPI.

role in the food web and are preferred by the local population: carnivorous jundia (*Pimelodema* sp.), cascudo (*Hypostomus* sp.), specialised in consuming periphyton, and omnivorous tambaqui (*Colosoma macropomum*). This polyculture system, with a moderate level of intensification, reaches yields of 5–8 tonnes/ha/year (Jorge de Matos Casaca, personal communication). It is generally integrated into family farming systems, which produce multiple products (e.g. pigs, poultry, soya bean, maize). Pond sediment is used as organic fertiliser for farm crops.

The other system developed, called "Modelo de Alto Vale de Itajai de Pisicultura Integrada" (MAVIPI), is a more intensive system based only on carp polyculture, producing 80–100 pigs/ha of pond/year (Figure 6). An aerator is used to distribute nutrients in the pond water better and decrease the risk of deoxygenation. Additional formulated feed is added when fish growth, permitted by pig waste, plateaus. This more "industrial-oriented" system constructs the farms containing both pigs and ponds. Fish yield reaches 10–15 tonnes/ha/year (Sergio T.J. Tamassia, personal communication). Pond sediment is planned for use as fertiliser by onion producers.

Development of these two production systems benefit from scientific and development knowledge and from the effort made in building and operating a network, not only dedicated to fish farming techniques, but also to related knowledge and skills (e.g. value chain, quality management, product image, cooking, fee-fishing).

These two models are based on agroecological characteristics. They are productive and based on nutrient recycling and species diversity. Integrating fish production helps to maintain the farmers' income when income from pigs decreases. Integration into a family farming system means that it relies more on local resources. The quality of the water the system releases is still under debate, but it does not seem to differ greatly from river water. The use of additional lagoons with floating plants (water hyacinth (*Eichhornia crassipes*) and water lettuce (*Pistia stratiotes*)) will be tested. The most critical point is the natural value of the system, which has not been studied, especially its interaction with natural biodiversity, especially in the more intensive model. Supplying exogenous species (carp) is debatable. It is difficult to deny the environmental risk of potential escapees (albeit lower for tilapia); however, such an efficient polyculture is yet to be built with local species. Overall, identifying and promoting the agroecological characteristics and associated ecosystem services associated with this activity helps integrate it into rural areas and encourages adaptations in response to environmental regulations.

5.1.3. Indonesia

Indonesia is the fourth largest international producer of aquaculture at 3.07 million tonnes, 68% of which are freshwater finfish (FAO 2014). The transition toward a more agroecological pond system was studied in the eastern Sumatra area in Indonesia (Aubin *et al.* 2014). Development of fish farming in this area was associated with the settlement of migrants from other parts of Indonesia in the lowlands of Sumatra. They cleared the rain forest to produce crops. Its high groundwater level and acidic soils failed to sustain crop production, except pineapple, which was produced on a large scale. In the 1990s, economic crisis pushed farmers to diversify, and they started producing fish as well. In 2010, 9,600 fish farms were reported, producing 17,500 tonnes of fish, 45% of it striped catfish (*P. hypophthalmus*) on farms with less than 1 ha. Production of striped catfish increased at a high rate (+59% from 2008 to 2010), but then the growth ceased due to disease outbreaks, fish-price volatility and feed-price increases. After the success of striped catfish production in the area, fish farmers' organisations searched for a new basis of more sustainable and reliable fish production. A new production system was designed and tested with help from Indonesian and French scientists and advisory service. It is based on existing trials of rearing striped catfish in association with other species. The giant gourami (*Osphronemus goramy*) was selected as it is a popular and herbivorous species. Farmers aimed to improve the performance of striped catfish (feed conversion ratio) to produce high-price fish as a secondary source of income and to better manage the pond environment, especially water quality. The production system developed is based on small traditional ponds for striped catfish (0.7 ha) in which settled cages of fed striped catfish are placed (Figure 7). Duck weed (*Lemna* sp.) was introduced to capture excess nutrients. As this species is high in protein, it is a beneficial feed for giant gourami. The gouramis are settled around the cages at low density and are fed only duck weed. It is possible to produce two cycles of striped catfish during one cycle of gourami as the species have different growth rates.

When draining the ponds, the output water is redirected towards palm trees and paddy fields to recycle nutrients. This practice supplements the use of pond sediments as fertiliser for banana and vegetable cultures on the pond banks.

(a) (b) (c)

Figure 7. (a) Cage in a pond system in Muara Jambi (Sumatra, Indonesia) combining striped catfish (b) and giant gourami (c) picture from Marc Legendre, IRD).

Trends show better management and efficiency of striped catfish production and the additional financial benefit of gourami production. Although yield of striped catfish decreases and gourami mortality occurs due to handling stress, the overall performance is economically and technically better than that of standard production (Legendre, J. Slembrouck, Ediwarman, and D. Caruso, personal communication). After the experimental phase, fish farmer organisations decided to adopt the model. They discussed how to integrate the system with existing agricultural activities. Integrated crop-aquaculture systems exist. For instance, crops on pond banks benefit from pond water and nutrients from pond sediments. In return, the crops help maintain bank structure and provide leaves for herbivorous fish species (i.e. tarot leaves for gourami) or beef cattle. The need for better integration was identified, and using water drained from ponds for, palm-oil plantations, banana plantations, and paddy fields was presented as a solution to increase crop yields and avoid potential pollution problems.

This system applies several of the agroecological principles. Less intensive production around the cages allows biodiversity to increase, which benefits natural regulation of the system. Recycling occurs within the polyculture and between it and terrestrial crops. The system depends less on external feeds (sometimes produced on-farm). The production of two fish species supplies two separate markets, a mass market (low-price striped catfish) and a market for festive traditional food (higher-price gourami), which improves economic stability. This new system needs to mature. Nevertheless, it shows that is possible to combine economic and agroecological performances.

5.2. *Agroecological practices for fish pond systems*

The literature has not yet defined agroecological practices for fish pond systems. We fill the gap here by presenting several of them as they are applied to the pond systems in France, Brazil, and Indonesia (Table 5). We present how these practices correspond with the five agroecological principles of fish pond systems described previously and how they are used in our systems studied.

Table 5. Agroecological practices used in the fish pond systems studied.

Agroecological practice	Description	Systems in which it is applied
Adjusting nitrogen fertilisation	Water in fish ponds can be low in nitrate in the spring due to intense consumption by aquatic plants and algae during periods of high photosynthetic activity. This can lead to a low mineral nitrogen phosphorous ratio, facilitating growth of N_2-fixing cyanobacteria. High proliferation can lead to cyanobacteria blooms, causing severe deoxygenation and fish mortality. Inputs of mineral nitrogen, but only up to a certain threshold (to avoid excess nitrate) can increase nitrate concentration in water, which rebalances the nitrogen phosphorous ratio and favours development of diverse algal communities. Input of 70–100 g of nitrate/100 m^2 of pond area in a liquid solution of ammonitrate is adequate.	Dombes, Lorraine (F); Brazil; Indonesia
Organic fertilisation	Manure can be added to enhance both primary and bacterial productivity and increase production of the natural food organisms of fish (phytoplankton, zooplankton, insects, and other invertebrates). Nitrogen, phosphorus, potassium, and calcium content should be measured to calculate the total amount of manure required (10–50 kg/100 m^2 pond area, depending on the type of manure).	Dombes, Lorraine (F); Brazil; Indonesia
Calcium liming	Adding calcium oxide (CaO) to water oxidises organic matter by increasing mineralisation at the sediment surface and favouring short term flocculation of dissolved organic matter. The amount of CaO required is about 2.5–3 kg/100 m^2 pond area.	Dombes (F)
	During the drying period, addition of CaO (same rate as given previously) or calcium carbonate (10 kg/100 m^2) enhances system productivity when water alkalinity is low. Calcium regulates pH during the season due to its buffering capacity.	Dombes (F)

Adding barley straw	Decomposition of barley straw in water produces and releases many compounds, some of which may control algal populations via allelopathy. Adding 30–50 g/m^3 of barley straw to the fish pond limits development of high phytoplankton biomass (e.g. cyanobacteria blooms).	Forez*, Lorraine (F)
Managing the water level	Variations in water levels in the drawdown zone provide diverse habitats in grass beds along the banks. Alternative moderate flooding (when the level is low) or drying of grass beds helps provide shelters, reproduction areas, and food for fish (and birds).	Lorraine (F)
Integrating a dry year in the production cycle	After 3–4 years, ponds are left to dry out for one year and then planted with oats, maize or sorghum, or are left fallow. This drying period allows (i) mineralisation at the sediment surface (nutrient recycling), (ii) recovery of plant species richness and water cover after the dry year, (iii) control of phytoplankton blooms due to intense competition with aquatic plants after pond refilling, and (iv) additional cropland for local agriculture.	Dombes (F)
Combining fish species	In polyculture systems, judicious combination of species allows multiple food resources and habitats to be exploited. The main objective is to exploit the trophic complementarity (i.e. no competition for resources) of several species; fish pond polyculture requires combining omnivorous fish (carp) with other fish with specific diets (benthivorous, herbivorous, zooplanktivorous). A moderate stocking density of only one carnivorous fish species is required to limit predation and ensure a good harvest. A related objective is to produce fish with different degrees of sensitivity to environmental changes.	Brazil; Indonesia; Dombes, Lorraine (F)

(Continued)

Table 5. (*Continued*)

Agroecological practice	Description	Systems in which it is applied
Managing vegetation along banks (including shallow zones <0.5 m deep)	Adapted management of vegetation along pond banks (i) supports biodiversity (refuge, nesting, or over-wintering habitat for birds, amphibians, dragonflies, etc.); (ii) supports food resources for animals; (iii) provides a water purification zone which captures pollutants, nutrients or organic matter; and (iv) provides shelter for fish reproduction. Management includes planting if no vegetation exists or creating openings in vegetation if it is too dense.	Dombes, Lorraine (F)
Using ponds as a filter	A pond can act as a buffer zone and a biological filter for its downstream catchment by capturing excessive nutrients, pesticides, and suspended matter from the catchment or other upstream ponds during the draining period. Development and management of reed-community species such as *Typha latifolia* and *Phragmites australis* is one of the most important practices for enhancing purification (phytoremediation). Ponds also promote other processes that degrade xenobiotic substances, such as photolysis at the air-water interface or microbial degradation in the sediment.	Dombes, Lorraine (F)

Introducing natural substrates	Adding substrates to the water (e.g. branches, bamboo,) significantly increases periphyton biomass, which is a natural food for some fish species. It also provides shelters and reproduction zones for fish. Periphyton biomass improves water quality as it enhances nitrogen-related processes, captures organic detritus, and controls dissolved oxygen concentration and pH.	Indonesia; Brazil
Irrigation and cattle watering	Water used for crop irrigation and cattle watering.	Dombes, Lorraine (F); Indonesia
Using sediments during draining/ drying period	Pond sediments are rich in nutrients and organic matter and can be used as fertiliser for nearby cropped fields. A maximum water input of 1 m^3/100 m^2 is required.	Indonesia; Brazil; Dombes (F)
Using aquatic plants as food for animals	Many aquatic plants can be used as fodder for cattle, sheep, goats, pigs, or other livestock.	Indonesia

*other fish pond areas in France near the Dombes study area.

Notes: (F) Indicates a French region

Many agroecological practices were identified in these fish pond systems. The same practice can have direct or indirect links with one or more agroecological principles (Figure 8). As the primary objective of these systems is fish production, it is evident that many different practices are intended to support Principle 1 "*Enhancing productivity*". We observed eight practices in connection with this agroecological principle, among which five were directly linked. It should be noted that Principle 4, "*Supporting environmental friendliness*", is also directly linked to five agroecological practices. The fish farmers' concern is to maintain the

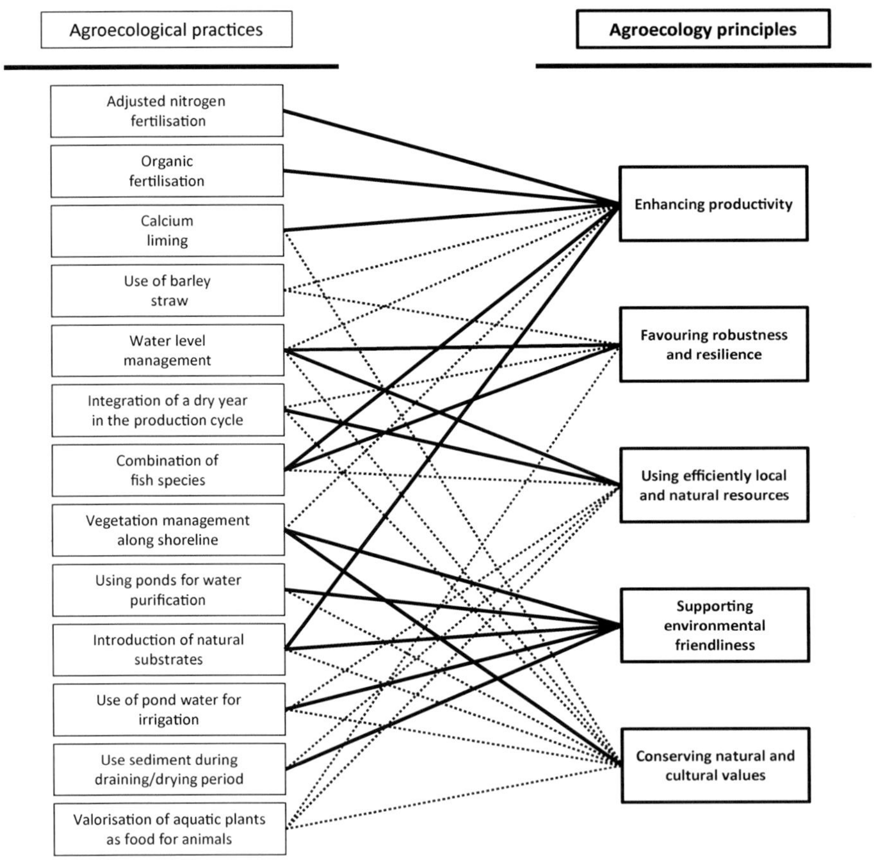

Figure 8. Links between agroecological practices and agroecological principles in fish pond systems. Solid lines indicate direct links, dotted lines indicate indirect links.

overall quality of their production units and the surrounding area to create a more sustainable fish production system, as well as to respond to regulatory obligations and social expectations. Another main objective of fish farmers is to effectively exploit the natural resources of the food web for fish production. Thus, up to six practices contributed to agroecological Principle 3, "*Using local and natural resources efficiently*", usually indirectly. Principle 2, "*Favouring robustness and resilience*", was directly linked to two agroecological practices: combining fish species and managing water level. These practices aim to ensure diversity of aquatic species (cultured and uncultured) and habitats in the system. This forms part of the complexity which is considered a factor of stability. Finally, Principle 5, "C*onserving natural and cultural values*", is often indirectly linked to practices. Maintaining traditions or conserving biodiversity associated with fish ponds are underlying principles generated by many agroecological practices. This means that the cumulative application of several practices in a fish pond, notably conservation of biodiversity, is important to encourage implementation of this principle.

6. Conclusions

Worldwide, aquaculture practices have evolved towards monocultures with a high dependence on inputs such as energy and feed. They increasingly demonstrate their limitations, especially from an environmental viewpoint, as they regularly push natural ecosystems beyond their carrying capacity. A new perspective on agroecological practices of fish pond systems is needed to foster development of aquaculture systems which respond to consumer demand for fish products and societal demand for conservation of ecosystem services.

This chapter on fish pond systems offers perspectives based on a wide range of agroecological practices. For instance, practices of ancestral pond systems using natural nutrient sources to feed fish are intrinsically agroecological. Practices of traditional polyculture systems applied to pond aquaculture are based on manipulating ecosystems by combining species with complementary trophic levels and behaviours. These practices are designed to enhance pond productivity and provide a range of products, but also to offer good ecological and economic robustness.

In this chapter, we describe and define agroecological practices used in extensive and intensive pond systems in Europe, Brazil, and Indonesia. Depending on environmental, economic and social contexts, different pathways are chosen to apply agroecological principles. We defined five main principles that drive agroecological practices in fish pond systems. They represent a range of objectives which guide the choice of agroecological practices: productivity, robustness and resilience, natural and cultural value, environmental friendliness, and use of local and natural resources. Agroecological fish pond practices lie at the interfaces between principles, emphasising that agroecology is a science of interactions.

The agroecology approach, applying agroecological principles to develop agroecological practices, is indispensable for future sustainable development of fish pond systems and overall aquaculture. However, there is no single formula. The wide range of existing practices shows the complexity and the need to adapt practices to local situations, constraints, opportunities, and objectives of fish farmers.

Acknowledgements

We wish to acknowledge the funding of research projects which allowed us to work together and create a synthesis of our work:

- The PISCEnLIT project was funded by the Agence Nationale de la Recherche (ANR), France, programme Systerra ANR-09-STRA-08.
- The TIPPINGPOND project was funded by the ERA-Net BiodivERsA, with national funds from the ANR, part of the 2012 BiodivERsA call for research proposals.
- The RESOTROPHIC project was funded by the Rhône-Alpes Region, France, and the district council of the Department of Ain, France.

The authors would like to acknowledge J.D.M. Casaca and S.T.J. Tamassia for providing useful information about the Brazilian case study, and M. Legendre, J. Slembrouck, D. Caruso (IRD) and Ediwarman (BBAT-Jambi) for information about the Indonesian case study.

References

Anh, P.T., Kroeze, C., Bush, S.R., Mol, A.P.J., 2010. Water pollution by *Pangasius* production in the Mekong Delta, Vietnam: causes and options for control. *Aquaculture Research* **42**, 108–128.

Altieri, M.A., 2002. Agroecology: the science of natural resource management for poor farmers in marginal environments. *Agriculture, Ecosystems and Environment* **93**, 1–24.

Altieri, M.A., Nicolls, C., 2005. *Agroecology and the Search for a Truly Sustainable Agriculture*. United Nations Environment Programme, Mexico, 290 p. www.agroeco.org/doc/agroecology-engl-PNUMA.pdf (accessed July 2015).

Aubin, J., Rey-Valette, H., Mathé, S., Wilfart, A., Legendre, M., Slembrouck, J., Chia, E., Masson, G., Callier, M., Blancheton, J.P., Tocqueville, A., Caruso, D., Fontaine, P., 2014. *Guide for Implementing Ecological Intensification of Aquaculture Systems*. INRA, Rennes, France, 131 p.

Avocat, C., 1975. La Dombes: milieu naturel ou milieu en équilibre. *Revue de Géographie de Lyon* **50**, 35–58.

Banas, D., Masson, G., Leglize, L., Pihan, J.C., 2002. Discharge of sediments, nitrogen (N) and phosphorus (P) during the emptying of extensive fishponds: effect of rainfall and management practices. *Hydrobiologia* **472**, 29–38.

Banas, D., Masson, G., Leglize, L., Usseglio-Polatera, P., Boyd, C.E., 2008. Assessment of sediment concentration and nutrient loads in effluents drained from extensively managed fishponds in France. *Environmental Pollution* **152**, 679–685.

Bätzing, W., 2013. Nutzungskonflikte zwischen Teichwirtschaft, Naturschutz und Freizeitinteressen im Aischgrund. Probleme und Potenziale bei der Aufwertung des Aischgründer Karpfens zum Qualitätsregionalprodukt. *Mitteilungen der Fränkischen Geographischen Gesellschaft* **59**, 81–100.

Bauer, C., 2014. Waldviertler Teiche. Denisia **33**, 157–166.

Bekefi, E., Varadi, L., 2007. Multifunctional pond fish farmers in Hungary. *Aquaculture International* **15**, 227–233.

Bernard, A., Lebreton, P., 2007. *Les oiseaux de la Dombes: une mise à jour*. Académie de la Dombes, Fondation Pierre Vérots. N°27 de la revue "Dombes", 168 p.

Bernard, C., 2008. *L'étang, l'homme et l'oiseau. Incidences des modes de gestion des étangs piscicoles sur les ceintures de végétation et l'avifaune nicheuse en Sologne*, Bresse, Territoire de Belfort et Champagne humide. Thesis, Ecole Normale Supérieure des Lettres et Sciences Humaines de Lyon, 632 p.

Bezirk Oberpfalz, 2015. *Gewässer, Teichwirtschaft und Fischerei in der Oberpfalz.* http://www.bezirk-oberpfalz.de/desktopdefault.aspx/tabid-5/140_read-178/ (accessed March 2015).

Billard, A., 1979. L'étang de Dombes. I. Insertion dans le paysage agraire. *Bulletin Français de la Pisciculture et de la Pêche* **272**, 71–79.

Billard A., 2010. *Derrière chez moi, y'a un étang.* Les étangs, textes d'hier, regards d'aujourd'hui et de demain. Editions Quae, 304 p.

Blayac, T., Mathe, S., Rey-Valette, H., Fontaine, P., 2014. Perceptions of the services provided by pond fish farming in Lorraine (France). *Ecological Economics* **108**, 115–123.

BMLFUW *Bundesministerium für Land- und Forstwirtschaft, Umwelt und Wasserwirtschaft* (2014). Teichwirtschaft in Österreich. http://www.bmlfuw.gv.at/wasser/nutzung-wasser/teichwirtschaft.html. (accessed March 2015).

Boyd, C.E., Wood, C.W., Chaney, P.L., Queiroz, J.F., 2010. Role of aquaculture pond sediments in sequestration of annual global carbon emissions. *Environmental Pollution* **158**, 2537–2540.

Bukacińska, M. D., Bukaciński, J. P., Cygan, K., Dobrowolski, K., 1995. The importance of fish ponds to waterfowl in Poland. *Acta Hydrobiologica*, Supplement **37**, 57–73.

Céréghino, R., Biggs, J., Oertli, B., Declerck, S., 2008. The ecology of European ponds: defining the characteristics of a neglected freshwater habitat. *Hydrobiologia* **597**, 1–6.

Chambre Régionale d'Agriculture de Lorraine (CRAL), 2005. Pisciculture d'étang, Référentiel Diversification. Available at: http://cra-lorraine.fr/fichiers/div-pisciculture-etang.pdf

Chary, K., 2013. *Analyse écosystémique des étangs, cas de la Lorraine.* Mémoire de fin d'étude de l'ISARA de Lyon, 85 p.

Conservatoire des Espaces Naturels de Franche-Comté, 2007. *Les étangs de Bresse comtoise.* http://www.cen-franchecomte.org/fich_59521/azure5.pdf (accessed March 2015).

Dumont, B., Fortun-Lamothe, L., Jouven, M., Thomas, M., Tichit, M., 2013. Prospects from agroecology and industrial ecology for animal production in the 21st century. *Animal* **7**(6), 1028–1043.

European Union, 2000. Directive 2000/60/CE du Parlement européen et du Conseil du 23 octobre 2000 établissant un cadre pour une politique communautaire dans le domaine de l'eau. Journal officiel de l'Union Européenne. http://eur-lex.europa.eu/legal-content/FR/TXT/?uri=CELEX%3A32000L0060 (accessed March 2016).

European Union, 2008. Directive 2008/105/CE du Parlement Européen et du Conseil du 16 décembre 2008 établissant des normes de qualité environnementale dans le domaine de l'eau. *Journal Officiel de l'Union Européenne*. http://eur-lex.europa.eu/legal-content/FR/TXT/?uri=celex%3A32008L0105 (accessed March 2016).

European Union, 2013. Directive 2013/39/EU du Parlement Européen et du Conseil du 12 Août 2013 modifiant les directives 2000/60/CE et 2008/105/CE en ce qui concerne les substances prioritaires pour la politique dans le domaine de l'eau. *Journal Officiel de l'Union Européenne*. http://eur-lex.europa.eu/legal-content/FR/TXT/?uri=celex%3A32013L0039 (accessed March 2016).

FAO (Food and Agriculture Organization of the United Nations), 2014. *The State of World Fisheries and Aquaculture. Opportunities and Challenges*. FAO, Rome, Italy, 243 p.

Fontaine, P., Banas, D., 2012. *Etat des lieux et approche prospective de la filière lorraine d'aquaculture continentale*. Rapport Université de Lorraine, INRA, 32 p.

Gaillard, J., Thomas, M., Lazartigues, A., Bonnefille, B., Pallez, C., Dauchy, X., Feidt, C., Banas, D., 2016a. Potential of barrage fishponds for the mitigation of pesticide pollution in streams. *Environmental Science and Pollution Research* **23**, 23–35.

Gaillard J., Thomas M., Iuretig A., Pallez C., Feidt C., Dauchy X., Banas D., 2016b. Barrage fishponds: reduction of pesticide concentration peaks and associated risk of adverse ecological effects in headwater streams. *Journal of Environmental Management* **169**, 261–271.

Horváth, L., Tamas, G., Seagrave, C., 2002. *Carp and Pond Fish Culture*. Fishing News Books, Blackwell Science, England, 170 p.

Kluts, I.N., Potting, J., Bosma, R.H., Phong, L.T., Udo, H.M.J., 2012. Environmental comparison of intensive and integrated agriculture-aquaculture systems for striped catfish production in the Mekong Delta, Vietnam, based on two existing case studies using life cycle assessment. *Reviews in Aquaculture* **4**, 195–208.

Korinek, V., Flott, J., Fuska, J., Lellak, J., Prazakova, M., 1987. Carp Pond of Central Europe. In: Michael, R.G. (Ed.), *Managed aquatic ecosystems. Ecosystems of the World* **29**, Elsevier, Amsterdam, pp. 29–62.

Lazartigues, A., Banas, D., Feidt, C., Brun-Bellut, J., Thomas, M., 2012. Pesticide pressure and fish farming in barrage pond in North-eastern France. Part I: Site characterization and water quality. *Environmental Science and Pollution Research* **19**, 2802–2812.

Lazartigues, A., Thomas, M., Cren-Olivé, C., Brun-Bellut, J., Le Roux, Y., Banas, D., Feidt, C., 2013. Pesticide pressure and fish farming in barrage pond in North-eastern France. Part II: Residues of 13 pesticides in water, sediments, edible fish and their relationships. *Environmental Science and Pollution Research* **20**, 117–125.

Lemmens, P. 2014. *Fish and the Management of Shallow Lakes*. PhD thesis, University of Leuven, Belgium, 182 p.

Le Bihan, J., Font, M., 2008. *Les étangs. Synthèse sur les zones humides françaises à destination des gestionnaires, élus et acteurs de terrain*. Rapport Fédération des Parcs Naturels Régionaux de France, 64 p.

Le Quéré G., Marcel J., 1999. *La pisciculture d'étang française*. ITAVI, Rapport de l'Institut Technique de l'Aviculture, Paris, 57 p.

Lhotský, R., 2010. The role of historical fishpond systems during recent flood events. *Journal of Water and Land Development* **14**, 49–65.

Little, D.C., Edwards, P., 2005. *Integrated Livestock-fish Farming Systems*. FAO, Rome, Italy, 197 p.

Little, D.C., Muir, J.F., 1987. *A Guide to Integrated Warm Water Aquaculture*. Institute of Aquaculture, University of Stirling, Stirling, Scotland, 238 p.

López-Ridaura, S., Masera, O., Astier, M., 2002. Evaluating the sustainability of complex socio-environmental systems: the MESMIS framework. *Ecological Indicators* **2**, 135–148.

Lorraine Chamber of Agriculture, 2005. *Pisciculture d'étang*. Référentiel Diversification. http://cra-lorraine.fr/fichiers/div-pisciculture-etang.pdf

Matzinger, T.M.E., 2014. *Teiche in der Landschaft - Bedeutung, Funktionen and Gefährdung*. Schriftenreihe des Bundesamtes für Wasserwirtschaft, Band 36, Vienna, Austria, 59 p.

Milstein, A., 2005. *Polyculture in aquaculture*. Animal Breeding Abstracts. CABI Publishing, pp. 15N–41N.

MEDD (Ministère de l'Ecologie et du Développement Durable), 2007. *Le réseau Natura 2000: Etangs du Forez*. http://natura2000.environnement.gouv.fr/sites/FR8201755.html (accessed March 2011).

Neori, A., Chopin, T., Troell, M., Buschmann, A.H., Kraemer, G.P., Halling, C., Shpigel, M., Yarish, C., 2004. Integrated aquaculture: rationale, evolution and state of the art emphasizing seaweed biofiltration in modem mariculture. *Aquaculture* **231**, 361–391.

Nhan, D.K., Phong, L.T., Verdegem, M.J.C., Doung, L.T., Bosma, R.H., Little, D.C., 2007. Integrated freshwater aquaculture, crop and livestock production in the Mekong delta, Vietnam: Determinants and the role of the pond. *Agricultural Systems* **94**, 445–458.

Oliveira Santos, A.A., Aubin, J., Corson, M.S., Valenti, W.C., Camargo, A.F.M., 2015. Comparing environmental impacts of native and introduced freshwater prawn farming in Brazil and the influence of better effluent management using LCA. *Aquaculture* **444**, 151–159.

PNR de la Brenne, 1998. *Document d'objectifs Natura 2000*. Site "Grande Brenne". http://www.parc-naturel-brenne.fr/info-actu/doc-telechargeables/zip/docob_Grande_Brenne_1998.pdf (accessed January 2009).

Pokorný, J., Hauser, V., 2002. The restoration of fish ponds in agricultural landscapes. *Ecological Engineering* **18**, 555–574.

Reinjntjes, C.B., Haverkort, Waters-Bayer, A., 1992. *Farming for the Future. An Introduction to Low-external Input and Sustainable Agriculture*. MacMillan Press, London. 250 p.

Robin J., Wezel,A., Bornette,G., Arthaud, F., Angélibert,S., Rosset,V., Oertli, B., 2014. Biodiversity in eutrophicated shallow lakes: determination of tipping points and tools for monitoring. *Hydrobiologia* **723**, 63–75.

Sceau, R., 1980. Les étangs de la Dombes, fondements socio-économiques d'un système hydraulique. *Revue de Géographie de Lyon* **55**, 129–159.

Steinbronn, S. 2009. *A Case Study*: *Fish Production in the Integrated Farming System of the Black Thai in Yen Chau District* (Son La province) *in Mountainous North-Western Vietnam — Current State and Potential*. Ph.D. thesis, University of Hohenheim, Stuttgart, Germany, 222 p.

Strehle, H.M., Trautmann, A., 2011. *Leitfaden für die Sanierung oberschwäbischer* Seen, PRO REGIO Oberschwaben, Ravensburg, Germany, 69 p. http://www.seenprogramm.de/fileadmin/pdf/Leitfaden_Vierte_Auflage_homepage.pdf (accessed March 2015).

Szumiec, M.A., 1996. The role of carp pond ecosystem management in wetlands conservation. *Acta Hydrobiologica* **37**, 13–20.

Touristische Gebietsgemeinschaft "Heide und Teiche im Bautzener Land", 2014. Die *Oberlausitzer Heide- und Teichlandschaft*. http://www.oberlausitz-heide.de/heide-und-teichlandschaft.html (accessed March 2015).

Troell, M., Joyce, A., Chopin, T., Neori, A., Buschmann, A.H., Fang, J.G., 2009. Ecological engineering in aquaculture - Potential for integrated multi-trophic aquaculture (IMTA) in marine offshore systems. *Aquaculture* **297**, 1–9.

UNESCO, 2015. UNESCO-*Biosphärenreservat Oberlausitzer Heide- und Teichlandschaft*. www.unesco.de/wissenschaft/biosphaerenreservate/deutsche-biosphaerenreservate/oberlausitz.html (accessed March 2015).

Vallod, D., Sarrazin, B., 2010. Caractérisation de l'effluent de vidange d'un étang de pisciculture extensive. *Hydrological Sciences Journal* **55**, 394–402.

Wezel, A., Bellon, S., Doré, T., Francis, C., Vallod, D., David, C., 2009. Agroecology as a science, a movement and a practice. A review. *Agronomy for Sustainable Development* **29**, 503–515.

Wezel, A., Luibrand, A., Le Quoc Thanh, 2002. Temporal changes of resource use, soil fertility and economic situation in upland north-west Vietnam. *Land Degradation and Development* **13**, 33–44.

Wezel, A., Oertli, B., Rosset, V., Arthaud, F., Leroy, B., Smith, R., Angelibert, S., Bornette, G., Vallod, D., Robin, J., 2014. Biodiversity patterns of nutrient-rich fish ponds and implications for conservation. *Limnology* **15**, 213–223.

Wezel, A., Robin, J., Guerin, M., Arthaud, F., Vallod, D., 2013. Management effects on water quality, sediments and fish production in extensive fish ponds in the Dombes region, France. *Limnologica* **43**, 210–218.

Chapter 13

Education in Agroecological Learning: Holistic Context for Learning Farming and Food Systems

Charles Francis*,†, Mary Wiedenhoeft‡,
Robert Dehaan§ and Paul Porter¶

*Department of Agronomy and Horticulture, University of Nebraska, Lincoln, Nebraska, USA
†Department of Plant Science, Norwegian University of Life Sciences, Ås, Norway
‡Department of Agronomy, Iowa State University, Ames, Iowa, USA
§Department of Environmental Studies, Dordt College, Sioux Center, Iowa, USA
¶Department of Agronomy and Plant Genetics, University of Minnesota, St. Paul, Minnesota, USA

1. Introduction

Education in alternative agricultural systems has evolved to embrace resource efficiency, chemical pesticide and fertiliser reduction, and economically viable strategies often called "agroecological practices" (Wezel *et al.* 2014). Holistic focus in agroecology education presents a contrast to much of the current training in the USA landgrant universities, where primary attention continues to focus on single solutions designed to solve specific problems. Yet initial ventures in agroecology education that search for environmentally benign practices often employ a similar reductionist approach that is used in conventional agricultural education. This is not surprising since most of us began our training in discipline-oriented courses in specific undergraduate departments, then continued to specialise through graduate training. In retrospect, we ask today if the early approaches in agroecology really represent education, or are they in fact simply a renamed focus on training to meet the specialised degree credit requirements already in place in current departments? Discipline-based education too often provides a narrow view of the world, which is too often devoid of an essential systems context. In a summer travel course called *Agroecosystems Analysis* we help students appreciate the complexity of farm enterprises, delve into integration efficiencies and emergent properties, explore how these ecological concepts can inform decisions on the farm, and examine multiple alternatives that will help farmers think about options to meet the long-term goals of farming and food systems (Wiedenhoeft *et al.* 2003). Farmers' decisions in aggregate also impact the character and resilience of each watershed, each local rural community, and each region.

In contrast to a new agroecology systems orientation, most landgrant scientists were trained to follow a compartmentalised problem-solving approach that is consistent with our normal institutional response to meet farmers' current challenges. Following research tradition using single- or two-factor experiments, most situations are viewed in a linear cause-and-effect manner, leading to solutions of such challenges one by one. To the credit of our institutions, much of the research agenda in the landgrant universities is designed by professionals with joint research/extension appointments who often have close working relationships with farmers. Research experiences under controlled conditions are combined with what is learned on farms, and from this are developed well-informed and documented "science-based" recommendations for cropping practices and rotation systems. For those who are also instructors, the experiences are translated into materials for classroom teaching. In many ways, methods of education in agroecology, organic farming, small-scale production, and sustainable agriculture mirror the mainstream in their reductionist approaches.

This is essentially how many of us learned as students, and we respected those elders who went before and demonstrated to us how good teaching should be done. With the marvels of Power Point and instant access to stimulating videos on the web, we are becoming reasonably good entertainers in the classroom, keeping students' attention, and maintaining above-average teacher evaluations at the end of each semester. So we are increasingly *getting better at doing things right*, at least in terms of helping students delve deeply into specific areas of interest and successfully taking exams. The real question is, *are we doing the right things*? Should we challenge some of our own assumptions, thus become students ourselves who explore what education should be, as compared to training? As learning facilitators, can we embrace a broader focus on whole systems and empower students to become responsible, autonomous, and lifelong learners? A practical immersion approach to understanding content, and use of an active learning process, is growing today in agricultural education under the inclusive umbrella of *agroecology* (Francis *et al.* 2011).

Ecologists frequently use the concept of hierarchy of spatial scale to organise understanding of structure and function of natural systems (Ahl and Allen 1996), and agroecologists have adopted the same perspective to look at components at the plant and soil, field, farm, watershed, and regional levels (Gliessman 2015). Over a decade ago, it was proposed that the term *agroecology* be expanded to embrace multiple perspectives by introducing an inclusive definition of "agroecology as the ecology of food systems" that embraces the dimensions of production, economics, environmental, and social issues (Francis *et al.* 2003). In his third edition of the seminal text in agroecology, Gliessman (2015) introduced a subtitle, *the ecology of sustainable food systems*. To further understand applications of this term, Wezel *et al.* (2009) observed that research and educational activities using the term *agroecology* could be recognised as "agroecology as science, agroecology as practices, and agroecology as a movement". They further found these definitions to predominate in certain geographic regions, for example "science" in Germany and France, "practices" in

France and the USA, and "movement" in USA and Brazil. After working with USA Midwest students for over two decades in the systems arena, we suggest as educators that the term may productively embrace all three.

In teaching agroecology and integrated systems, we have reflected on our own past classroom experience and concluded that multiple hierarchies could help shape the design of education. Certainly an expanded time horizon is essential if the goal is to discuss, design, and operationalize farming and food systems for sustainability, which by definition must be long-term rather than single season. Thus enters the hierarchy of time, with short-term, medium-term, and long-term challenges all included in analysis of agroecosystems. We introduce a hierarchy of thinking that could move a learning community from the current "monoculture mentality" about crops, cover crops, and systems into new dimensions of agriculture, food systems, and society that today are constrained by thinking primarily of the here and now. Such an evolution in educational focus on the spatial, the temporal, and the conceptual dimensions of learning has compelled us to reflect on what we are teaching in the classroom and field. Lower order issues include the crop varieties and fertiliser treatments we use, while higher order questions explore the interactions among components, for example crop rotations or crop/animal integration, and the long-term rotation and cover crops effects on soil physical properties and soil fertility (Blanco-Canqui *et al.* 2015). Moving up this hierarchy of priorities leads to economic and social issues, and we begin to ask about why we are involved with certain types of crops and production systems, and who benefits from them. The answers to these questions may lead our inquiry and learning in quite different directions than the current near-singular emphasis on increasing commodity crop yields. The all-too-common paradigm that the future will likely be an extrapolation of the past is challenged by humanist and Pulitzer prize-winning author Rene Dubos (1971), who stated, "Whenever human beings are involved, trend is not destiny". This challenge is passed on to our students in agroecology.

We could perhaps borrow a chapter from the business community. In the book *Breakthrough: How Great Companies Set Outrageous Goals and Achieve Them* (Davidson 2004), the author describes the early competitive edge of Japanese companies in the small-scale electronic market and how this edge was likely to continue into the computer age. He presents an intriguing story of the rise and fall of several major USA electronic companies that first rode the crest of a growing market for computers, and some that were able to foresee the emergence of personal computers as the real future in the new century. At the risk of "comparing education to a business" and "considering our students as clients" — which is all too common in academia today — we ask whether similar bold thinking, over the horizon, into the future could be what is needed in educational reform to move from training to education, from teaching to learning, and from passive observing to active participation?

2. Midwest summer course in agroecosystem analysis

In a Midwest summer course in *agroecosystem analysis* we have captured the broad concepts of change and involvement by setting up a learning agenda to emphasise whole systems and farm-level evaluation of sustainability, and urged students to consider single practices in the context of holistic farm systems as well as how they reflect the world views and goals of farmers they visit. We are currently evaluating the need to expand the scope of this education to encompass a spatial hierarchy and include more issues at the landscape level as well as explore broader themes such as the relevance of farming to society. This chapter summarises our reflections on nearly two decades of this course where student teams interview farmers, process and discuss observations and new knowledge, and conduct comparisons of sustainability among farms. The course has received consistent and highly positive evaluations, and our goal is to continue to improve on its relevance to contemporary society and food systems and provide useful educational experiences for today's students. An early report on class structure and outcomes provided an overview, and these early ideas still provide the structural foundation for the course (Wiedenhoeft *et al.* 2003). An evaluation of behavioural changes in students as a result of the course was the basis of a doctoral dissertation that found an increased appreciation of a systems focus in learning and broader perspectives on the complexity of farmer decision making (Harms *et al.* 2009).

We first explore how students *learn about agroecological practices at the field level* when farmers describe their specific management decisions on principal field crops of maize, soybean (Figure 1), and other Midwest crops (Figure 2). In the interviews conducted by students, farmers describe how their practices have changed in response to new crop genetics, increased knowledge about fertiliser use efficiency, integrated pest management, and potential changes to no-till. Use of crop rotations, crop/animal integration, and cover crops have been adopted by the farmers we visit in a four-state region to reduce costs and negative environmental impacts. The section borrows from ecological practices described by Francis and Porter (2011), and the examples illustrate first-order, field-level thinking about the components and combinations with specific enterprises and fields on the farm.

Next we describe our student teams' task to look beyond practices at the field level and to *explore on each farm the integration of enterprises* and how these contribute to farm efficiency and sustainability. Most of the farms visited include both crops and livestock, and their integration efficiencies are examined by students as they interview farmers. After reflection as a team, the students move the responses into a broader context. In this step, the students consider several models for farm analysis. Whole-farm exploration and assessment could be considered a second-order, farm-level approach to an agroecology-driven evaluation of production and economic challenges in the system. This learning method has been called a phenomenological approach to understanding a farm (Østergaard *et al.* 2010).

During the week of farm interviews, we also *focus on the broader landscape* and imagine what this prairie region was like in pre-agricultural times and how it has evolved since farmer settlement. We visit the Ocheyeden Mound, a *kame* near the western edge of the last advance of glaciers about 10,000 years BCE (12,000 years ago). A kame is a small hill made up of sand, gravel and till left behind by a retreating glacier such as the Des Moines Lobe in Iowa. We explore the Steele Prairie to experience the vegetation on a virgin prairie, noting the rich diversity of plant life and the multiple unique niches and speciation within 130 ha of virgin prairie land. The goal is to examine the structure and discuss the functions of a highly diverse prairie ecosystem, and to search for potential lessons from the prairie that could inform design of diverse agroecosystems for the future. Study of where systems fit into the larger rural

Figure 1. Prevailing maize — soybean system in the Midwest of the USA.

Figure 2. Biodiverse vegetable farm in the Midwest of the USA.

landscape, how farms interact among each other and with nearby rural communities, and what emergent properties of political and economic decisions impact long-term success could be considered a third-order of complexity in food systems.

Finally we consider *long-term perspectives and innovative ideas useful in designing a future educational agenda* for using agroecology to inform learning about farming and food systems. This fourth-order level of educational design includes both spatial and temporal components, informed by the larger world food situation and the complexities of a global industry, and today framed in the context of climate change. We recognise that combining all these factors is a daunting challenge. Yet with innovative learning methods, cooperation among instructors, visits with enthusiastic farmers, and open-minded exploration by students, it appears that we are making progress toward a new order of understanding. We introduce the current controversy about transgenic technologies now widely used in Midwest maize and soybean systems as a case study that involves biology, economics, regulation, and social acceptance in this region. In the following sections we share more details on learning at the field, farm, and landscape, and community levels, as well as how we project this learning into the future through the students we involve in the course. The discussion includes what we think has been achieved, based on student evaluations, mixed with evolving methods and goals that we hope will be pursued and accomplished in the future.

3. Integrative agroecological practices at field level

Because the majority of our students are involved in agricultural majors, farming practices are high on the agenda for students asking interview questions of farmers. In fact, these lower-order issues often represent the comfort zone for farmers to describe their types of crops and specific hybrids, row widths and planting dates, and the makes and models of equipment observed on the farm. One of our goals is to help lead students and the overall discussion to higher order issues such as why these crops are grown, how farmers decide among multiple options of management, and how their world views and long-term objectives are reflected in the choice of crops and practices. This is accomplished by discussion with students ahead of the interviews, and with our initial contacts with farmers about what types of students are coming to visit and what their interests will be. We are partially successful in meeting this goal. Methods of orientation and practice for interviews have been described (Francis and Salomonsson 2012, Østergaard *et al.* 2013).

Initial questions in most interviews centre on farm size, principal crops and animal species produced, and how successful these enterprises have been in recent years. These are background topics that provide each student team with an overview of the operation, and also the data to calculate rough economic estimates when data on production costs and recent prices are available from state statistical sources. It is one way to get estimates about economics without asking direct questions about costs, production levels, and profits. In many cases, asking about yield levels and other data is acceptable, then students are able to conduct a preliminary economic analysis. We urge students to move beyond initial descriptive facts to learn about owned versus rented acres, who does the labour, and what equipment is owned versus what operations may be contracted. Farmers generally are forthcoming with most of this general information, and we discuss later why some issues are sensitive and how to learn as much as possible while asking tactful questions within the rural culture of the area. In the early years of the course, much of the focus was on this type of questions about enterprises, specific choices of practices, and schedules through the cropping cycle. More recently we plan discussions about the interview process for preparation prior to farm visits (citations above). Teams are asked to review their pre-course readings (Bland and Bell 2007, Conway 1985, Gomez *et al.* 1996, Visser 2005) and discuss possible models with which to evaluate the sustainability of farms, and what questions to ask to be able to make these models as robust as possible.

Of greater interest in an agroecological analysis of the farm is examining each enterprise and its practices as components and sub-sets of a complex and sustainable whole farm system, including the perspective of holons (Bland and Bell 2007). For student teams to achieve this we need to explore, even at the field level, a deeper set of questions such as how fertility is maintained using a range of strategies, how pest control is a function of system design and integrated management rather than the product of applying a menu of "conventional best management practices", and how these management practices impact the immediate farm environment as well as the surrounding landscape and watershed. Some specific topics include questions about:

- Holistic strategies for maintaining adequate soil nutrient status;
- Pest management without chemical pesticides: beyond conventional IPM;
- Value of cover crops in annual cropping systems;
- Rotations to enhance yield stability in resilient production systems;
- Tillage options to increase water use efficiency, including zero-till planting and strip tillage.

We observe that for some farmers their goals have changed over time from maximum yields to greater focus on yield stability and production resilience in a changing climate. And we hope it becomes obvious to all students that what they observe is one snapshot in time, and that the specific practices at field level are imbedded in whole farms, in landscapes and watersheds, and ultimately in food systems.

4. Integrative evaluation of farm sustainability

Analysis of sustainability at the farm level is in fact the focus of the agroecological analysis course, where teams aggregate the information collected during the farmer interviews about specific enterprises and attempt to evaluate the long-term resilience and adaptability of the whole farm operation in a time of uncertain climate and markets. Among the most obvious types of integration and diversity are appropriate crop rotations, complementary enterprises on the farm, and crop/animal integration. These are encountered in the context of farms visited, and farmer interviews provide essential information for student teams to undertake multifunctional analyses of farms from an agroecological point of view (Rickerl and Francis 2004).

Temporal diversity through rotations is the norm on most farms in the region. Although there are exceptions with a few monoculture and continuous maize-cropped fields, especially when commodity prices are high for this favoured crop. Students encounter the expected maize-soybean rotation on most conventional farms. In discussions by teams and in whole-class sessions the students examine the factors that would contribute to the sustainability of this dominant two-year rotation pattern, recognising that farmers who choose such a predominant system must have many logical production and economic reasons for this decision year after year. Some students who come from farms where this is the norm may become defensive when their team scores a maize-soybean rotation as low compared to other more diverse rotations. The discussion is further enriched by visits to organic farms where such a narrow rotation is not allowed under certification rules and where other options are introduced (Porter 2009). Rotations include alfalfa, cover crops, short-term perennial pastures, winter grains, and other alternative and specialty crops. Such temporal diversity provides resilience to fluctuations in markets and the vagaries of unpredictable weather events that are becoming more the norm in the Midwest region (Peterson and Janke 2009).

Innovative farmers we visit during the week have moved beyond traditional row crops and cereals to include specialty crops and products in their rotations. Among the enterprises that we encounter are turnips, winter peas, and radishes mixed or added to summer row crop fields for cover and grazing; cover crops such as oats or rye to provide nutrient capture, fodder for grazing, and erosion prevention; and vegetable crops to meet local demand through farmers markets or community supported agriculture (CSA) operations.

Another step toward production and economic diversity, plus a strategy to add value to raw commodities on the farm and sell a higher value product, is to integrate crops with livestock or poultry (Entz and Thiessen-Martens 2009). Grain, silage, or hay produced on the farm can be fed to ruminant livestock, and a nutrient cycle completed by returning manure to fields by spreading or grazing forages or crop residues. Students learn from organic farmers that they are hesitant to sell any forages off the farm, as this represents a loss of organic matter and nutrients that are better cycled within the system (Clark 2009). The principle is to sell the highest value product possible with the lowest possible export of nutrients off the farm, since these in some way must be regenerated or imported to maintain soil quality and nutrients for succeeding crops.

Students quickly recognise the many trade-offs among often conflicting and even incommensurate goals the farmer is trying to meet. They have been able to identify these and evaluate their importance using several tools for analysis and visual description listed above: the holon model (Bland and Bell 2007), the amoeba or spider diagram (Gomez *et al.* 1996), the wealth wheel (Visser 2005), and the multidisciplinary analysis model of agroecosystems (Conway 1985). Often students manipulate these models and derive their own hybrid approach to evaluating sustainability and creating visual representations of their findings. One common shortcoming in the use of most models is the simplistic assignment of equal weightings to all factors included. Students and faculty appreciate *a priori* that some factors are much more important than others, but for lack of time or perceptive interpretation of possible relative weightings students (and others) most often draw their diagrams or construct mathematical representations of systems that consider all factors equal. A more convincing and accessible method to approach this educational hurdle is needed.

Some of the most important systems and interactions that students have identified in their whole-farm evaluation of interview results include:

- Complementary rotations on whole farm basis to provide temporal diversity;
- Enhanced biodiversity in farming systems with new specialty crops;
- Perennial crops and pastures to enhance nutrient cycling;
- Crop/animal integrated systems to build farm resilience to climate change;
- Value-added activities on farms and in local communities.

These systems when viewed from the holistic perspective of the entire farm provide new insights for students as they look beyond the narrow spatial scale of individual fields and enterprises, and gain an appreciation of the importance of evaluating integration at the farm level. It is obvious at this point that farms are imbedded in a rural landscape that includes watershed and ecological features larger than a single farm, as well as nearby communities that are vital to social sustainability. It is also important to appreciate that the farms visited are on land where tallgrass prairie dominated for centuries, and this knowledge can perhaps help inform the design of future sustainable farming systems.

5. Evaluation of farms as components of rural landscapes

After guiding students through an analysis of farming practices and enterprises, then seeking to build appreciation of how farmers make decisions using multiple criteria at the whole farm level, we encourage the students to explore farms as components of rural landscapes and how they are dependent on the social infrastructure of rural communities (Flora *et al.* 2015). One important foundation for this search is a better understanding of the natural environment and the history of the Great Plains, including how geology and glaciers, precipitation, wind, and historical plant communities have shaped the landscape that is farmed today. More than just an interesting history lesson, an appreciation of soil formation helps students to understand how and why these soils and local topographies came to be. Based on this knowledge we can together envision future management options and integrated enterprises that can best make use of available resources while preserving production capacity for the future.

In a seminal morning learning activity, the class scales to the top of Ocheyeden Mound, the second highest point in Iowa. From here we can see the results of glaciation some 12,000 years ago. We are visiting the edge of the Des Moines Lobe of the last glacier to retreat from the area, imagining what it would be like to stand on a mountain of ice 1 km high, with a sheer drop to the glacial till plain to the southwest that slopes to what is now the Missouri River and carries water through the Mississippi River to the Gulf of Mexico. Standing on this kame, or gravel deposit left when the glacier retreated, is a heady experience. We are told by an Iowa State University pedologist that most of the other kames in the area have been removed and used as sources of gravel for local roads, and to create more level fields to grow maize and soybeans.

To the northeast are prairie potholes that were left by glacial retreat, often now with drainage tiles to allow farming virtually every acre in this part of Iowa. A few have been preserved as wetlands, because of new federal programs to promote their restoration as wildlife habitat. To the southwest are rolling hills in the glacial moraine, where maize and soybeans also predominate and gravel is mined along the small stream corridors. These are geological realities that impact soil quality, thus the farming practices and crops that are grown today. Students and faculty are entranced by this glimpse back into both geological and more recent agricultural history.

Further insight into the past is gained by taking short journeys through several native and restored prairies, with extended walking transects through areas of native plants. Students learn to appreciate the diversity of an intact prairie ecosystem (Figure 3) that appears to be rather uniform from a distance, but actually contains more than 100 grass and forb species found in unique assemblages corresponding to subtle differences in topography, drainage, and nutrient availability. Students begin to appreciate how this landscape looked, felt and functioned before intervention by European settlement, and to further understand how humans have simplified the original diverse assemblage of ecological niches by installing subsurface drainage tile and applying synthetic fertiliser. These modifications reduce the underlying environmental variability, but do not remove it entirely. After taking a close look at a native prairie, some students begin to consider the potential advantages of using ecological information and principles to inform selection of agricultural species, and placement of crops and livestock on the landscape. They are also encouraged to explore

Figure 3. Native biodiverse prairie of the Midwest of the USA.

ways in which we all might learn to work with the environment instead of trying to force a given production system onto the landscape, often at great economic and environmental cost.

The students learn first-hand the meaning of resilience based on native prairie diversity, and then easily contrast this with the familiar monoculture soybeans and maize that dominate the surrounding farming landscape. To be sure, well-crafted soil maps, variable fertiliser and seeding rates, and global positioning system (GPS) guidance systems provide new technological tools for farmers to do a better job of using nutrients and crop cultivars that are well-suited for specific areas in each field. But students also begin to appreciate the cost of this technology and potential futility of total dominance of the production environment, in a small spatial sense, and how it might be more valuable as agroecologists to ask other questions, in a larger spatial and temporal context. Why are farmers growing maize and soybeans across a landscape that was engineered, at great cost, in a large area that might be better suited to perennial forages, to crop/livestock integrated systems, or to multiple cropping or perennial polycultures where species are put in those places where they are best suited to the natural landscape? Why are Midwest farmers not providing the majority of the vegetables consumed by their urban neighbours? We inform students of innovative research at The Land Institute in Salina, Kansas (https://landinstitute.org/), where the prairie is considered a prototype for future biodiverse perennial polycultures on the plains that could provide grain crops for people and forage for livestock. This is a real stretch of the imagination for most.

Prairie walk activities are among the methods we employ to meet goals in the educational process, where we use experiential learning on the ground to bring important large lessons to our students. Are we always successful? Of course some students with life-long (20–25 years) experiences are not sure we are really asking the right questions, and would prefer to continue their conventional education about how to grow more maize, hopefully profitably, across the landscape. But other students are thoughtful about what might be possible.

Our field experience with students in viewing details of production, of whole systems at the farm level, of economics involved in current enterprises and farming strategies, and of where these activities fit in the landscape leads us to explore together the social dimensions of agriculture. In the interview process, students learn from farmers about the histories of their farms, plus the current human involvement of family, employees, and friends in each operation. Students explore the markets and multiple options for current enterprises, including how to improve them. At the same time, they ask about succession on the farm, mindful that the average age of Midwest farmers is about 58 years and that major changes in ownership and management within the next decade are expected. Students learn that one farmer we visit in South Dakota allows his retired friends to come hunt deer and pheasants in exchange for driving grain trucks in the fall during harvest. They learn about shared use of equipment, and some farmers who have neighbours who bring in livestock for grazing crop residues. On one organic dairy farm, managed by two brothers and their father, the owners decided to forego their feeding of grain from certified organic acres in order to sell this valuable product to organic dairies in the area, and to buy conventional feed for their own herd. On the same farm, they are certified organic for crops but not for the dairy. They may soon add value to their milk product by making ice cream and cheese, and selling milk directly to consumers. These are options and details that the students could read about on the web or discuss in class, but we have found that lessons are much more real when standing in the dairy barn or in the maize field, and building on the credibility of farmers whose lives are invested in the results of both immediate production decisions and long-term planning. Students learn first-hand the involvement of people and institutions, and the value of social capital in the rural community.

Thus among many other lessons learned from farmers in the landscape and others from rural communities, the students and instructors are seeing the potential of agroecology to achieve expanded understanding of:

- Farming systems that are informed by topography, soil type, and native prairies;
- Structure and function of natural systems that contribute to resilience in farming;
- Perennial options for Midwest systems including permaculture and polycultures;

- Social factors that contribute to long-term evaluation and success of sustainable systems;
- Production, economics, environmental principles, regulations, and social issues integral to sustainability.

As a learning community each year, we all share a common experience on farms and across the prairie landscape that can inform our analyses of sustainability of agroecosystems, and can help catalyse meaningful discussion of options for the future of Midwest agriculture, as explored in the next section on education.

6. Education on farming practices, farms, landscapes, and food systems

As described in the above sections, choice of farming practices and types of technologies are primarily informed by factors at the field and farm levels, and to some extent by the rural landscape location of the farm in the watershed, and by current economics and policy. These range from regulations on input use imposed at the natural resource district, state, or national level to others related to marketing such as product identity, for example certified organic foods. Thus a farmer's decisions on adopting agroecological practices will depend on a host of production, economic, environmental, regulatory, and social issues. Briefly, one could think about such issues in categories such as:

- *Production*: will the practice work within current systems, including existing equipment, are the inputs available, and is knowledge accessible to make a change?
- *Economic*: will agroecological practices result in reduction of expensive inputs, equal or acceptable profits, and provide a special product with ready local or regional market?
- *Environmental*: is the practice positive within the farm and the larger ecosystem and watershed, and does it solve a current problem or one anticipated in the future?
- *Regulatory*: is the practice required by the local natural resource district, or by state/national environmental regulations that must be met?

- *Social*: will the practice be accepted by neighbours, by nearby urban residents, by bankers who provide annual operating loans, and will they bring positive recognition to the farmer and family?

In the section on field level decisions and practices, we explored the importance of both temporal and spatial biodiversity, with extended and more diverse crop rotations described as valuable alternatives to the dominant maize-soybean rotation. For farmers long-invested in monoculture cropping and large equipment to maximise returns to labour, often decisions are made to upgrade and enlarge farms and access the equipment to cover more ground, sometimes described as "get bigger or get out" (Hightower 1973).

An intriguing element in the discussion of farm size and integration in the input industry is recent adoption of (GMOs) in the Midwest. Many argue that the transgenic (GMO) technologies available for weed and insect control have helped drive this trend to larger scale farms, less biodiversity, and resulting reduction in resilience and sustainability of farming operations (Chappell and LaValle 2009). Some question whether these new hybrids and varieties provide any advantage in yield, and where widely deployed may add to convenience for farmers but also promote expansion of farm and field size (Rótolo *et al.* 2014). One negative biological consequence of the wide adoption of transgenic technologies is input-resistance, e.g. herbicide-resistant weeds and by Bt-resistant insects, and methods are needed to assess this risk (Morris 2011). The challenge of resistance is countered by industry through development of other modes of action that will introduce "chemical diversity". An agroecologist would prefer thoughtful redesign of systems to avoid the problem in the first place. Although site-specific variable-rate applications of fertiliser and herbicide may be advantageous, and even include the options of planting different cultivars in unique parts of a field, the large industrial farmer has less time to adapt practices to variations in fields and farms that agroecologists would call agronomic niches deserving unique management. Further, transgenic technologies are not acceptable in organic certified production, and limit crop export to E.U., Japan, and some other markets.

When systems are examined at the field, farm, and landscape levels, we must also evaluate our current fixation on maize, soybeans, and wheat and consider the need to increase priority on alternative crops, biodiverse and complex systems, and exploration of emerging issues of importance to society. When we look at a hierarchy of needs to satisfy human nutrition, it is valuable to consider broader issues in the food systems such as food waste, equitable access to food, and production of grains for food versus fuel. Thoughtful farmers today ask how agriculture will move to:

- Go beyond a "monoculture mentality" and face the challenges of adopting more biodiversity;
- Assess environmental criteria for farming systems evaluation and providing ecosystem services;
- Evaluating importance of the structure of farming and its impact on rural communities;
- Integrated systems that add value by processing raw commodities and enhancing local economies;
- Addressing food waste to reduce costs and improve food access for people with limited resources;
- Thoughtfully examine the real distribution of benefits from agriculture and food systems;
- Embrace long-term evaluation of farming systems and their sustainability in rural landscapes;
- Objectively explore the future of mid-sized, small, peri-urban, and urban farms in a future of limited land and production resources.

We raise these questions with students as we evaluate learning each year in our summer travel course, and when we consider how the course can evolve towards an even more holistic approach that is accessible to a wide range of participants with different backgrounds.

7. Conclusions on future emphasis in holistic education in agroecology

Based on our long experience as educators in the landgrant system and colleges of agriculture, we have enjoyed the challenges as well as experienced the frustrations of moving toward study of holistic strategies in dealing with systems. In a university and college environment that is organised around academic disciplines, and reflecting on our own preparation for educational careers, we recognise both the need for broad thinking and the burden this places on everyone who is accustomed to dealing with single issues, one detail at a time. It has been liberating to find that agroecology provides an academic as well as practical field platform to study holism in agriculture and how we can begin to appreciate the ecology of food systems in a rigorous manner.

Agroecology in future education and research agendas is now increasingly recognised as a method to approach learning through:

- Informing the development of sustainable and resilient food systems that are inclusive in promoting participation of players who are active in all steps from production through processing, marketing, consumption, and recycling at all stages;
- Evaluating potentials for local and regional food systems, their benefits and shortcomings including the distribution of benefits throughout the community;
- Quantifying the energy use efficiency of alternative systems through life cycle and emergy analyses that begin to evaluate the real long-term costs and benefits of farming and food systems in real energy terms;
- Applying system design principles at different levels of the spatial hierarchy to achieve food sustainability in a time of uncertainty with climate and scarce resources.

How do we achieve or at least begin to build awareness of these goals in a one-week course in agroecosystems analysis, plus the time that student teams work collaboratively on their projects and then reflect individually about the experiences? It is clear that students thrive on first-hand experience, walking across farmlands and pastures to experience the context of a farm using all their senses. In our visits to rural areas in four states and interviews with farmers, students begin to appreciate the complexity of challenges faced in running a farming operation in times of uncertainty. They inquire about crop and animal enterprises, how well suited these are to the particular soils and environment of each farm, and how various activities on the farm are linked in ways that can help the farmer meet their income, lifestyle, and long-term objectives. It is also useful to evaluate how the practices, systems, and plans fit with the worldview of each farmer.

Students respond well to an open-ended case approach (Francis *et al.* 2009) that gives them responsibility for their learning, although this is challenging for people who have most often been told in previous classes what to do and how to do it. We urge them to break out of a "monoculture learning culture" to embrace multiple perspectives in how to evaluate current systems and to envision more productive, profitable, and sustainable farming systems in the future. Students are encouraged to explore higher order issues such as succession on the farm and how each farm fits into the larger landscape and watershed, as well as how it relates to neighbours, the rural community, and all of society. As instructors, we do respond with our own ideas when asked, but leave the responsibility to students to wrestle with the most important issues and the methods they might use for analysis.

The course provides safe space to explore non-conventional ideas, to think outside the field or farm, and to bounce creative ideas off their peers and faculty mentors. We encourage students to include assessment of environmental impacts of future systems as part of their evaluation. In the social sphere, we challenge students to look at distribution of benefits among people of different income levels and the potential viability of rural communities that depend on agriculture and natural resources. Based on feedback in evaluations from students each year, this course appears to meet their need for a practical, comprehensive, and forward-thinking approach to evaluation of farming and food systems.

This course in agroecosystems analysis is in constant evolution as we respond to students' evaluations each year, and we assess and reassess which of the activities appear to be most effective in helping us study together as a learning community to meet the course objectives. We consider evaluation to be an integral and ongoing component of the learning process, and thus we dedicate quality time to looking at learning using one-page daily evaluation sheets and group sessions each evening to take stock of what has happened and how we are building together on shared experiences. In nearly two decades of conducting this course each summer, we think that much has been achieved in furthering our own understanding of the learning process in agroecosystem analysis, and students confirm in their evaluations that they share this optimism in better understanding not only farming systems but the future of communities and food systems.

References

Ahl, V., Allen, T.F.H., 1996. *Hierarchy Theory, A Vision, Vocabulary and Epistemology*. Columbia University Press, New York, 215 p.

Blanco-Canqui, H., Shaver, T., Lindquist, J., Elmore, R., Shapiro, C., Francis, C., 2015. Cover crops and soil ecosystem services. *Agronomy Journal* **107**, 2449–2474.

Bland, W.L., Bell, M.M., 2007. A holon approach to agroecology. *International Journal of Agricultural Sustainability* **5**(4), 280–294.

Chappell, M.J., LaValle, L.A., 2011. Food security and biodiversity: can we have both? An agroecological analysis. *Agriculture and Human Values* **28**, 3–26.

Clark, E.A., 2009. Forages in organic crop-livestock systems. In: Francis, C. (Ed.), Organic Farming: the ecological system. *American Society of Agronomy*, Madison, Wisconsin, pp. 85–112.

Conway, G.R., 1985. Agroecosystem analysis. *Agricultural Administration* **20**(1), 31–55.

Davidson, B., 2004. *Breakthrough: How Great Companies Set Outrageous Goals and Achieve Them*. John Wiley & Sons, New York, USA 244 p.

Dubos, R., 1971. Trend is not destiny. *Engineering and Science* **34**(3), 5–10.

Entz, M.H., Thiessen-Martens, J.R., 2009. Organic crop-livestock systems. In: Francis, C. (Ed.), *Organic Farming: the Ecological System. American Society of Agronomy*, Madison, Wisconsin, pp. 69–84.

Flora, C.B., Flora, J.L., Gasteyer, S.P., 2015. *Rural communities: Legacy and Change*, fifth edition. Westview Press, Boulder, Colorado, USA 487 p.

Francis, C.A., Jordan, N., Porter, P., Breland, T.A., Lieblein, G., Salomonsson, L., Sriskandarajah, N., Wiedenhoeft, W., DeHaan, R., Braden, I, Langer, V., 2011. Innovative education in agroecology: experiential learning for a sustainable agriculture. CRC Crit. Rev. *Plant Sciences* **30**(1&2), 226–237.

Francis, C., King, J., Lieblein, G., Breland, T.A., Salomonsson, L., Sriskandarajah, N., Porter, P., Wiedenhoeft, M., 2009. Open-ended cases in agroecology: farming and food systems in the Nordic Region and the U.S. Midwest. *Journal* of *Agricultural Education and Extension* **15**(4), 385–400.

Francis, C., Lieblein, G., Gliessman, S., Breland, T.A., Creamer, N., Harwood, R., Salomonsson, L., Helenius, J., Rickerl, D., Salvador, R., Wiedenhoeft, M., Simmons, S., Allen, P., Altieri, M., Flora, C., Poincelot, R., 2003. Agroecology: the ecology of food systems. *Journal* of *Sustainable Agriculture* **22**(3), 99–118.

Francis, C.A., Porter, P., 2011. Ecology in sustainable agriculture practices and systems. CRC Crit. Rev. *Plant Science* **30**(1&2), 64–73.

Francis, C., Salomonsson, L., 2012. Farmer interview role play exercise. *NACTA Journal* **56**(4), 87–88.

Gliessman, S.R., 2015. *Agroecology: The Ecology of Sustainable Food Systems*, 3rd edn. CRC Press, Boca Raton, Florida, USA 405 p.

Gomez, A.A., Swete Kelly, D.E., Syers, K.J., Coughlan, K.J., 1996. Measuring sustainability of agricultural systems at the farm level. In: Doran, J.W., Jones, A.J. (Eds.), *SSSA Special Publication* 49, *Soil Sciences Society of America*, Madison, Wisconsin, USA, pp. 401–409.

Harms, K., King, J., Francis, C., 2009. Behavioural changes based on a course in agroecology: A mixed methods study. *Journal of Natural Recourses and Life Sciences Education* **38**, 183–194.

Hightower, J., 1973. *Hard tomatoes, Hard Times*: *Failure of the Landgrant College Complex*. Agribusiness Accountability Project. Schenkman Publ. Co., Cambridge, UK, 22 p.

Morris, E.J., 2011. A semi-quantitative approach to GMO risk-benefit analysis. *Transgenic Research* **20**, 1055–1071.

Østergaard, E., Francis, C., Lieblein, G., 2013. Practicing and preparing for stakeholder interviews. *NACTA Journal* **57**(1), 97–99.

Østergaard, E., Lieblein, G., Breland, T.A., Francis, C., 2010. Students learning agroecology: phenomenon-based education for responsible action. *Journal* of *Agricultural Education and Extension* **16**(1), 23–37.

Peterson, H.H., Janke, R.R., 2009. Organic marketing. In: Francis, C. (Ed.), *Organic Farming: The Ecological System. American Society of Agronomy*, Madison, Wisconsin, USA, pp. 217–234.

Porter, P., 2009. Crop rotations in organic production systems. In: Francis, C. (Ed.), *Organic Farming: The Ecological System. American Society of Agronomy*, Madison, Wisconsin, USA, pp. 51–68.

Rickerl, D., Francis, C. (Eds.), 2004. *Agroecosystems Analysis*. Monograph Series No. 43, *American Society of Agronomy*, Madison, Wisconsin, USA, 207 p.

Rótolo, G.C., Francis, C., Craviotto, R.M., Viglia, S., Pereyra, A., Ulgiati, S., 2014. Time to re-think the GMO revolution in agriculture. *Ecological Informatics* **26**, 35–49.

Visser, J.R., 2005. Worldview and global leadership. In: Sims, R.R., Quantro, S.A. (Eds.), *Leadership: Succeeding in the Private, Public, and Not-for-profit Sectors*. Routledge, Taylor and Francis Publ., New York, pp. 321–347.

Wezel, A., Bellon, S., Dore, T., Francis, C., Vallod, D., David, C., 2009. Agroecology as a science, a movement or a practice. *Agronomy for Sustainability Development* **29**(4), 503–516.

Wezel, A., Casagrande, M., Celette, F., Vian, J.F., Ferrer, A., Peigné, J., 2014. Agroecological practices for sustainable agriculture: a review. *Agronomy for Sustainability Development* **34**(1), 1–20.

Wiedenhoeft, M., Simmons, S., Salvador, R., McAndrews, G., Francis, C., King, J., Hole, D., 2003. Agroecosystems analysis from the grass roots: a multidimensional experiential learning course. *Journal of Natural Recourses and Life Sciences Education* **32**, 73–79.

Chapter 14

Teaching Agroecological Practices to Higher Education Students, Farmers, and Other Stakeholders: Examples from France

Aurélie Ferrer, Florian Celette, Joséphine Peigné, Marion Casagrande, Jean-François Vian and Alexander Wezel

ISARA Lyon, Department of Agroecology and Environment, Lyon, France

1. Introduction

In the coming decades, humankind faces the major challenge of feeding an increasing world population while avoiding negative impacts on climate change and environmental quality. The current world food production system has become increasingly scrutinised, often described as neither environmental friendly nor sustainable, and considered to be unable to ensure long-term food security. More and more voices are calling for a revisioning of the present food production system, and even undertaking its deep transformation.

Producing a large diversity of commodities world-wide, agriculture, while serving as the foundation of our global food system, is also responsible for related negative environmental impacts. For example, agriculture is a major contributor to climate change, generating 10–12% of the global anthropogenic greenhouse gas (GHG) emissions (Smith *et al.* 2014). Modern agricultural systems are characterised by a high dependency to fossil fuel-based energy via both direct utilisation of fossil fuels (e.g. oil, fuel, natural gas for agricultural practices), and indirect consumption related to the manufacturing of synthetic fertilisers and pesticides. The use — and particularly overuse — of these inputs contributes to the release of GHG and other air pollutants, as well as other serious negative environmental consequences such as soil fertility erosion, water pollution, and biodiversity loss (Millennium Ecosystem Assessment 2005). In addition, simplified cropping systems are increasingly common across the globe, including crop rotations with limited diversity and large-scale monoculture production which are more vulnerable to abiotic stresses and less resilient to variability in environmental conditions (Olesen and Bindi 2002, Jackson *et al.* 2007, Kremen and Miles 2012). To ensure sustainable food production, agriculture must reduce the negative externalities of its practices, decrease its dependence on fossil-based energy, and increase its resilience to variation in environmental conditions.

In this context, the education and training of current or future agricultural practitioners, professionals, and other agriculture-related stakeholders have leading roles to play in adaptation and transformation of agricultural systems. The transition to more sustainable production systems will require people who not only are highly technically skilled and specialised, but also experienced in working with multi-disciplinary perspectives, allowing for the successful implementation of the holistic approach necessary for such a marked transition. Students engaged in agricultural sciences must be provided with an education that challenges them to address concrete real-life situations (Francis *et al.* 2011, Lieblein *et al.* 2012), where the productive, economic, environmental, and social dimensions of farming and food systems constantly interact. Future protagonists in farming and food systems must be prepared to address the complexities associated with the acute transition of agricultural approaches, both with respect to technical and social aspects, and be able to adopt a multiperspective lens through which to analyse the world. Their education will therefore require a multidisciplinary teaching strategy, allowing the biophysical, economic, social, and political components of the systems to be evaluated in concurrent and related perspectives.

The teaching of agricultural science within the larger framework of agroecology — including its three dimensions of practice, science, and movement (Wezel *et al.* 2009) — can significantly contribute to the attainment of these education goals. Agroecological practices aim to create productively yielding systems while valorising and integrating ecosystem services as fundamental elements in the production of food (Wezel *et al.* 2014). This enhancement of ecosystem services within the agroecosystem can help to increase agricultural autonomy from external inputs and to decrease the negative side effects of agricultural activities on the environment. Agroecological practices therefore not only serve as a tool to increase the sustainability of production systems, but also serve as the foundation for sustainable food systems more broadly.

In France, agricultural sciences are taught in higher education colleges specialising in agricultural and food sciences teaching. The curriculum is designed to be completed within 5 years, combining undergraduate and graduate level instruction, leading to an engineer diploma. ISARA-Lyon represents such a French engineering school specialised in agriculture, food and environment, where agroecology and its related practices are taught throughout the whole five-year curriculum in the general framework of agricultural sciences. Typically, the five-year program is split in 10 academic semesters, including internships, with the final semester dedicated to the completion of a Master's thesis. In addition, in collaboration with universities in Europe, USA and South America, ISARA-Lyon also offers as two-year graduate MSc program in Agroecology, consisting of three academic semesters and a fourth semester for the master thesis work.

While this chapter does not provide a general instruction manual on teaching agroecology, it provides an example of agroecological instruction in France. Teaching elements addressing agroecological practices are integrated in the educational pathway throughout the five year programme. We describe the successive modules of the ISARA-Lyon curriculum, progressively building upon agroecological practices with increasing complexity. We present module objectives, content and structure, as well as some pedagogical methods used. Professional training offered by ISARA-Lyon is also addressed at the end of the chapter as an additional example of teaching agroecological practices.

2. Progression principle of the curriculum: How to teach agroecological practices in an integrative way

In our view, agroecological practices are not restricted to the technical farming methods that can readily be implemented in agroecosystems to increase the sustainability and decrease the negative environmental impact of farming activities. Although some more simple agroecological practices can directly substitute for less environment friendly agricultural practices (Hill and MacRae 1995), many involve greater complexity in order to be integrated in the farming system, with their impact on agroecosystems expressed only upon long-term management through modifications of ecological processes (Wezel *et al.* 2014). Some agroecological practices may require a full redesign of the farming system and, more broadly, a redesign of the food system.

Therefore, to master agroecological practices, students must simultaneously achieve:

— a thorough understanding of the agroecosystem as a whole, including its extension from the natural ecosystem and the impacts of imposed design, including human management and interactions of technical, social, and economic components.
— a sound knowledge of common agricultural production practices and their impacts on ecological processes and ecosystem services.
— a view of the agroecosystem as a dynamic system extending beyond the field and farm in both time and space.
— a comprehension of the food system on a regional scale as an essential framework to assess agroecosystem design and functioning.

The agricultural studies curriculum at ISARA-Lyon approaches these student learning goals by presenting agroecological practices in the broader framework of agricultural, environmental and social sciences. During the five years of study, instructional progression is designed with an hourglass principle (Figure 1).

Throughout the first year of education, "basic" sciences (biology, ecology, geology, physics, and chemistry) are taught to students in order to provide with the context necessary to understand mechanisms, processes, and functioning of ecosystems.

During the second year (see Section 3), teaching goals include a greater emphasis on the students' comprehension of the agroecosystem and its main components (Figure 2). Second-year curriculum focuses on agroecosystem structure and functioning, as well as the underlying ecological and biological processes supporting ecosystem services. Students

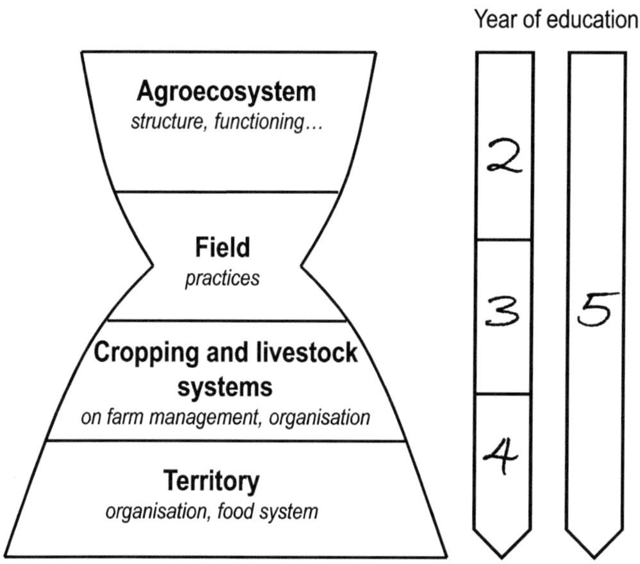

Figure 1. Instructional progression across successive modules contributing to agroecological practices teaching in the ISARA curriculum (in year one only basics are taught with no specific relation to agroecological practices). Each teaching focuses on presenting agroecological principles at various scales; the thickness of the water glass indicates the magnitude of the scale addressed. Instruction regarding agroecological practices begins in the second year at the agroecosystems scale, narrowing it down to the field scale, and then broadening to the cropping systems and territory scale in the next 2 years. Instruction during year 5 integrates each of the scales that were previously introduced individually.

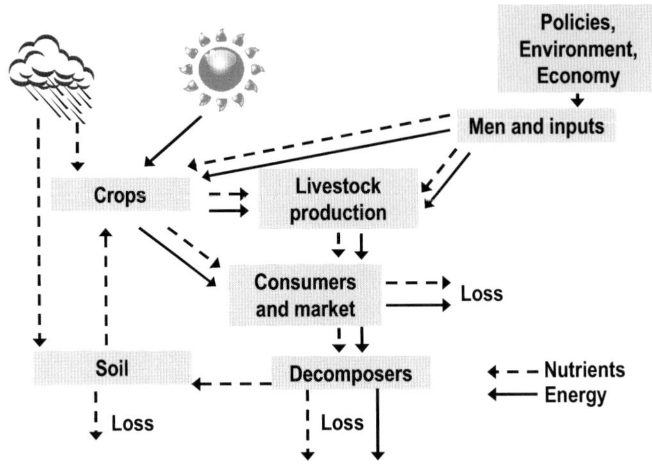

Figure 2. The agroecosystem and its main components (after Gliessman 1997).

are provided with a holistic view of an agroecosystem, becoming aware of the different biophysical, technical drivers and of some of the socio-economic drivers vital to the design of an agroecosystem (farm and adjacent landscape). During field visits, they learn to perform a diagnosis of an agroecosystem, identifying the assets and constraints of the farm, natural environment, abiotic conditions, and available resources, and discussing crop and livestock production choices of the farmer accordingly. During the summer semester break between years two and three, students must complete an internship on a farm, applying diagnostic methods to their individual situation.

In the third year (Section 4), instruction is primarily dedicated to further the students' knowledge of agricultural practices implementation (particularly at the cropping and livestock systems scale) and how socio-economic drivers impact farmer decisions regarding these practices. The curriculum therefore allows students to more deeply explore the socio-economic drivers of the farm agroecosystem, including farmer management decisions. As during the second year, students employ a holistic approach in their assessment of the farm agroecosystem, but they also engage in a more in-depth analysis concerning socio-economic drivers of agroecosystem management. Then, students apply this knowledge through group projects, completing a three week-long diagnosis of working farms producing crops or livestock. As the culminating activity of this teaching, they engage in a discussion of their findings with the farmer.

During year four (Section 5), the larger scales of the territory and the food system are addressed. Food system stakeholders beyond the farmers are included in the assessment of the farming and food systems, with an emphasis on human and social sciences. The connection between farm and agricultural enterprises and industries are now taken into account, including an investigation of the link between farmers, consumers, and food markets.

Finally, during a fifth year module, students fully integrate the different scales and disciplinary approaches (Section 6) of the curriculum. They synthesise the different dimensions and stakeholders introduced during the previous years of the curriculum through the study of a specific territory and the assessment of its potential to be defined as an "Agroecology-Territory". Through the concurrent consideration of each element and complexity of agroecosystems and territories, this phase of study provides a solid preparation for the master thesis work that students must complete during the last semester of their fifth year. Additionally, during this fifth year, students participating in the MSc Agroecology complete a module focused on the last advances of research related to agroecological cropping practices and associated ecosystem services in this field (Section 7).

The following sections highlight the main pedagogic activities and provide greater detail of the teaching modules that significantly contribute to the teaching of agroecological practices and agroecology across the ISARA curriculum.

3. Understanding an agroecosystem, its different components, and the drivers of its design — An initial introduction of the concepts of ecosystem services and agroecological practices (year 2)

The goal of this teaching module is to raise student awareness of the complexity of an agroecosystem (farm within a landscape) and of the diversity of the components, particularly the biophysical and socio-economic components, that determine its design and functioning. This pedagogic activity, which is the first holistic approach to an agroecosystem analysis in the curriculum, is built on a diagnosis of a case study farm carried out by students (Box 1). During this diagnosis, students first describe the biophysical system by investigating its different agroecosystem components: climate, topography, soils, water circulation network, and land use. In a second phase, they examine the agricultural and other land management practices, including agroecological strategies, implemented on the farm. During a third phase, they assess some of the socio-economic drivers that determine the production choices of the farmer in his farm agroecosystem. In the final phase, they engage in a group brainstorming to synthetise the information that they have collected, building an overview of agroecosystem function and discussing the relationships between the biophysical agroecosystem components and the farmers' choices of management and agricultural practices.

To begin this work, students first receive a brief classroom presentation of the key components studied within an agroecosystem (environment and human activities) and the interaction of these components at farm and agroecosystem scale. Following this introduction in the classroom, students attend a field trip to the case study farm (Box 1). During the field trip, students first attend three different workshops led by instructors (Step 1 in Figure 3) to become introduced to the methods that will be employed to study the different farm-scale biophysical components:

— the soil component: students are introduced to the characterisation of soil types (soil classification, origin) and the influence of these soil characteristics on crop growth and ecosystem services. As part of this instruction, the components of the soil profile (Peigné *et al.* 2013) are defined by the instructor and students learn to perform an auger test to describe the soil profile.
— the water component: students' attention is brought to the fish ponds present on the studied farm and on infrastructures installed to manage water flow (e.g. ditches, rills, and drains). Also, a short introduction about the traditional fish production in the territory is provided, with specific production practices described.

Box 1. Description of the case study farm

The case study farm has been chosen in a very specific territory of Rhône Alpes region: the Dombes. The Dombes is a plateau of about 1,000 km^2 flanked in the east, south and west by fluvial valleys with the rivers Ain, Rhône and Saône. This results in a landscape with a water supply of the plateau originating exclusively from rainfall (annual precipitation from 900 to 1,300 mm).

During the late Würm glaciation, substantial amounts of loess were mainly deposited in the depressions between the drumlins (Williams 2006). Post-glacial rain leached much of the loess creating decalcified clayey soils (about 60% of clay and silt) in the depressions that induce water stagnation when soils are wet (Avocat 1975). Four main soil types can be found on farms, from silty-clay soil (on plateaus) to very degraded acidic silty soil (in depressions). Today, the Dombes region is characterised by about 1,100 shallow fish ponds with about 12,000 ha, located in an agricultural area with pastures, cropped fields and forests (Bernard and Lebreton 2007) and sustaining a significant fish production activity with a net fish yield average of 250 kg/ha (Wezel *et al.* 2013, Robin *et al.* 2014). Agriculture and fish farming are therefore intimately intertwined in this territory. A high degree of connectivity among most ponds is ensured by a dense network of ditches which collect rainwater from the catchment. The farm used in this study has been held by the same family since more than 200 years. The farmer manages 100 ha (Figure 4(a)) leading different activities: crop production (maize, cereals), livestock production (mainly sheep), fish production, and firewood production. He also rents some of the land for private hunting. The strong territorial identity, the important biophysical constraints, the diversity of the productions, of management and agricultural practices of the farmer makes this farm particularly exciting and challenging to study in the framework of agroecology.

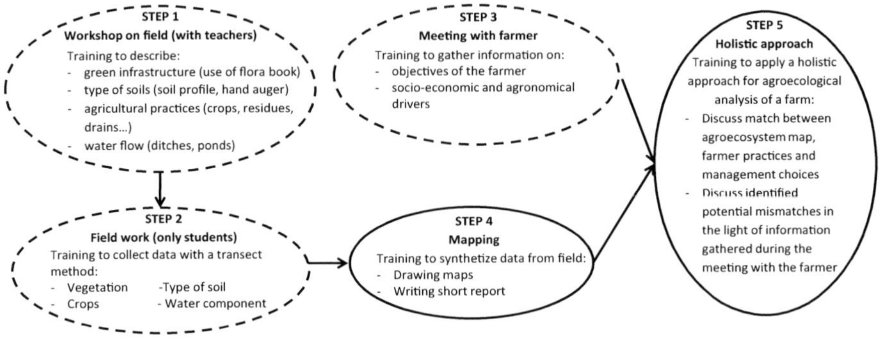

Figure 3. The different steps of the second year teaching module, the methods, and skills that students are expected to develop in each step. Activities are performed on the field (dotted line) or in classroom (solid line).

— the vegetation component: students learn to assess the structure and composition of the green infrastructure (woods, hedges, and grass strips) on the farm and to recognise some specific indicator plant species related to soil fertility or water parameters.

Following these instructional sessions, students explore the farm (Figure 4(a)) in groups using a transect method, investigating the organisation of the agroecosystem and noting pertinent information on raw maps outlining the farm and land parcels (Step 2 in Figure 3). In addition, during the field trip, students conduct an interview with the farmer (Step 3), during which the farmer presents his farm history, his production choices and purposes, his prioritisation of production decisions, and the rationale thereof. Students are encouraged to ask further questions on crop and livestock production, agricultural practices, management of green infrastructure, and other production issues in order to better understand the farming systems and the agroecosystem organisation. After the field trip, students use this information to draw two maps (Step 4): one including information on the occurrence of different soil types, water flow and water circulation network (see Figure 4(b)), and the second including information on land use, green infrastructure, and areas of indicator plants (see example of one year in Figure 4(c)).

In addition to the map, students must integrate all of information that they have collected into a global diagnosis of the organisation of the farm

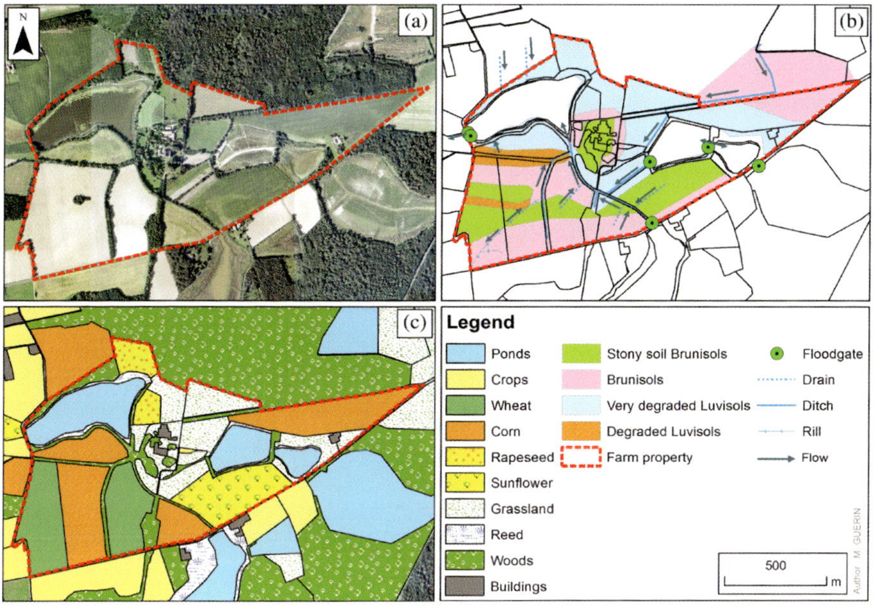

Figure 4. The case study farm: (a) aerial photograph view, (b) map of the water circulation network and the type of soils, and (c) map of the land cover.

agroecosystem (Step 5). Creating a short report, they are encouraged to discuss the relationships and different interactions between the soil and water components, crops and livestock production, and green infrastructure elements. During this exercise, they also must comment on farmer choices of land use and practices. For example, students observe on the farm that the crop rotation and cropping practices are largely determined by the soil structure. Because of the hydromorphic soil in some areas, wheat is only cropped on the better quality soil on a bit higher located land of the farm, whereas hydromorphic areas are mainly used as pasture or meadow. As soils are sensitive to compaction and erosion, agricultural practices, especially tillage, have to be adapted. In this way, students can highlight the assets and constraints of the farm agroecosystem and anticipate potential mismanagement.

Further using the information collected through the interview with the farmer, students are able to reassess the links between the biophysical components and the agricultural practices and production choices in the context of the technical and socio-economic dimensions of the farm. For example, on the case study farm, the green infrastructure (Figure 4(a)), particularly wood areas and hedges, is very dense, highly connected, and less impacted by agricultural activities, creating favourable habitats for large wild fauna such as wild boars and birds. Yet, this situation is not beneficial for the crops that can suffer from shading, high weed pressure from the field surroundings, and even direct predation by the wild fauna. At first glance, students define this as mismanagement of the production system. However, the interview with the farmer reveals that he can get more income from the hunting activity than from the crop production, and is therefore prioritising to a certain degree practices and management promoting wild fauna rather than practices improving crop yield. Some crops are even sometimes deliberately left unharvested to serve as food for the wild fauna. This highlights for the students the importance of an integrative study of agroecosystem in all its components, including economic.

After this module, students put into practice their acquired skills, methods, and tools by engaging in the same kind of integrative analysis during their internship on a farm at the end of their second year.

➔ During this pedagogic activity, students understand the specificity of an agroecosystem and the variety of drivers in a real farming situation that determine the design of the system. They learn to adopt for their analysis a multiscale approach from the crop field to the farm and the surrounding landscape. This highlights to students that the integration of agroecological practices within an agroecosystem must be considered in a multidimensional approach that is not restricted to the technical/biophysical system, but also considers the socio-economic realities and components of the landscape.

4. Practices as a cog in the wheel of cropping systems: Integration in time and space (year 3)

During the third year, students further progress in their comprehension of the functioning of different cropping systems (orchards, annual grains, etc.) and livestock systems and associated agricultural practices, including agroecological practices. The third-year objective is to provide students with a thorough understanding of crop and livestock systems in their spatial and temporal dimensions, the management logic at field/grassland and farm scales, and the derivation of farmer choices made among a given suite of practices. Instruction also addresses different decision tools developed for farmers to improve their system management and the indicators to evaluate the sustainability of different cropping or livestock systems. The framework proposed to students is largely based on the definition of systemic agronomy as defined by Sebillotte (1977): "Study in time and space of the relationships within the plant–soil–climate system and their respective chemical, physical and biological components on which farmers act to obtain a production". This definition highlights that a transdisciplinary approach is necessary for the study of agronomic systems, stressing the numerous parameters required during the consideration of system management and the complexity of their integration in a systemic approach, which demands a high degree of technical skills, expertise and know-how on the part of the farmers (Doré *et al.* 2006).

If we take the example of teaching cropping systems, the first step of the approach consists of a broad overview of the different cropping practices applied by farmers in the management of soil quality and crop yield and quality. Students receive a presentation on the concepts needed to describe and understand cropping systems and crop management. Subsequently, different practices used to maintain or enhance soil quality, such as tillage regimes, organic amendments, crop rotation, intercropping, and cover-cropping, are presented. Students learn the rationale of the fertilisation strategies for nitrogen, phosphorus, potassium, and other elements, and become familiar with the different decision tools available for farmers to manage fertilisation. The different strategies for crop protection are also presented (e.g. integrated pest management, biological control, chemical strategies of pesticides, herbicides, and fungicides), as well as water and irrigation management strategies. Each of these components of agricultural production practices is then integrated into study at the cropping system scale.

This approach enables students to integrate the spatial dimension with crop allocation on the farm territory and the temporal dimension with crop rotation, succession of practices, and levels of the farm management. It also considers the socio-economical aspects of farmer decision-making and their specific farming objectives, e.g. to produce higher yielding crops versus employing less extensive management, obtaining a higher quality of the production, or decreasing their workload. Each of the implemented cropping practices, as well as the succession of practices at the cropping system scale, are evaluated regarding: (i) agronomical performances in terms of yield and quality of the production; (ii) economic and social performances (e.g. cost of production, profitability, and workload); and (iii) environmental impacts such as soil, water and air pollution, soil protection, and biodiversity conservation. Teams of students complete these evaluations during tutorials with the different application tools in order to assess the multiperformances of cropping systems (e.g. calculation of nitrogen balance or evaluation of crop protection strategy).

The second step involves different case studies based on working farms in grain cropping systems. The goal is to understand farmers' objectives and logic, as well as to describe and to analyse the multiperformances of the systems in relation to production in quantity and quality, profitability and environmental impacts. For this activity, students work in groups of 5–6 individuals for three weeks. Groups are provided access to 15 years of production and financial data from different grain cropping farms located near Lyon, France, including indicators of performances such as yield, prices, and inputs. They visit the farms several times, interviewing the farmers. With this information, groups are expected to analyse the systems and explain the reasoning developed by each farmer for their farming decisions, including the different evolutions observed on their farms over the 15 years. After using the various indicators to assess the sustainability of the different cropping systems, students present their results to the farmers and propose modifications to improve the sustainability of their systems, using the principles of agroecological systems. These proposed modifications are then critically discussed with the farmer.

➔ The approach in this module enables students to be confronted to the challenge and complexity a farmer has to face when managing his farm. It helps them to put into practice the concepts and tools that were presented during lectures. They calculate different indicators and carry out system assessment tools to integrate the organisation dimension of the production processes at the farm level (i.e. considering available resources and machinery on farm, adapting system objectives to these resources and to the socio-economic environment of the farm).

5. Agroecological practices as elementary components of organic farming at territorial scale (year 4)

During year four of the curriculum, a three-week module is proposed to students interested in organic farming and with backgrounds in agronomy, sociology and economics (Casagrande *et al.* 2012). It has four learning objectives: (i) provide a general background on organic farming, (ii) distinguish and analyse the main components of organic farming systems, including organic practices, (iii) analyse and understand the social expectations towards organic farming, and (iv) carry out a diagnosis in order to assess the potential development of organic farming in a French case study area and the contribution of organic farming to local development. As organic farming is based on agroecological practices such as organic fertilisation or biological control (Wezel *et al.* 2014), agroecological practices are thus at the core of organic systems that are embedded within agroecological territories.

The module combines different types of learning activities such as farm visits, lectures, and case studies. Agroecological practices are taught in several ways. On the first day of the module, the students visit a mixed crop–livestock organic farm, with an opportunity to discuss with the farmers their transition process, their change in practices during this transition period, and their current organic practices. Most organic farmers employ some degree of agroecological practices, such as the intercropping of cereals with pulses. Professionals from the organic farming sector (such as those providing advisory services) provide lectures on different farming systems including viticulture, arable cropping systems, and livestock systems, focusing on the agroecological practices on which organic systems rely.

Instructors organise practical work on the transition from conventional to organic farming, with the objective of redesigning a virtual conventional farm in order to meet organic agriculture requirements. For example, students must identify the elements that require redesign, such as the modification of the crop rotation to include the benefit of legume crops or the substitution of chemical inputs with organic inputs, biological control or mechanical control. Students are also expected to reflect on the magnitude of each proposed change in practice, and whether the practice impacts crop management, the cropping system, and/or the whole farming system.

Case study work is also presented to students groups with the generic question "How can organic farming contribute to local food system development?" In order to answer this question, students typically interview farmers and advisory services to understand organic farmers' practices and to identify the diversity of farming systems within a certain territory.

During the module, students are also invited to take part in a half-daylong game, with the objective of involving students in a virtual project aiming to develop organic farming in a specific territory. During this game, each student is given the role of a stakeholder potentially involved in the process, such as farmer, local authorities, extension services, and citizens. Stakeholders have different, sometimes even contradictory, opinions and visions about organic farming. Each student has then to defend the interest and the point of view of the stakeholder she/he represents in virtual meetings organised to implement the project. This game provides students with the understanding that the development of organic farming, and more generally agroecological practices, can be hindered or fostered by factors that are external to the farming system. They can foresee that lock-in effects (Lamine 2011) can strongly limit the uptake of agroecological practices at farm level.

➔ This module provides students with the opportunity for in-depth study of organic farming, which is highly reliant on agroecological principles. Based on inductive pedagogy as well as lectures and practical exercises, students discover different facets of organic farming and reflect on the role of organic farming for the development of territories.

6. Agroecological practices: From field to territory (year 5)

During the semester before the master thesis in the fifth year of the curriculum, a teaching module focused on agroecology and agroecological practices at a regional scale is presented. The objective of the module is to develop student awareness of specific environmental and agricultural issues that are intertwined in a given region. Students have to study a selected area and to discuss whether this case study area is a good candidate for being qualified as an "Agroecology Territory" (Wezel *et al.* 2016), i.e. a place where a transition process toward sustainable agriculture and food systems is engaged (Figure 5). Three major domains must to be considered for the transition to take place: adaptation of agricultural practices, conservation of biodiversity and natural resources, and development of embedded food systems. The existence or the development of strategies of stakeholder groups who engage in these three domains and act for the transition is also a critical component.

In order to answer this question, the students engage with a French small region where they (i) meet and interview various stakeholders including institutions, (ii) observe and analyse landscapes, farming systems, and selling channels, and (iii) work in groups on specific topics encompassing the different dimensions of an agroecology-territory. They learn to use different methodological tools to perform their diagnosis of the region, e.g. landscape analysis, agricultural production systems analysis, and interviewing techniques. The areas chosen are usually characterised by a strong identity in agricultural productions and a specific

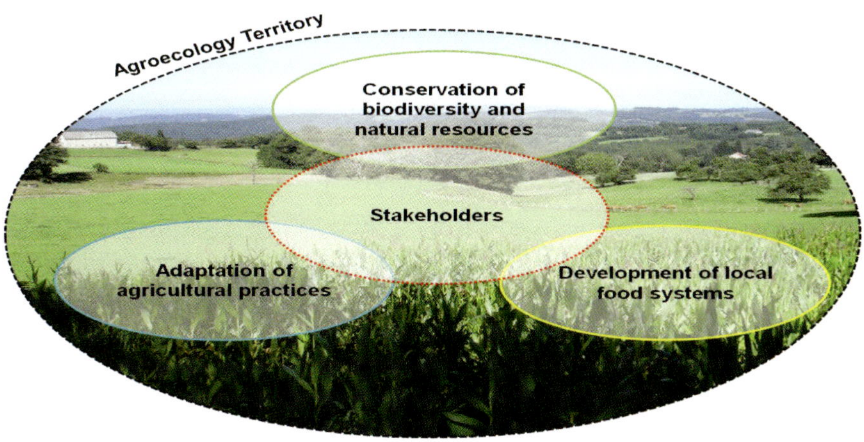

Figure 5. Definition of agroecology-territories: places where a transition process toward sustainable agriculture and food systems is engaged (adapted from Wezel *et al.* 2016).

territorial context such as presence of natural protected areas, for example a regional natural park or/and biosphere reserve.

The module is organised around a one-week excursion in a selected region characterised by different environmental and agricultural production issues and proceeds in three steps. First, students undertake a preparatory period with introductory lectures related to the area, explanation of the different methodological tools for the diagnosis, and presentation of the different group topics. This preparatory time allows for familiarisation with the area and their group topics, and for preparation for their field work (e.g. targeting relevant stakeholders or institutions for their specific topic and scheduling interviews and/or visits).

The second step is the excursion in the area. At the beginning of the excursion, meetings are organised with the entire class and key stakeholders in order for students to understand the agricultural, economic, and environmental characteristics and major issues within the study area. This provides an initial overview to students prior to beginning their fieldwork. Then, the class is divided according to their group topic. The specific topics addressed encompass the different dimensions of an agroecology-territory as, for example:

(i) The food system: students are expected to identify and describe "Terroir" products, local and regional product distribution.
(ii) The agricultural and agroecological practices: student work on past, present, and future land use in the area, or work on agroecological transition of agricultural productions in the territory.
(iii) The biodiversity: students study nature conservation, and the importance of different types of policies for the management nature and more specifically biodiversity of the study area.

As students meet and interview farmers, stakeholders from cooperatives, extension service, and protected areas managers, they enquire about the use of agroecological practices and stakeholder viewpoints concerning these practices. In addition, all students participate in a land use study consisting of surveying land use and creating maps while driving across the study area (for the car transect method please see Wezel *et al.* 2010). Students thus observe the occurrence of agroecological practices, such as cover cropping, intercropping, agroforestry, and integration of semi-natural landscape elements such as hedges and grass strips around fields.

The third step is a general presentation and a written report by the student groups describing their findings and analyses, followed by a discussion with the instructional staff and the other students. This final presentation provides an opportunity to further discuss the concept of agroecology-territories applied to the case study area.

➔ With this relative complex question about the concept of agroecology-territory, students are encouraged to take into account, in a holistic approach, different components, topics, and scales in their analysis of a study area. The final discussion invites them to consider their more specific findings of the group topic within the larger setting of a territory and food system perspective and to consider agroecological practices as a piece in the larger frame of an agroecology-territory.

7. Agroecological practices as research forefronts (year 5)

In their fifth year of higher education, students are expected to master the technical, economic, and sociological components in relation to agroecological practices, and their implementation process at different scales: field, agroecosystem, and/or territory (Figure 1). For students participating in the instructional programme of the MSc Agroecology, we established an additional fifth year module specifically addressing Agroecological Cropping Practices. There, students' understanding of agroecological practices and associated ecosystem services are further deepened through a focus on recent advances of research in this field. This module aims at presenting the state of the art and latest findings concerning different agroecological practices and their impact on ecosystem services and agroecosystem functioning.

Three main pedagogic activities are presented to the students: lectures given by international researchers, the completion of a literature review, and a field trip to analyse the field-based implementation of agroecological practices.

In the lectures, the set of different agroecological practices, their degree of integration in today's agricultural systems, and the consequences of their implementation on agricultural system structure are presented and discussed with the students. Also, different international researchers present their latest research findings in class. They focus on specific ecosystem services and their supporting ecological processes, highlighting the potential impact of agroecological practice implementation on these processes and therefore on the service provision. Lectures cover topics like IPM and biocontrol practices, intercropping and relay cropping practices for weed and nitrogen management, and reduced tillage practices and their effects on soil fertility.

From the beginning of the module, students are invited to choose one specific agroecological practice and to perform a thorough literature review on it. Through this exercise, they are expected to search for recent scientific and technical literature, to read the scientific articles on the last research findings, and to finally present during a seminar the state of the art on the chosen practice to other students and instructors. This seminar provides an opportunity to share the latest knowledge on a large set of agroecological practices and the degree of implementation of these practices worldwide in today's agriculture.

As an illustration of agroecological practices implementation in real farming systems, a field trip is carried out to a vegetable production near Lyon. The case study farm is characterised by very low use of inputs and zero use of pesticides. The farmer's management strategy is to integrate high tunnels and the farm's semi-natural landscape elements in order to achieve pest regulation through a conservation biological control approach. The farmer himself produces and collects natural enemies for release on pest infested vegetables when necessary. In addition, students exchange with the farmer on farm design, cropping practices, and management of semi-natural elements. They discuss together the pros and cons of the implemented practices for the potential of pest regulation on vegetables. This visit allows students to engage in a real farming situation and become aware of the difficulties and uncertainties linked to the transfer of research/theoretical findings into practical situation, in this case conservation of biological control.

➔ With this module, students receive information concerning a large set of agroecological practices, up-to-date scientific knowledge, and an illustration of the difficulties to implement them in real farming situations. They learn about the limits in our understanding of the impact of these practices on ecological processes and ecosystem services provided.

8. Professional training sessions in agroecology with focus on soil management

For more than 10 years, ISARA-Lyon has developed specific training sessions for farmers and advisors on agroecology and soil management. Sessions focus on different topics, depending on target audience. In most cases, the training sessions are developed based on an external stakeholder demand. In the last few years, advisors or other agriculture stakeholders requested sessions focused on the different dimensions, interpretations and concepts of agroecology (practices, movement, and science) and how to implement and develop this approach at the local scale, e.g. territory, or water catchment area.

As a response to farmer interest in the practical applications of soil management, trainings were developed to deepen farmer knowledge of the soil system, i.e. the global soil functioning and the different indicators that can be implemented on their fields to make tactical and strategical decisions. First, ISARA researchers and instructors provide a half-day training presenting the theoretical bases of global soil functioning, with a specific focus on the soil biological functioning and with adaptation to the specificities of the farmer group (e.g. pedoclimatic conditions of their territory, farming systems, and productions). The group attending the training then engages in a discussion of their cropping practices and soil characteristics in the light of the last scientific findings on soil quality management. During the second part of the day, researchers and farmers visit an agricultural field for a soil observation training session. Together, they perform a diagnosis on the soil structure and the global soil functioning using different methods, discussing the agricultural practices responsible for the observed soil state. Then, farmers are trained to perform a soil profile investigation and to observe different indicators of soil quality (Batey *et al.* 2015); for example, a spade test combined with the extraction of earthworms to assess both soil structure and biological activity of earthworms. Farmers are also taught to observe the density-activity of the epigeic soil macrofauna using pit-fall traps. The field visit provides a critical opportunity for farmers to exchange information and to observe and characterise together the results of their cropping practices applied in the last years. These field sessions also allow for the transfer of farmer decision tools developed in research programs to end-users. During this training session, researchers and farmers also exchange ideas on other different practices that farmers can test in their farm by integrating their own constraints and objectives.

9. Conclusions

The education and training of current or future agricultural practitioners, professionals, and other agriculture-related stakeholders has a determining role to play in the transition of agricultural systems towards more sustainable food production. A successful transition of the agricultural systems will strongly depend on the ability of stakeholders to adopt a multidisciplinary approach allowing the production systems to be examined and evaluated in their biophysical, economic, social, and political components and perspectives.

Teaching of agroecological practices within the larger framework of agroecology can strongly contribute to reaching these goals. At ISARA-Lyon, a French engineering school specialised in agriculture, food and environment, practical agroecological elements are integrated throughout related disciplines (agricultural, environmental, and social sciences) and considered at different scales (field, farm, territory) throughout the curriculum. The pedagogic strategy is based on the constant combination of theoretical inputs and practical application with case studies and field trips.

Students begin their instruction during the second year of their five-year curriculum at the field and agroecosystem scales, primarily addressing agricultural and environmental issues. Student learning is focused on the comprehension of an agroecosystem, its main components, its functioning, and the underlying ecological and biological processes supporting ecosystem services. This instruction provides students with a first occasion to adopt a holistic approach in their agricultural system analysis. During the third year, students are encouraged to go deeper in the understanding of the biophysical and socio-economic drivers of the farm agroecosystem and its management in time and space by the farmer. The teaching is dedicated to deepen knowledge about farmers' decisions and practices at the cropping and livestock systems scale. During year four, the larger scales of the territory and the food system are tackled. Other non-farmer stakeholders are included in the assessment of the farming and food systems. At this point, a greater emphasis is placed on human and social sciences. During the last year of the curriculum, students return to the synthesis of the different dimensions and stakeholders perspectives considered during the first four years, through an integrative study of a specific territory and the assessment of its potential for a transition towards sustainable agriculture and food systems.

In addition, within the teaching programme of the MSc Agroecology, a module is fully devoted to agroecological cropping practices and associated ecosystem services with a focus on the latest advances of research in this field. Moreover, specific training sessions on agroecology and soil management targeting farmers and advisors are developed to contribute to the training of professionals of agricultural systems in France.

The successful dissemination of agroecological practices, their integration in modern agricultural systems, and their efficiency to increase ecosystem service provision, depends upon their successful integration in the larger framework of the global food production system. For this reason, we believe that the teaching of agroecological practices as particular elements within the larger scheme of agricultural, environmental, and social sciences, taking into account different scale approaches, is the best way to promote their current and future use in the process of transition towards more sustainable food systems.

Acknowledgement

We thank Mathieu Guerin for preparation of maps and Audrey Vincent for her contributions. We warmly thank Erin Silva and Charles Francis for reviewing this chapter.

References

Avocat, C., 1975. La Dombes: milieu naturel ou milieu en équilibre. *Revue de Géographie de Lyon* **50**, 35–58.

Batey, T., Guimaraes, R., Peigné, J., Boizard, H., 2015. Assessing structural quality for crop performance and for agronomy (VESS, VSA, SOILpak, Profil Cultural, SubVESS). In: Ball, B., Munkholm, L.J. (Eds.), *Visual Soil Evaluation: Realizing Potential Crop Production with Minimum Environmental Impact*. CAB International, pp. 15–28.

Bernard, A., Lebreton, P., 2007. *Les oiseaux de la Dombes: une mise à jour*. Académie de la Dombes, Fondation Pierre Vérots, N°27 de la revue "Dombes", 168 p.

Casagrande, M., Fleury, P., Peigné, J., 2012. A teaching module in organic agriculture: Territorial development and market trends, In: Wezel, A., Rembiałkowska, E., Francis, C. (Eds.), *Learning in Future Farming and Food Systems: European Education in Organic Agriculture and Agroecology*. European Network of Organic Agriculture Teachers (ENOAT) Workshop, Lyon, France, pp. 48–51.

Doré, T., Le Bail, M., Martin, P., Ney, B., Roger-Estrade, J. (Eds.) 2006. *L'agronomie aujourd'hui*. Quae edition, Paris, France, 367 p.

Francis, C.A., Jordan, N., Porter, P., Breland, T.A., Lieblein, G., Salomonsson, L., Sriskandarajah, N., Wiedenhoeft, M., DeHaan, R., Braden, I., Langer. V., 2011. Innovative education in agroecology: experiential learning for a sustainable agriculture. *CRC Crit Review Plant Science* **30**(1&2), 226–237.

Gliessman, S.R., 1997. *Agroecology: Ecological Processes in Sustainable Agriculture*. CRC Press, 384 p.

Hill, S.B., MacRae, R.J., 1995. Conceptual framework for the transition from conventional to sustainable agriculture. *Journal of Sustainable Agriculture* **7**, 81–87.

Jackson, L.E., Pascual, U., Hodgkin, T., 2007. Utilizing and conserving agrobiodiversity in agricultural landscapes. *Agriculture, Ecosystems and Environment* **121**, 196–210.

Kremen, C., Miles, A., 2012. Ecosystem services in biologically diversified versus conventional farming systems: Benefits, externalities, and trade-Offs. *Ecology and Society* **17**, 40.

Lamine, C., 2011. Transition pathways towards a robust ecologization of agriculture and the need for system redesign. Cases from organic farming and IPM. *Journal of Rural Studies* **27**, 209–219.

Lieblein, G., Breland, T.A., Francis, C., Østergaard, E., 2012. Agroecology education: action-oriented learning and research. *Journal of Agricultural Education and Extension* **18**(1), 27–40.

Millennium Ecosystem Assessment, 2005. *Ecosystems and Human Well-being. Current State and Trends Assessment*. Island Press, Washington, D.C, USA. 917 p.

Olesen, J., Bindi, M., 2002. Consequences of climate change for European agricultural productivity, land use and policy. *European Journal of Agronomy* **16**(4), 239–262.

Peigné, J., Vian, J.-F., Gautronneau, Y., Cannavaccuiolo, M., Lefevre, V., Boizard, H., 2013. Assessment of soil structure in the transition layer between topsoil and subsoil using the profil cultural method. *Soil and Tillage Research* **127**, 13–25.

Robin, J., Wezel, A., Bornette, G., Arthaud, F., Angélibert, S., Rosset, V. Oertli, B., 2014. Biodiversity in eutrophicated shallow lakes: determination of tipping points and tools for monitoring. *Hydrobiologia* **723**, 63–75.

Sebillotte, M., 1977. Jachère, système de culture, système de productions: méthodologie d'étude. *JATBA* **24**(2&3), 241–264.

Smith, P., Bustamante, M., Ahammad, H., Clark, H., Dong, H., Elsiddig, E.A., Haberl, H., Harper, R., House, J., Jafari, M,, Masera, O., Mbow, C., Ravindranath, N.H., Rice, C.W., Robledo Abad, C., Romanovskaya, A., Sperling, F., Tubiello, F., 2014. Agriculture, Forestry and other land use (AFOLU). In: Edenhofer, O., Pichs-Madruga, R., Sokona, Y., Farahani, E., Kadner, S., Seyboth, K., Adler, A., Baum, I., Brunner, S., Eickemeier, P., Kriemann, B., Savolainen, J., Schlömer, S., von Stechow, C., Zwickel, T., Minx, J.C. (Eds.), *Climate Change 2014*: *Mitigation of Climate Change*. Contribution of Working Group III to the Fifth Assessment Report of the Intergovernmental Panel on Climate Change. Cambridge University Press, Cambridge, United Kingdom and New York, NY, USA, pp. 811–922.

Wezel, A., Bellon, S., Dore, T., Francis, C., Vallod, D., David, C., 2009. Agroecology as a science, a movement and a practice. A review. *Agronomy for Sustainable Development* **29**, 503–515.

Wezel, A., Brives, H., Casagrande, M., Clément, C., Dufour, A., Vandenbroucke, P., 2016. Agroecology Territories: places for sustainable agricultural and food systems and biodiversity conservation. *Agroecology and Sustainable Food Systems* **40**(2), 132–144.

Wezel, A., Casagrande, M., Celette, F., Vian, J.-F., Ferrer, A., Peigné, J., 2014. Agroecological practices for sustainable agriculture. A review. *Agronomy for Sustainable Development* **34**(1), 1–20.

Wezel, A., Dubois, P., Lagaise, B., Boissière, C., 2010. Land use transects with cars — a potential tool for students to explore an agricultural region. In: Lieblein, G., von Fragstein, P., Francis, C. (Eds), *E-Learning and Experiential Learning in Organic Agriculture and Agroecology*. Proceedings European Network of Organic Agriculture Teachers (ENOAT), Nitra, Slovakia, August 31–September 2, 2008, pp. 28–41. http://vega.isara.fr/awezel/Car%20transects%20Wezel%20ENOAT%202008%20.pdf

Wezel, A., Robin, J., Guerin, M., Arthaud, F., Vallod, D., 2013. Management effects on water quality, sediments and fish production in extensive fish ponds in the Dombes region, France. *Limnologica* **43**(3), 210–218.

Williams, D.D., 2006. *The Biology of Temporary Waters*. Oxford University Press, New York, USA, 337 p.

Chapter 15

Learning Agroecological Practices as Components of Complex Farming Systems

Charles Andrew Francis*,†, Tor Arvid Breland*, Geir Lieblein*, Lennart Salomonsson‡, Suzanne Morse*,§ and Anna Marie Nicolaysen*

**Department of Plant Science, Norwegian University of Life Sciences, Ås, Norway*
†Department of Agronomy and Horticulture, University of Nebraska, Lincoln, Nebraska, USA
‡Department of Urban and Rural Development, Swedish Agricultural University, Uppsala, Sweden
§College of the Atlantic, Bar Harbor, Maine, USA

1. Introduction

Students and many teachers in agriculture are intrigued with details of farming systems, perhaps because of years of training as specialists in discrete disciplines, a desire to simplify questions to make them more understandable, or the immediacy of decisions on farming practices in the field. It has become predictable each year in our agroecology class in Norway that many students come to learn practices. We have learned that a useful way to connect with practically oriented students, as well as those motivated by whole-system thinking and long-term planning for sustainability in agroecosystems, is to observe and participate in as many practical applications as possible within the context of the whole farm in its ecological and societal environment. We explore in this chapter how to involve students in hands-on learning of practices as components of complex farming systems rather than how to attempt to teach and contextualise practices at a distance.

One primary goal in agroecology education is to emphasise the overriding importance of looking at whole systems. This has roots in the widely-cited book *Seeing Things Whole* by naturalist John Wesley Powell (deBuys 2001). But in fact, the concept goes back at least to *The Allegory of the Cave* which was a provoking story in Plato's book *The Republic* (Griffith 2000) where people chained to the wall of a cave could only see images projected by light from a fire behind them onto the opposite cave wall. The people thus confined could only see and interpret the images without benefit of knowing from where they came, how they were connected to anything real, or what they signified in the larger context. Perhaps we are all "chained to the wall" and unable to see a wider vision of reality, when we focus only on details, see images in two dimensions, and in farming systems relate only to monocultures with singular purpose to produce maximum yields and short-term economic returns.

A similar lesson can be learned from the fable about "The blind men and the elephant", (https://en.wikipedia.org/wiki/Blind_men_and_an_elephant) where each man draws conclusions from feeling one small part of the whole, and no one perceives that this is really an elephant. Our goal as educators is to break these chains of disciplinary and narrow thinking and remove "blinders from horses" that are content to move along without the confusion and complexity that would emerge from peripheral vision, and thus help students to move beyond the "monoculture mentality" of menu-driven systems that were promoted as part of the Green Revolution. Practices are important, but are better understood and remembered as part of the whole system.

We have found that specific key competencies are needed by students to become agroecologists who understand how practices are essential parts of and influence whole farming system (Lieblein *et al.* 2015). One of these is *observation*. In learning to read the context of a farm, students engage in observation walks, physical work, interviews with farmers and others who work there, and generally employ all their senses to observe as much as possible about the biological and human activities on the farm. It is important to observe and take notes to later explore complexity and consider possibilities for dealing with both immediate and long-term challenges. Some changes mean a modification of current practices. Cultivating the competency of observation is key to become aware of the site specifics of any farm, and it also forms the basis for inductive and experiential learning about farming systems. All observation starts with the parts, as we cannot observe the farm as a whole. Different observations of singularities are connected through reflection, described below, to enable understanding of the interactions and complexities of the whole farm, where the farmer lives and makes decisions.

A vital but undervalued component of learning is *reflection* about what has been observed or accomplished, how things were done, and what it all means. Dewey (1938) states that "we do not learn from experience… we learn from reflecting on experience". According to Kolb (2015), reflection is what transforms experience into knowledge. Although we often reflect on what we have been doing, it is not frequent that this activity is explicitly identified as an important step in the learning process. For this reason, we have scheduled specific times each week during the course in the autumn to have a group activity in reflecting on the lectures, field visits, or discussions just completed. The students find this activity unusual and somewhat strange early in the semester, as most have never participated in a formal reflection session nor can they see the immediate value to doing this so frequently. Later in the term, we find that people appreciate the chance to sit back and reflect, first on their personal experience on a farm or in group meetings, and then to share these reflections with others in a small group. Our sessions often round out the activity with a search for relevant action steps to further support learning, and a report out to the entire class. This step-wise process has worked well, and the results often appear in a final "learner document" that the students individually write at the end of the semester, where they reflect on all of their different experiences throughout the whole course. Thus, we are convinced at the end of the semester that most students have internalised the process of reflection. If the goal of agroecological techniques is that they are site-specific applications of ecological principles within a societal frame, then farmers must rely on their own observations of their specific site and have the ability to reflect on their experiences in order to continuously improve their choice and application of specific techniques. As such, the competencies of observation and reflection are generic skills of importance to all actors within the agroecological spectrum.

Another key competency is *dialogue*, going beyond articulating ideas in a group and includes active and engaged listening to others. We conceive of dialogue as a carefully structured but open way of communicating, in agreement with Bohm (1996). It is the opposite of a debate, "a conversation with a centre, not sides" (Isacs 1999). The ability to listen and the willingness to be influenced is key to dialogue. It is a move from thinking alone to thinking together. We joke about the importance of designing an "elevator speech" if you want to communicate an idea quickly to someone who is marginally interested in what you have to say. But in serious communication about farming systems, we can engage our stakeholders by posing questions about details and issues we know are important to them. This will often start with specific details about the farm, and often comes down to practices. Thus another way to introduce a discussion of practices is to start at the farm level so that the specific component technologies are considered within the context of the field or farm. Dialogue is important with farmers during interviews and also in the meetings of student teams as they begin to summarise and evaluate data. It is essential that all voices are being heard both to obtain a comprehensive understanding of the whole situation, which is usually much more complex than what one mind can capture, and to maintain good dynamics among those whose contributions are essential for improving it. An example often invoked is that to be a true listener, it is important that the mind is engaged in what others are saying and the mental motor is not running full speed just waiting for the chance to jump in with the next intervention without carefully considering what is going on from others in the moment. Another basic rationale for dialogue as a key competency is that the complexity of a farm and the larger food system exceeds what one mind can capture. In dialogue, we can listen to different observations and reflections by several people, then build a collective understanding of the whole situation. To integrate these capacities, observation is observing carefully the parts, reflection is when each individual generates connections, and dialogue is reflecting together. In dialogue between different minds the goal is to create an "ecosystem of thought" that can serve the development of the ecosystem of farming and food.

One of the most valued activities by students is *participation* on farms as a part of the overall learning experience. Several years ago, we introduced a one-day "working visit" to each project farm, where the farmer identified an important practical task that needed to be completed during a day in August, when students could make a real contribution. It was first viewed mainly as a chance for students to compensate the farmer for the time spent engaging in interviews with the team and mentoring the group, but now has resulted in a much richer and deeper appreciation by students of what farming is all about. A number of reports by students in their end-of-term learning documents confirm that this is a valuable day in the agroecology semester in Norway, and it is one concrete way that students learn specific practices on the farm through actually doing chores and appreciating their importance. Being an active participant in the agroecosystem also opens for a range of phenomena that would otherwise not surface. It represents an additional window into the agroecosystem, one that hardly can be replaced by other activities.

Lastly, a key skill to be developed is the capacity for *visioning* a future desired situation (Parker 1990), not exactly the crystal ball gazing and invention of some mystical situation but a serious venture into examining potential future realities that could provide a direction for a farmer in seeking the best route to improvement. Rather than training students to be keen diagnosticians whose task is to come up with specific recommendations for solving a certain challenge, we urge them to first suspend judgment as much as possible about a problematic situation and then seek and reflect on a range of potential ways to approach it. Next they can practice visioning, both individually and as a student team group that may include the stakeholder farmer to imagine a desired future for the farm. This activity will be followed up by an exploration of a number of possible ways that challenges met on the road towards the vision could be overcome. Within a range of possible action steps, the farmer will often pick and choose components of several ideas for what needs to be done and chart her or his own most indicated path. Thus visioning is a key capacity that rounds out the education of future agroecologists. We can learn many things from our experiences, from the past, but today's complex challenges require in addition that we can also learn from visioning the desired future. Learning to create shared visions represents such a shift from learning from the past to learning from visioning the desired future.

Each of these capacities (Figure 1) that is a part of what we design in agroecology education is now described, using examples of how each contributes to learning more about appropriate future farming practices within a systems perspective. We conclude the chapter with observations and reflections about the learning process by students over our 15 years teaching in Norway, and comment on how these skills have been applied in their next steps in thesis research as well as in future careers.

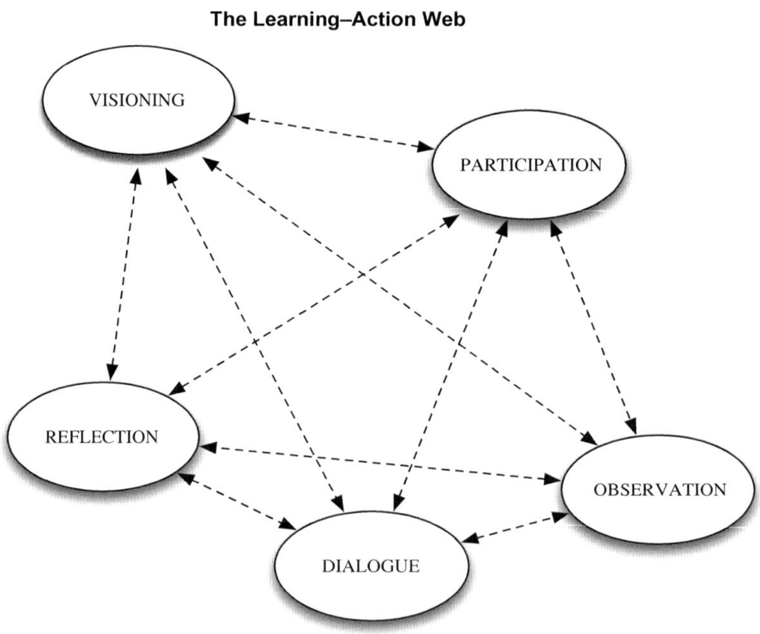

Figure 1. Competencies developed in agroecology courses and degree programmes, including the multiple interactions in a "learning–action web" of educational activities (from Francis *et al.* 2015).

2. Observation skills as a key capacity

Learning is an experiential activity. As we guide students through a series of learning activities they are gaining skills in reading each unique context, using observation involving all the senses, and connecting this to prior experiences which are essential to the learning process. On observation walks across the rural landscape near the university (Francis *et al.* 2012), our students have observed crops new to them in this high-latitude environment, well-tended farms, and a patchwork of small monoculture fields interspersed with forests and grazing lands. Similar walks across their project farms provide student teams with first-hand observations of integrated crop-animal systems, the deployment of these enterprises within the farm, and where they appear to fit into the local rural landscape. It is this shared experience that provides the foundation for creating rich pictures of the farm and its enterprises, practices, and many internal interactions. All these inform the process of assessing what the students observe and learn from the farmer during interviews, as well as their eventual design of potential scenarios to help the farmer meet her or his goals. So how do we introduce new ways of observing and evaluating practices, and thus help students to gain new skills for this important capacity?

A key point in the process of "observation", as in "dialogue" is to have an open mind and to not jump into conclusions and judgement. A common human challenge is not to squeeze phenomena we see and experience into our present worldview models to the extent that these models become a "filter" for new experiences and knowledge. Not surprising, it is also a challenge for students, after 10–14 years of formal school training, not only to look for what they have previously learned about and make interpretations strongly influenced by their pre-knowledge, which is often of disciplinary nature. Further, students have often been trained in an academic pedagogical environment where theorising an "object" is the point of departure for a learning process forming their "knowledge" of the phenomenon. Without sufficient openness they can miss the opportunity to challenge their present worldview models, broaden them and explore new ones to better understand the complex, constantly changing reality. Yet, an even more complicated challenge in guiding students to practice open-minded observation is that many teachers have the same problem. It is difficult for many who observe a specific practice to not make a snap judgment about how good or appropriate it is, so we urge students to just observe at this stage and not draw conclusions.

At the Swedish University of Agricultural Sciences (SLU) several educators developed an opportunity for creating master's programs in Agroecology, together with colleagues at Uganda Martyrs University (UMU) and Mekelle University (MU) in Ethiopia. During the planning process, with core staff members from the three universities, a theoretical entrance into the process of learning was used for students to practice open-minded observation. From experience, we know that both faculty and students find this a "comfort zone" to learn new things. Inspired by the analyses, knowledge and experiences from the "Hawkesbury model" (Bawden and Pakham 1993), Kolb's (2015) cycle of experiential learning was used as a theoretical model for framing open-minded observations and the subsequent steps needed to take informed and purposeful action. Kolb's theory describes learning as a process where the mind shifts in perceiving between "Concrete Experiences" and "Abstract Conceptualisation" and activities shift between "Reflective Observation" and "Active Experimentation". In everyday life, the learning process is, of course, a constant "jump" between these four phases, but we can use the theory to ask what kinds of attitude and practice are essential and effective in the different phases. Kolb describes the first phase — perceiving the world as "Concrete Experience" in a process of "Reflective Observation" — as a phase that needs an attitude of "Divergence". Some call this the phase of finding out "What is?", a process that is supported by curiosity, held-back judgment, and an open mind for new phenomena and experiences. Central to this open-minded approach is to reserve conclusions until the person gains more depth on how a specific component of a situation and a specific practice impacts the entire system and its outcomes. Experience from the agroecology education in Norway suggests that Kolb's theory, showing the observation phase as one of several in a full cycle, helps students to become more motivated for and committed to observation and fact gathering, as they see that analysis, conclusions, and actions will follow in due time and that the quality of these obviously will depend on the quality of their observations.

3. Learning and practicing dialogue

Effective learning is both an individual and a social interactive process, and students need dialogue skills for working with farmers. These skills also provide foundation for interactions within their project teams to complement other academic classroom activities including acquisition of new ideas and knowledge. Ability to articulate ideas and arguments and the capacity for listening are equally important components of dialogue (Vella 2002). Five students who walk a transect on a farm, conduct an interview with a farmer, and begin to evaluate and analyse their experiences, soon learn that even in a shared experience each one has seen that place through a slightly different lens, one that is shaped by the personality of each individual and their prior experience. The importance of discerning higher-order information from farmer interviews, for example, takes a particular concentration on what is said as well as on non-verbal cues (Hargie and Dickson 2004). Perspectives on the worldviews and ethical concerns of the farmer can emerge from an interview, and students who are listening need to put this into the context of that particular farm, its resource base and potentials, the multiplicity of goals and obligations facing the farmer, and the complexity of decision making that often reflects far more than what is on the surface (Driessen 2012). Thus, dialogue includes sharing observations and reflecting as well as careful and thoughtful listening, and a process that leads each student to broaden their understanding of farming practices by interviews with farmers and effective dialogue among those in the teams. According to Isaacs (1999), "Dialogue addresses problems farther upstream than conventional approaches. It attempts to bring about change at the sources of our thoughts and feelings, rather than at the level of results our ways of thinking produce", and further "Dialogue seeks to address the problem of fragmentation not by rearranging the physical components of a conversation but by uncovering and shifting the organic underlying structures that produce it".

Practicing in preparation for farmer interviews is a part of the pre-field visit activities of students, and we have used several methods for this: mock interviews, role play, practice in small groups (Østergaard *et al.* 2013). In some courses, the farmers have several years of experience with previous groups visiting the farms, and they become mentors in the process.

Often, their discussions of specific practices are well imbedded in the whole farm system and informed by their worldviews. In other situations, such as those in Uganda and Ethiopia the students and instructors need to be better prepared when they undertake the initial visits to key farmers to be interviewed. But all the preparation is not focused on students, and needs to include orienting the educators as well.

In the planning process for three master's programs in agroecology (at SLU, UMU, and MU) there were also "training of trainers" (ToT) courses. In these activities, with staff members from each university, we introduced the idea of Kolb's learning cycle and put the focus on how to practice the first phase (what is this?) by organising the teachers in heterogeneous groups from different specialty areas of agriculture and rural development to visit a complex small-holder's farm and household. To help academically trained people open up for new phenomena and preparing them for just observing and listening, and not judging, we provide some introduction to PRA (Participatory Rural Appraisal) methods and tools (Chambers 1994) The introduction of such concrete tools helped them feel more comfortable in doing the exercise, i.e. "how shall I do in this exercise?". Each group was directed to prepare and present a "Rich Picture" of the small-holder's case (Bell and Morse 2013). This design of the exercise to learn to observe with an open mind was highly successful when applied by our educator colleagues. Some students said they now entered the small-holders farm with another, more open attitude, compared to how they had previously visited small-holders to "teach" them new farming practice or apply new external inputs. One said with enthusiasm, "This is the first time I learned from the farmer!" The success of this design of the ToT courses for educators, was then incorporated as a guideline in designing the introductory course in the agroecology master's program at SLU. This is an essential step for discussing farm practices and how they fit into farming systems.

4. Education through participation

Since the early writings of John Dewey (1938) we have appreciated the importance of practical learning and integrating new knowledge into our existing experience base. Education through action or participation includes work on farms, in communities, and in extra-curricular activities that can generate knowledge and build confidence and competence needed for becoming responsible agents of change. The purposes of action of any learning may be grouped in three main categories, a concept not confined to agroecology. One category contains the generic pedagogic goal of motivating students, e.g. by immersion in situations where experience and relevant theory can be more effectively linked. This includes providing a rich and stimulating learning environment to foster learning that is more profound and durable than what is possible in a classroom with fewer stimuli. In agroecology learning, we emphasise the goal to expose phenomena not readily understood from a purely theoretical standpoint, such as complexity, dynamism and interactions leading to emergence of case-specific system properties. This reveals the need for transdisciplinary knowledge and agroecosystems thinking. Building on this experience is the goal of learning theory, approaches and methods in such an environment and in a manner that students will be able to apply their knowledge and competencies in similar situations in future professional activities. Beyond learning for the sake of knowing more things is the potential for future responsible participation and action (Lieblein and Francis 2007). We consistently observe on evaluations from students that they consider practical experience and participation on the farm as among the most important learning activities. We hope students will remember and incorporate methods into the toolbox they take away from the course to apply in new situations, so it is important that the value of context is fully understood. Thus any specific practice is not to be applied as a quick band-aid fix for a situation that is viewed only superficially, but should be considered carefully within the context of each farming system encountered. These practices are most often viewed through the lens of agroecology, and thus are considered important as they impact the whole system (Francis and Porter 2011).

We have accomplished this in several ways in the agroecology course in Norway, where students visit a number of farms, interview farmers, and work on farms for at least one day under supervision of the farmer. Visiting provides exposure to a range of different practices, in the context of whole farms, while farmer interviews add depth to the discussion about why certain practices are used in site- and resource-specific situations.

But the days of farm work have provided the most meaningful experience in terms of actual hands-on involvement, especially in the discussions that are promoted in the field and in later reflection on what this participation means. Several years ago, a student team was assigned the mundane task of stacking split firewood in an orderly way for storage through the winter. Although they learned the details of proper stacking within a short time, much of the rest of the work day was spent discussing the sources of the wood, how sustainable that was in the long term, the environmental implications of burning wood on the farm versus alternative energy sources, the opportunity costs of burning versus selling the wood, the potential to add value to this resource in creative ways, the carbon sequestration in trees and how this is released by burning, and a potential for life cycle analysis of the entire energy balance of the farm. This turned a routine task into a valuable learning opportunity. We have a cartoon of two men splitting stones, with one thinking "I am splitting stones" and the other thinking "I am building a cathedral". We observe with amazement what creative students will derive from participation on the farms.

5. Learning through reflection

Education through contemplation and reflection on applications in a real-world context complements and enriches the learning achieved in prior steps. Mezirow (1990) maintains that "Reflection enables us to correct distortions in our beliefs and errors in problem solving. Critical reflection involves a critique of the presuppositions on which our beliefs have been built". Although some degree of reflection is a part of our everyday activity, the formal structuring of reflection in the agroecology program in Norway is perhaps unique in agroecology education. In class meetings, we often begin with individual reflection through posing a question about recent shared experiences in the field, then ask students to discuss in pairs and then in groups of four their individual ideas about that experience. Each group is urged to come to some consensus about the most important ideas gained from the group experience, and thus pursue a process of social learning around that topic. Halfway through the semester, each student submits an individual learner document, one that is then refined and broadened to submit at the end of the term for a grade by faculty.

These learner documents, provide a rich trove of examples of farming practices that have been learned during the semester, how these specific practices fit into the larger farming system, links between practices and experiences on the one hand and agroecological theory on the other, and what these elements mean to the student as a life-long learner and while becoming a professional. We are currently doing an in-depth analysis of these documents to see how effective we are at meeting the learning objectives of the course, and this includes what they learn about farming practices.

Although students are hesitant at first to share frankly their feelings about the educational experience, this improves as they realise that both faculty and other students really care about what they think and that there is no judgment of them for frank expression of feelings. We build this confidence in the classroom through our own behaviour as faculty in welcoming a wide range of opinions, and also through group travel to farm sites, sharing meals, and informal gatherings off campus (Francis *et al.* 2011). Once a level of trust is established in the learning community, the individual reflection arising in quiet meditation and embodying deeply held values can emerge and contribute to better understanding of the complexity of current systems. These individual values can shape a team's communication with the farmer as they assess ideas for improvement.

Beyond the reflection on specific practices, the value of scheduled sessions became obvious one year in our final potluck lunch when we asked students for their last-minute reactions to the course as well as recommendations for the future. One Italian student immediately jumped into the discussion, saying "I have been in school for over 18 years and this is the first time any professor ever asked my opinion about the classes and learning process!" One must resist the urge to weep when realising there has been so little empowerment over his years in the educational system, and the dearth of concern about getting feedback from students about the process.

6. Education is projected into the future by visioning

Education through examining future scenarios builds the skill potential for visioning alternative solutions across the spatial spectrum from individual practices to systems design, and a priori evaluation of possible outcomes of action (Lieblein *et al.* 2011). Based on practical farm experiences in project work, student teams are exposed to a range of field, farm, marketing, and farm subsidy participation decisions made by farmers. They also have a clear understanding of the current challenges that farmers face in meeting their long-term goals. In order to design potential scenarios that farmers could consider to meet these goals, it is useful to envision an ideal future desired situation and then creatively explore ways of acting that could lead to that desirable state. After practice in a three-day visioning workshop in the classroom with mentors and peers, and after learning from interviews and in-depth visits to the farm, students have conducted on-site visioning sessions with their project farmers. This has proven useful to test their ideas with the reality of the current farm situation, as well as the potential for following one or more components of each proposed future scenario. Building a shared vision in the team together with the farmer client creates an alignment of values that can result in a compelling image so the group can move forward towards commitment, not merely compliance. Visioning a desired future definitely includes some specific farming practices, but this is accomplished within the larger framework of more sustainable production, processing, and marketing to add value to products from the farm. Consequently, with this as guiding principle we stimulate the students not to view relevant farming practices in isolation but to envision them in the perspectives of their contribution to the pursuit of key ecological principles and goals (Gliessman 2015) of (1) providing energy for decomposer organisms in the soil, e.g. relay cropping and growing of perennial crops and recycling of crop residues, animal manure and compost to improve soil structure and health, (2) managing nutrients within the farm unit with, for example, plant cover as continuous as possible, legumes, mycorrhizal species, inter and relay cropping, the use of green and animal manures and compost, and inclusion of animals for better balance between nutrient input and output across the farm boundary, and (3) diversification for maximum exploitation of available niches and ecosystem services, for better regulation of the balance between beneficial (crop plants, microbes, insect predators) and detrimental organisms (weeds, pests, diseases), and for increased interconnectedness of farm elements and process as well as ecological, economic, and social resilience. Examples of diversification include crop rotation and increase in the number of crops grown each year, intercropping, and introduction of animals. Students have also included more specific techniques in some of their visions, for example those common in permaculture (Mollison 1990) and holistic management (Savory 1988).

We have observed in many thesis research projects the value of students observing and reflecting on current practices and then putting suggested ideas into future potential scenarios for farmers to consider. The role of the systems researcher is transformed from the disciplinary problem solver to a holistic observer and analyst who applies principles of agroecology to envision action steps for the farmer to achieve her/his future wanted situation.

7. Summary and conclusions

A large expectation of many students who arrive to enrol in an Agroecology MSc program is to learn more agroecological/organic farming practices. Although we think that skills needed for many of these more sustainable practices are better acquired in a full-time internship on an appropriate farm, it is important to meet the students' needs to some degree in our course to maintain interest, enthusiasm about the assigned tasks, and credibility of the faculty as practical people who do appreciate the need to consider changes in practices as part of whole-system improvement. We accomplish this by having students spend substantial time on farms, observing activities and working on needed tasks under supervision of farmers. Furthermore, we maintain that practical skills should be obtained and farming practices should be understood in a context of agroecological systems thinking and systems practice, linking information from the literature with practical knowledge about the farm and its resources, including the ethics and values of the farmer and a clear idea of her or his goals, that makes our approach to agroecology learning especially relevant to a future that is complex and unpredictable.

A major change in the educational paradigm from training for quick-fix solutions to education for whole-system understanding is essential. Based on experience in the agroecology class at the Norwegian University of Life Sciences over 15 years we have identified several key competencies that appear valuable for students to acquire and practice in the "safe space" of the course. These competencies of observation, participation, dialogue, reflection, and visioning have each been described in the context of our autumn agroecology class in Norway, along with examples of specific practices that are discovered in each of the steps in the semester. Important for students is how each of these practices is conceptualised, evaluated, and integrated into whole-farm systems, and how each is articulated in terms that are relevant to the project farmer. These are perhaps specific useful practices that are carried forward by students in their thesis work and in later positions, but much more important is the process through which the practices are discovered, fine-tuned, discussed with farmer stakeholders, and communicated through effective dialogue. This process is at the core of the agroecology learning experience in the one-semester intensive course in Norway as well as a foundation for thesis research and effective performance in later work environments in agriculture.

References

Bawden, R.J., Packham, R.G., 1993. Systemic practice in the education of the agricultural practitioner. *Systems Practice* **6**(1), 7–19.

Bohm, D., 1996. *On Dialogue*, Ch. 2, Routledge, London, UK and New York, USA, pp. 6–54.

Bell, S., Morse, S., 2013. Rich pictures: a means to explore the 'sustainable mind'? *Sustainable Development* **21**(1), 30–47.

Chambers, R., 1994. The origins and practice of participatory rural appraisal. *World Development* **22**(7), 953–969.

deBuys, W. (Ed.) 2001. John Wesley Powell. *Seeing Things Whole: The Essential* Island Press, Washington D.C., USA.

Dewey, J., 1938. *Experience and Education.* Kappa Delta Pi Publishers, Indianapolis, Indiana, USA.

Driessen, C., 2012. Farmers engaged in deliberative practices: an ethnographic exploration of the mosaic of the concerns in livestock agriculture. *Journal of Agriculturally Environment and Ethics* **25**(2), 163–179.

Francis, C., Morse, S., Breland, T.A., Lieblein, G., 2012. Transect walks across farms and landscapes. *NACTA Journal* **56**(1), 92–93.

Francis, C., Morse, S., Lieblein, G., Breland, T.A., 2011. Building a social learning community. *NACTA Journal* **55**(3), 99–100.

Francis, C., Østergaard, E., Nicolaysen, A.M., Breland, T.A., Lieblein, G., Morse, S., 2015. Learning through involvement and reflection in agroecology. In: Mendez, V.E., Bacon, C.M., Cohen, R., Gliessman, S.R. (Eds.), *Agroecology: A Transdisciplinary, Participatory and Action-Oriented Approach*. CRC Press, Boca Raton, Florida, USA, pp. 73–99.

Francis, C.A., Porter, P., 2011. Ecology in sustainable agriculture practices and systems. CRC Critical Reviews in *Plant Science* **30**(1&2), 64–73.

Gliessman, S.R., 2015. *Agroecology: The Ecology of Sustainable Food Systems*. 3rd edn. CRC Press, Boca Raton, Florida, USA.

Griffith, T., (translator) 2000. Plato's The Republic: *The Allegory of the Cave*. Cambridge University Press, Cambridge, UK.

Hargie, O., Dickson, O., 2004. *Skilled Interpersonal Communication: Research, Theory and Practice*. Routledge Publication, London, England.

Isaacs, W., 1999. *Dialogue and the Art of Thinking Together*. Currency, New York, London.

Kolb, D.A., 2015. *Experiential Learning: Experience as the Source of Learning and Development*, 2nd edn. Pearson Education, Upper Saddle River, New Jersey, USA.

Lieblein, G., Breland, T.A., Morse, S., Francis, C., 2011. Visioning future scenarios. *NACTA Journal* **55**(4), 109–110.

Lieblein, G., Francis, C., 2007. Towards responsible action through agroecological education. *Italian Journal Agronomy/Riv. Agronomy* **2**, 79–86.

Lieblein, G., Nicolaysen, A.M., Morse, S.R., Østergaard, E., Breland, T.A., Francis C., 2015. Students learning agroecology: the case of agroecology MSc education in Norway. In: *IFOAM*, *Agroecology for Organic Agriculture in the Mediterranean*, Modena, Italy, 10–13 September, 5 p.

Mezirow, J., 1990. How critical reflection triggers transformative learning. In: Mezirow, J. (Ed.), *Fostering Critical Reflection in Adulthood*. Jossey Bass Publishers, John Wiley, Hoboken New Jersey, pp. 1–20.

Mollison, B., 1990. *Permaculture: A Design Manual*. Tagari Publications, Tasmania, Australia.

Østergaard, E., Francis, C., Lieblein, G., 2013. Practicing and Preparing for Stakeholder Interviews. *NACTA Journal* **57**(1), 97–99.

Parker, M., 1990. *Creating Shared Vision*. Dialog International Ltd, Sandown, Isle of Wight, UK.

Savory, A., 1988. *Holistic Resource Management*. Island Press, Washington, D.C., USA.

Vella, J., 2002. *Learning to Listen, Learning to Teach: The Power of Dialog In Educating Adults*. Barrett-Koehler Publications, San Francisco, California, USA.

Chapter 16

Agroecological Practices: Potentials and Policies

Alexander Wezel* and Charles Francis[†,‡]

*Department of Agroecology and Environment, ISARA, Lyon, France
[†]Department of Plant Science, Norwegian University of Life Sciences, Ås, Norway
[‡]Department of Agronomy and Horticulture, University of Nebraska, Lincoln, Nebraska, USA

1. Introduction

The search for sustainable development of agriculture is currently achieving high priority in many parts of the world. Among the indicators of change are an accelerated development of agroecological practices and growing attention to integrated systems design. Beyond our research organisations, there is increasing concern about improving, testing, and extending these advances to make meaningful impacts on nutrition and health. We are coming to recognise that many of our research results are not impacting the lives of most people on the planet whose need is greatest, and that gross inequities persist in levels of income and well-being. We understand that the global economic system as well as national and regional policies undercut many of our grassroots initiatives to improve the lives of people. As indicated by several authors in this book, research, education and development in agroecology provide an avenue to expose and evaluate the present situation in farming and food systems, and introduce strategies to improve the human condition by implementing new ideas presented in recent international publications (e.g. IAASTD 2009).

Multiple and diverse examples of research and development are presented in prior chapters of this book, including the rationale for the supreme importance in teaching of agroecological practices and systems in higher education. One key to success and an essential first step is to convince other professionals about the importance of not only agroecology, but of including a systems focus in all agricultural courses, professional training, and extension service. Our attention to all components and their integration into viable systems needs to consider production, economics, environment, and social implications to be relevant to comprehensive strategies for development. Agroecology has been defined variously as a science, as practices, and as a movement, depending on the country and organisation that embraces this vision (Wezel *et al.* 2009). One overarching definition that has guided our choice of topics in the educational arena in our universities has been to consider agroecology as *the ecology of food systems* (Francis *et al.* 2003). This umbrella approach has yet to gain general currency in academia and in the development of sustainable agriculture and food systems, and one reason may be that we, and others, have not given adequate attention to policy.

In the sections that follow, we address first the important issue of scale, and how we approach research and application of results when most farmers and other food systems professionals tend to address single issues at a time, and generally at the lowest levels of spatial and temporal scale. This is human nature, no doubt developed over millennia as a survival strategy. When faced with protecting one's family from a bear at the mouth of the cave, it is hardly the time to speculate on where we should have located the abode in a safer place or why the bear evolved to be a carnivore? So how do we address practices as components of whole systems? As agents for education and change, how do we address the challenges of potential adoption of new ideas, both practices and systems? And lastly, how does current agricultural policy in a specific country and context impact research decisions as well as possible adoption of improved management that will help farm families meet their goals. A larger issue is how communities can achieve desired nutrition for everyone, and food security and food sovereignty for the long term (Francis *et al.* 2013)? Examples from the EU, New Zealand, and the USA are used to illustrate ways that policy currently influences recommendations that emerge from agroecology, and how policy could be used as a creative tool in the future.

2. Scale of application and systems change

One frequent challenge faced by researchers is how to address broad and systemic constraints in farming systems. Most of our prior research experience has been with system components, informed by years of experience and success in designs and analysis methods to study one or a few variables and employ basic or multivariate statistics. Applied in agricultural research in development of new crop cultivars, appropriate levels of nutrients that give an economic return, and pesticide application rates, these methods have been highly successful in providing the foundation for improved practices. In large part, the Green Revolution technologies were successful due to focused effort on plant breeding and system components that confirmed their value in raising production (Conway 1997). Yet further examination of the obvious gains in productivity and contributions to solving short-term challenges of hunger for some people has revealed that a focus on only genetic technologies and immediate production advances may bring a host of other economic, environmental, and especially social emergent properties or challenges to long-term sustainability (Jordan 2002). It is within this context that we examine agroecology as an approach that can help guide the orientation of both research and education into future agricultural systems.

Research in soil fertility and regenerating soil health serves as a useful example. Ryan and Peigné (Chapter 3) describe various management practices for regenerating soil health, and then extend the discussion to invoke the importance of an agroecological approach that goes far beyond soil fertility management for the farmer and include profitability, resilience, and environmental impact on systems in the short and long terms. In a similar review of nutrient management in major cereal crops, Blesh and Galt (Chapter 4) explore ecological processes and social issues as part of the process of transitioning to holistic nutrient management. In each chapter, the authors move beyond the development of a single technique or practice, and examine soil fertility in the frame of how these impact whole systems and how such change should be evaluated. It is obvious that we are not starting from the beginning in building a foundation for sustainable nutrient management. Many appropriate practices have been developed over years of research and have already reached farmer testing and widespread implementation, whereas others reflect new innovations based on recent research and appear to have high potential. There already exists, specific recommendations derived from research which favour the adoption of agroecological practices in soil fertility, for example the use of early spring nitrogen tests in Iowa, USA that allowed maize farmers to reduce nitrogen application rates by 50 kg ha^{-1} without reduction in yields (Blackmer *et al.* 1989). They are reflected now in policies that regulate in some areas the application of nitrogen to reduce excess levels of this nutrient moving into the groundwater (Ribaudo *et al.* 2011). Similar examples are found in EU policies governing use of applied fertilisers (European Commission 2016a). These questions are best answered by research that takes into account the example of agroecological cropping practices, and policy needs to be informed by science that support transition to agroecological farming systems by regulating such practices.

Research on integrated pest management provides another clear example of the importance of limited success of single control measures for insects, pathogens, or weeds, and why a suite of methods is preferable to assure long-term economic management of pests. Smith and Mortensen (Chapter 5) describe an "expanded assembly framework" to inform the development of ecologically-based weed management. The concept of combining multiple strategies to reduce harmful weed populations, sometimes called the use of "many small hammers", was articulated by Liebman and Davis (2009) for application in organic production systems. Challenges of integrated pest management and genetically modified (GM) crops as one tool in the toolbox are discussed by Hokkanen and Menzler-Hokkanen (Chapter 6). This also includes specific provisions related to the introduction of GM crops such as the requirement for sowing refuge areas with non-Bt cultivars in Bt crops, or coexistence-related measures.

A large diversity of agroecological practices that improve agricultural production have existed for decades. Nevertheless, they are still applied to with various degrees of success in different parts of the world. There are also differences in applications in the prevailing regional or national farming systems. One of the relevant agroecological principles is that of site specificity of practices and production systems. There are no widely relevant "silver bullets" that can be adopted across a wide range of conditions, such as a spectacular new maize hybrid or miraculous soil amendment that could be widely marketed. In some ways, this is a reason why there has been less research on agroecological practices, since they most often do not involve a patented product that could be successfully commercialised. So far, agroecological practices are more generally accepted and applied by farmers practicing integrated agriculture, organic farming, or biodynamic systems. The concepts have been less interesting to farmers practicing monoculture methods and homogeneous practices in large fields that are found in conventional agriculture. In most of the highly industrialised large-scale cropping and livestock systems, the use of agroecological practices is still limited and poorly understood. In contrast, in less intensive conventional systems, e.g. in less-favoured hilly or mountainous areas, different agroecological practices have been more widely applied for decades as these areas have greater potential for small-scale, intensive production than for industrial monocultures.

Examples of different specific agroecological practices are presented by Wezel and Silva (Chapter 2), and their application often implies modifying the farming system, either at individual field crop management scale or at the larger farming system scale. In the case of a single practice, the level of change needed is usually low because only part of the crop management strategy needs to be changed or adapted to specific conditions by the farmer (Table 1). This is usually the case when considering practices designed to improve efficiency or which are only substituting a new practice for a current one. For example, applying split fertilisation or selecting a locally better adapted crop cultivar can be relatively easily understood, demonstrated, and implemented. In contrast, when the practices require modification of the cropping or farming system, the necessary level of

Table 1. Agroecological cropping practices, scale of application, level of system change, and integration in today's agriculture in Europe (adapted from Wezel *et al.* 2014b).

Agroecological practice	Scale of application[1]	Level of system change needed	Level of integration in today's agriculture	Potential for the next decade
Efficiency increase and substitution practices				
Cultivar choice	Practice	Low	High	High
Split fertilisation	Practice, system	Low	High	High
Biofertiliser	Practice	Low	Low	Medium
Organic fertilisation	Practice, system	Medium	Medium	Medium
Drip irrigation	Practice	High	Medium	High
Natural pesticides	Practice	Low	Low	Medium
Biological pest control	System	Medium	Medium	High
Cover crops	System	Medium	Medium	High
Redesign practices				
Crop choice and rotations	System	Medium	Low	High
Intercropping and relay intercropping	Practice, system	High	Low	Medium
Agroforestry with timber, fruit, or nut trees	System	High	Low	Low
Allelopathic plants	Practice, system	Low	Low	Medium
Direct seeding into living cover crops or mulch	System, practice	High	Low	Low/medium
Reduced tillage	System, practice	High	Medium	Medium/high
Integration of semi-natural landscape elements at field or farm scale	System, practice	Medium	Low	Medium
Management of landscape elements at landscape scale	Landscape	High	Low	Low

Note: [1] Practice = only the specific practice has to be changed or adapted. System = the cropping or farming system has to be changed or adapted. Landscape = multi-stakeholder agreement is necessary to apply management.

system change is normally medium or high because not only a single practice, but a much larger part of the system has to be redesigned. The example of direct seeding into living mulch demonstrates, where a strong system change is needed. New machinery will be necessary to prepare fields and carry out seeding, changes in mechanical weed management will have to be applied to avoid the need to apply high and costly amounts of herbicide, and crop rotations have to be redesigned to take into account the benefits and challenges of additional mulch. This high level of system changes may explain, for example, why this agroecological practice is not yet widely applied in current agriculture.

Another important dimension of complexity when implementing many agroecological practices is increased diversity (Wezel *et al.* 2014a). To successfully introduce agroecological practices based on diversification, a redesign of cropping systems is often necessary. This is very contrary to the general trend that has followed the green revolution, where farming systems and their practices were developed towards simplification and widespread deployment of homogeneous and menu-based or 'technical package' recommendations. This approach is contrary to the principles of agroecology. Useful applications of agroecological practices, their scale of use, needed level of change and integration, and probabilities of adoption of each are summarised in Table 1. These are but a few of the potential practices that are being used by farmers in the field, and they help summarise the ideas presented previously by Nicholls *et al.* (Chapter 1) and Wezel and Silva (Chapter 2).

3. Potentials for adoption of agroecological practices and systems

Most of the agroecological practices listed in Table 1 and described earlier in Chapter 2, have not been widely adopted in agricultural systems (Figure 1). Two practices that are already in broad use are split fertilisation and planting adapted cultivars. This can be explained because they have long been part of conventional technology packages, and thus part of the mainstream. In addition, there have been high levels of experience and knowledge developed through research and promoted by extension, and also they do not require high levels of system change. The use of organic fertilisation, cover crops, drip irrigation, and biological pest control have already reached a medium level of integration in today's agriculture. Along with split fertilisation and planting new cultivars, these practices have a high potential to be more broadly implemented in the next decade because they already benefit from good scientific knowledge as well as broad experience of farmers.

Moreover, legislative regulations and laws in Europe, such as the EU Nitrate Directive, Water Framework Directive, Pesticides Framework Directive, the greening of the second pillar of the Common Agricultural Policy, and agri-environment measures, will probably boost implementation of more environmentally friendly practices that are less polluting and less reliant on external inputs. Some regulations are more strict at national levels. These regulations could enhance an extended use of split fertilisation, cover crops, diversified crop rotation, and biological control, as public and political awareness grows about alternatives to current heavy use of chemical products in agricultural production.

Although some agroecological practices have good potential, many of them only have a medium potential for broad implementation in the next decade. For some there is limited scientific knowledge available on which to base recommendations, or there is limited practical on-farm experience with practices such as direct seeding and relay intercropping, and the use of natural pesticides and application of biopesticides in agriculture. A major constraint for direct seeding and intercropping is consistency in their agronomic performance, as yields are highly variable (Malézieux *et al.* 2009, Soane *et al.* 2012). Natural and botanical pesticides as well as biopesticide applications will probably not be used on larger areas in the next decade, due to limited research, long and costly legislative registrations processes for new products, and reluctance of industry to move away from currently profitable chemical products. In contrast, practices such as direct seeding into living mulch or cover crops and the integration of landscape elements around fields that are classified with medium or low potential for the next decade might even develop much faster than expected as substantial research is currently in progress on these topics. Growing awareness of their efficacy and efficiency is leading an increasing number of farmers or farmer groups to adoption, especially when they are involved and feel ownership of the development of these practices.

Integration of allelopathic plants, biofertilisers, agroforestry systems, and management of landscape elements at field scale currently have a low level of integration in today's agriculture and will probably not be broadly implemented in the near future. The main constraints for the development of practices based on allelopathy are limited scientific knowledge and practical experience, and challenges in managing allelopathic plants. Biofertilisers such as compost are used on a small scale due to high labour costs, and commercial bioproducts are too expensive for wide use. Broader watershed-based integration of landscape elements across farm boundaries will strongly depend on the regional and national general conditions, as well as cooperation among neighbours, as it involves a larger scale of agreement and management. Moreover, it is likely that this type of agroecological practice/approach can only be implemented in the framework of territorial development projects and planning. Although agroforestry systems exist in many European regions, the combined management of trees and meadows, for example, is considered an "old-fashioned" system without high yield potential, and also difficult to manage. It will require a change in farmer mentality for these practices to become viewed as progressive and futuristic, and several will need substantial additional research and monetary support to become more popular in agriculture.

Most practices shown in Figure 1, have a low level of integration in today's agriculture, and only low or medium potential for the next decade to be more broadly implemented. Further research is needed on many of these practices. In contrast, organic fertilisation, reduced tillage, drip irrigation, biological pest control, cultivar choice, and split fertilisation already have medium or high integration levels in today's agriculture, and medium or high potential for the future. A number of practices in the left-hand column of the figure involve changes in management and do not involve purchase of additional inputs from the commercial sector, and this reduces the incentives for industry to invest in research and development. Such practices as allelopathic plants, intercropping, and biofertilisers produced on the farm actually would replace purchased inputs, and research on these management elements will have to be financed by public agencies.

What can be generally expected from agroecological practices in the following years? Although some of them are already quite well known,

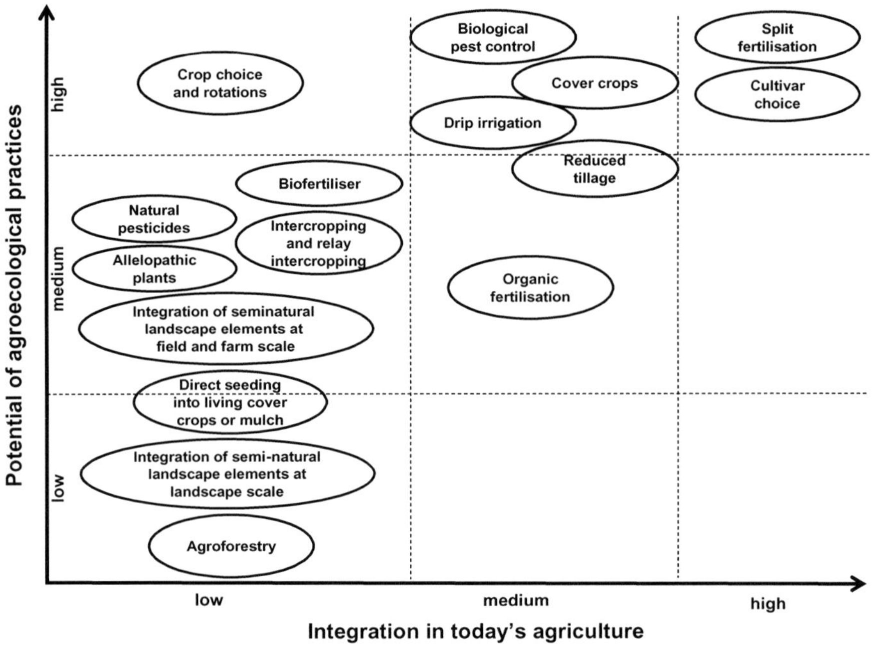

Figure 1. Potential of agroecological practices for the next decade in relation to their integration in today's agriculture (adapted from Wezel *et al.* 2014b).

most will probably not be applied in the near future at larger scales due to the reasons described, and unless there are strong regulatory or policy incentives for farmers to adopt. In the near future, it seems more realistic that we will continue to focus on current highly fertile production areas with high inputs to guarantee high yields, but at the same time move to set clear rules against environmental degradation and pollution as well as biodiversity loss. A broader dissemination of agroecological practices will probably happen first in less-favoured agricultural areas or low potential production zones.

In contrast to temperate zone agriculture, potentials for yield increases with agroecological practices are much higher in developing countries as they are currently most often not at yield maximums, except for irrigated rice systems. Pretty *et al.* (2003) and De Schutter (2011, 2012) summarised noteworthy examples from tropical and subtropical countries.

Nevertheless, as in the case for temperate zone agriculture, many of these agroecological practices with high potential in developing countries are not yet widely applied and are unlikely to be developed by private sector research. This conclusion leads us to recommend a strong involvement of the public sector in research and extension on agroecological practices.

4. Agroecological practices in European, USA and other national agricultural policies

Thus far there has been no clear EU strategy for agroecological practices and sustainable agriculture, and national action plans and political will on this topic still remain both marginal and varied. France is the sole country among the 28 member states to have set up an explicit "Agroecological Project for France" strategy in December 2012 (Ministre de l'Agriculture, de l'Agroalimentaire et de la Forêt 2016). However, the newly defined Common Agricultural Policy (CAP) for 2014–2020 includes valuable elements, in addition to already existing agri-environment measures, which are oriented towards some agroecological practices.

The major novelty of the new CAP is a new financial sub-heading named "Green Payment" representing 30% of direct farm supports. A green component based on compulsory practices to be followed by farmers addressing both climate and environment policy goals is set up in Pillar 1 (direct payments), while previously the trend was only to reinforce environmental measures within Pillar 2 (rural development). Greening practices take the form of simple, generalised, non-contractual, and annual actions that go beyond the common requirements and the regulatory cross-compliance which is the EU directive for good agricultural and environmental practices. The three proposed compulsory practices of greening include (i) crop diversification, (ii) maintenance of permanent grasslands, and (iii) establishment or maintenance of ecological focus areas (European Commission 2016b). More explicitly, these include:

- Crop diversification: minimum of two different crops for between 10 and 30 ha of arable land and three different crops for a holding exceeding 30 ha of arable land. The main crop may cover at most

75% of arable land, and the two main crops a maximum of 95% of the arable area.

- Maintenance of permanent grasslands in and outside environmentally sensitive areas, including permanent grasslands on carbon rich soils.
- Establishment or maintenance of ecological focus areas for holdings where arable land covers more than 15 ha corresponding to at least 5% of the arable land of the holding (and planned to be increased to 7% from 2017 onwards). Ecological focus areas may include fallow land, terraces, landscape features, buffer strips, afforested areas, hectares of agroforestry, strips of eligible hectares along forest edges, but also areas with short rotation coppice where no use is made of mineral fertiliser and/or plant protection products, areas with catch crops and green cover, and areas with nitrogen-fixing crops.

In reality, those practices are not a novelty for some member states, as they are already included to some degree to meet the cross-compliance requirements in one way or another. Furthermore, the CAP reform established in spring 2014, allows Member States to apply for exemptions and equivalences with current farming practices. In that respect, the impact of greening might be much lower than expected at the beginning.

Under the second pillar of the CAP also different agri-environment measures are proposed by the different member countries (European Commission 2016c). Examples which are covered by national/regional schemes are:

- Environmentally favourable extensification of farming;
- Management of low-intensity pasture systems;
- Integrated farm management and organic agriculture;
- Preservation of landscape and historical features such as hedgerows, ditches and woods;
- Conservation of high-value habitats and their associated biodiversity.

An example of an agri-environment measure recently developed for cereal-dominated systems in France includes different elements concerning crop diversification, use of inputs, and maintenance of woody infrastructure (Ministère de l'Agriculture, de l'Agroalimentaire et de la Forêt 2015), and can be simply described with:

- Crop diversity:
  - Major crop less than 60% in year 2 and 50% from year 3 onwards;
  - Minimum of 4 different crops in year 2 and five from year 3 onwards;
  - Minimum of 5% of legumes from year 2 onwards;
- Limitation to not follow the same crop on the same field;
- Limitation of pesticide applications: two different levels of treatment frequencies;
- Control of nitrogen fertilisation: less than +50 kg/ha in nitrogen balance, prohibition of fertilisation on legumes;
- Maintenance of woody infrastructure such as hedges, isolated trees, other permanent planting.

Payments depend on the level of engagement the farmer wants to take as well as on regional parameters decided by provincial governments.

In the USA, farm programmes, there is less focus on environmental concerns, with a continuing emphasis on maximising production of a small number of commodity crops through various subsidy payments. Notable among conservation initiatives are these programmes summarised on the Natural Resource Conservation Service (NRCS 2014) web site:

- Conservation Reserve Program (CRP): both financial and technical assistance is available to eligible farmers and ranchers to address soil, water, and other natural resource concerns in a cost-effective and environmentally beneficial way. The program reduces soil erosion, protects the country's potential to produce food and fiber, reduces sedimentation in streams and lakes while improving water quality, establishes wildlife habitat, and encourages conversion of erodible cropland into permanent cover (www.nrcs.usda.gov/wps/portal/nrcs/detail/national/programs/?cid=stelprdb1041269).

- Environmental Quality Incentives Program (EQIP): both technical and financial assistance is provided to farmers to help deal with natural resource concerns that will provide improved air and water quality, conservation of both surface and groundwater, reduced soil erosion and sedimentation, or create and maintain wildlife habitat (www.nrcs.usda.gov/wps/portal/nrcs/detail/national/programs/farmbill/?cid=stelprdb1237774).
- Conservation Stewardship Program (CSP): this program helps farmers maintain and improve existing conservation activities to address priority natural resource concerns. Farmers are required to improve soil, water, and habitat quality, as well as improve water quantity stored and energy conserved. There are additional payments for use of crop rotations, and contracts may be renewed after an initial five years (www.nrcs.usda.gov/wps/portal/nrcs/detail/national/programs/farmbill/?cid=stelprdb1237774).
- Agricultural Management Assistance Program (AMA): the programme helps farmers use conservation to manage risk and solve natural resource issues through conservation. Farmers can construct or improve water management structures or irrigation structures plant trees for windbreaks or to improve water quality and mitigate risk through production diversification or resource conservation practices, including soil erosion control, integrated pest management, or transition to organic farming (www.nrcs.usda.gov/wps/portal/nrcs/detail/national/programs/farmbill/?cid=stelprdb1237774).
- A number of other programmes include those for water enhancement, air quality, cooperative conservation partnerships, conservation innovations, emergency watershed protection, and wildlife habitat incentives (www.nrcs.usda.gov/wps/portal/nrcs/main/national/programs/financial/ama/).

It is generally assumed in the USA that the farmers in the EU are more highly subsidised and controlled, but it is difficult to compare the two systems with such a wide array of programmes that can be accessed by agricultural producers in each system. It has been widely suggested in both regions that about half of net farm income is due to subsidies from national or region-wide supports, but data are difficult to access and compare. According to Kohl (2016), subsidies in the USA have ranged from

16% in 1996 to 77% in 2000; more recently they dropped to 70% in 2002 and to 5% in 2014, then increased to 33% in 2015. This fluctuation is due to subsidies connected to market prices of commodity crops.

New Zealand provides a unique model in today's economy, as they eliminated most agricultural subsidies in 1984 when a conservative, free-market supporting government came into power. The rationale for this major policy shift was that most traditional subsidy programmes allow inefficient farmers to produce unprofitable crops, and to produce them beyond what consumer demand dictates, and then reduce their risk of loss due to government protection. The decision in New Zealand was to "level the playing field" for farmers in all countries to produce only for market demand. This appears to work in part because of a small domestic market and large reliance on exports. How widely applicable the model will be is yet to be determined, but it apparently is working well for New Zealand (Arnold 2007).

5. Conclusions

Many agroecological practices already are employed in unique production situations around the world, and they are applied to different degrees in different economies and systems. What many applications have in common is that, they include higher biological diversity either at the field, farm, or landscape scale, and this results in a corresponding economic diversity of products available to the market. However, many agroecological practices so far have a low level of integration in today's agricultural systems, and their potential to be more broadly implemented in the next decade will depend on education and training and especially on policies and incentives provided to support their inclusion in a whole set of agro-ecological practices.

There is a large need for research on agroecological practices that will improve water and soil quality, reduce soil and nutrient loss through erosion, increase water capture and storage for crop and pasture growth, provide pollinating insects and other species to maintain production of critical crops, and generally enhance ecosystem services for maintaining a healthy environment. When these practices are not recognised nor rewarded in the marketplace, it is essential to have policies and incentives for farmers to adopt them, as they impact the well-being of people and other inhabitants of our complex global ecosystem. When development of new practices and systems will not include commercialisable products, thus attracting investment from the private sector, it is essential that financial support for public sector research and development be identified and provided. This is critical for long-term, sustainable development of the agriculture and food sectors and a key element in promoting more agroecological practices.

We find examples of different policies for further promotion of agroecological practices in both the EU and the USA, and a unique model where subsidies have been eliminated in New Zealand. Each of these models can function at different spatial and temporal scales, but overall, the adoption of sustainable agroecological models is still limited. Nevertheless, from ongoing discussions on the Common Agricultural Policy at the EU level and also new research, the funding for programmes is still at a low level. However, this area is attracting increasing interest and support, and there is growing potential to integrate more elements of agroecology. Continued development and outreach of agroecological practices can be expected. Moreover, European initiatives such as the creation of pan-European associations promoting agroecology, e.g. Agroecology Europe (www.agroecology-europe.org), and the expansion of regional FAO agroecology conferences such as those already organised in different continents since 2014, will continue to support the expansion of agroecological systems and practices. We find the initiatives described in several chapters in this book to be potentially valuable contributions in that important direction.

Acknowledgements

The evaluation of potentials of agroecological practices and their implication into today's agriculture is based on joint discussion and earlier writings with colleagues M. Casagrande, F. Celette, A. Ferrer, J. Peigné, and J.F. Vian, for which we are very grateful.

References

Arnold, W., 2007. *Surviving Without Subsidies*. New York Times, 2 August 2007. www.nytimes.com/2007/08/02/business/worldbusiness/02farm.html?_r=0 (accessed June 2016).

Blackmer, A.M., Pottker, D., Cerrato, M.E., Webb, J., 1989. Correlations between soil nitrate concentrations in late spring and corn yields in Iowa. *Journal of Production Agriculture* **2**(2), 103–109.

Conway, G., 1997. *The Doubly Green Revolution*. Cornell Univ. Press, Ithaca, New York.

De Schutter, O., 2011. *Agroecology and the Right to Food*. Report of the Special Rapporteur on the right to food. United Nations. www.srfood.org/images/stories/pdf/officialreports/20110308_a-hrc-16-49_agroecology_en.pdf (accessed May 2016).

De Schutter, O., 2012. Agroecology, a tool for the realization of the right to food. In: Lichtfouse, E. (Ed.), *Agroecology and Strategies for Climate Change*, Sustainable Agriculture Reviews Vol. 8. Springer, Dordrecht, The Netherlands, pp. 1–16.

European Commission 2016a. *The Nitrate Directive*. http://ec.europa.eu/environment/water/water-nitrates/index_en.html (accessed June 2016).

European Commission, 2016b. *Greening*. http://ec.europa.eu/agriculture/direct-support/greening/index_en.htm (accessed May 2016).

European Commission, 2016c. *Agri-environment Measures*. http://ec.europa.eu/agriculture/envir/measures/index_en.htm (accessed May 2016).

Francis, C., Lieblein, G., Gliessman, S., Breland, T.A., Creamer, N., Harwood, R., Salomonsson, L., Helenius, J., Rickerl, D., Salvador, R., Wiedenhoeft, M., Simmons, S., Allen, P., Altieri, M., Flora, C., Poincelot, R., 2003. Agroecology: The ecology of food systems. *Journal of Sustainable Agriculture* **22**(3), 99–118.

Francis, C., Miller, M., Anderson, M., Creamer, N., Wander, M., Park, J., Green, T., McCown, B., 2013. Food webs and food sovereignty: research agenda for sustainability. *Journal of Agriculture, Food Science and Community Development* **3**(4), 95–101.

IAASTD, 2009. *International Assessment of Agricultural Science, Technology and Development*. United Nations, Rome.

Jordan, C.F., 2002. Genetic engineering, the farm crisis and world hunger. *Bioscience* **52**(6), 523–529.

Kohl, D., 2016. *Are U.S. Farmers Supported by Government Programs*? Corn and Soybean Digest, June 14, 2016. http://cornandsoybeandigest.com/blog/are-us-farmers-supported-government-programs (accessed June 2016).

Liebman, M., Davis, A.S., 2009. Managing weeds in organic farming systems: an ecological approach. In: Francis, C. (Ed.), *Organic Farming: The Ecological System*. Agronomy Monograph 54, American Society of Agronmy, Madison, Wisconsin, pp. 173–196.

Malézieux, E., Crozat, Y., Dupraz, C., Laurans, M., Makowski, D., Ozier-Lafontaine, H. Rapidel, B., de Tourdonnet, S., Valantin-Morison, M., 2009. Mixing plant species in cropping systems: concepts, tools and models. A review. *Agronomy for Sustainable Development* **29**, 43–62.

Ministère de l'Agriculture, de l'Agroalimentaire et de la Forêt, 2015. *MAEC: les nouvelles mesures agro-environnementales et climatiques de la PAC 2015.* http://agriculture.gouv.fr/maec-les-nouvelles-mesures-agro-environnementales-et-climatiques-de-la-pac-2015 (accessed June 2016).

Ministre de l'Agriculture, de l'Agroalimentaire et de la Forêt, 2016. *Le projet agro-écologique en France*. http://agriculture.gouv.fr/agriculture-et-foret/projet-agro-ecologique (accessed June 2016).

Pretty, J., Morison, J.I.L., Hine, R.E., 2003. Reducing food poverty by increasing agricultural sustainability in developing countries. *Agriculture, Ecosystems and Environment* **95**, 217–234.

Natural Resources Conservation Service (NRCS), 2014. *Natural Resource Conservation Programmes*. U.S. Department of Agriculture, Washington, D.C. http://www.nrcs.usda.gov/wps/portal/nrcs/detail/national/programs/farmbill/?cid=stelprdb1237774 (accessed June 2016).

Ribaudo, M., Hansen, L., Livingston, M., Mosheim, R., Williamson, J., Delgado, J., 2011. *Nitrogen in Agricultural Systems: Implications for Conservation Policy*. USDA-ERS Economic Research Report No. 127. Washington D.C. http://ssrn.com/abstract=2115532 or http://dx.doi.org/10.2139/ssrn.2115532

Soane, B.D, Ball, B.C., Arvidsson, J., Basch, G., Moreno, F., Roger-Estrade, J., 2012. No-till in northern, western and south-western Europe: a review of problems and opportunities for crop production and the environment. *Soil and Tillage Research* **118**, 66–87.

Wezel, A., Bellon, S., Doré, T., Francis, C., Vallod, D., David, C., 2009. Agroecology as a science, a movement or a practice. A review. *Agronomy for Sustainable Development* **29**, 503–515.

Wezel, A., David, C., Ferrer, A., Letort, A., Féret, S., Peigné, J., Vian, J.V., Celette, F., 2014a. *Agroecological Practices Supporting Provision of Goods and Services in Agriculture: Examples from France and Europe*. ISARA Lyon, France, 67 p.

Wezel, A., Casagrande, M., Celette, F., Vian, J.F., Ferrer, A., Peigné, J., 2014b. Agroecological practices for sustainable agriculture. A review. *Agronomy for Sustainable Development* **34**(1), 1–20.

Index